T5-CVD-917

Brune, Dag, 1931-
Nuclear analytical
 chemistry

DATE			

Nuclear Analytical Chemistry

Dag Brune Bengt Forkman Bertil Persson

 Studentlitteratur

 verlag chemie international

Distributed in the Americas by:

VERLAG CHEMIE INTERNATIONAL, INC.
1020 N. W. 6th Street, Deerfield Beach, Florida 33 44 1.

ISBN 0-89573-300-5

© Dag Brune, Bengt Forkman, Bertil Persson 1984
Printed in Sweden
Studentlitteratur
Lund 1984

ISBN 91-44-19561-3   (Studentlitteratur)
ISBN 0-86238-047-2   (Chartwell-Bratt)

Contents

Part C. Activation analysis

Part D. Applications

Preface

A review is given of the application of the nuclear techniques to chemical analysis, particularly to elements at low and trace concentrations. The nuclear reactions involved are outlined, and brief consideration is given to the sources of activation methods, such as reactors, accelerators, neutron generators and isotope neutron sources.

Following three decades of varied use, it is evident that nuclear activation techniques have a wide scope of application. These range from the mapping of trace element occurrence in living tissues, monitoring effluent discharges for pollution control purposes and charting the occurence of metallic elements on the sea floor to in vivo analysis of intact human organs.

In the early days of activation analysis the thermal neutron played the most predominant role as the bombarding particle responsible for inducing radioactivity. Today the thermal neutron is probably still the particle most used for such purposes. However, epithermal and fast neutrons are also important as projectiles in nuclear analytical chemistry.

Thermal and epithermal neutrons are usually produced in reactors, whereas fast neutrons are emitted from radionuclide sources (e.g. ^{252}Cf-sources) or generated in 14-MeV neutron generators.

For photon activation analysis, various electron accelerators producing bremsstrahlung in heavy materials are used. In some instances hospital accelerators can be used for such studies.

Neutron and photon activation analysis is mainly used for bulk analysis, inasmuch as the intensity or energy of the bombarding particles or radiations is usually not essentially changed when passing the sample. On the other hand, charged particles at low or intermediate energies are stopped in a metal surface. For this reason charged particles are mainly used for surface analysis, i.e. the determination of an element in a surface or of its concentration profile just below a surface. In charged particle activation analysis, low energy accelerators of various types are used, e.g. the Van de Graaff type.

In order to extract the most useful information from the textbook it should be used together with the nuclear data information obtained from the well-known Table of Isotopes, Eds. C.M. Lederer and V.S. Shirley, J. Wiley & Sons (1978), as well as the Handbook on Nuclear Activation Cross Sections, TR-156, IAEA 1974, the Activation and Decay Tables of Radiosotopes by Bujdosó et al., Elsevier Publ. Co 1973, and the Gamma-ray Spectrum Catalogue for Ge(Li) and Si(Li) Spectrometry by R.L. Heath, TID 4500, 1974 (ANCR-1000-2). The specific analytical methods to be chosen for the activation analysis procedure may be found in bibliographies edited by Lutz et al. at the National Bureau of Standards, Gaithersburg, USA. Also a bibliographic and data section is found in the periodical Radioanalytica Chemistry.

Non-nuclear methods based on back-scattering and particle-induced X-ray emission (PIXE) techniques, which are often accomplished simultaneously with charged particle activation analysis have been briefly outlined.

In the description of the various entities in the present text the SI-system has been widely used. However, in the literature of nuclear chemical analysis other units often appear. For the purpose of convenience such units have been introduced in the applied parts of the book facilitating comparisons etc.

At the present time several analytical methods, such as flameless atomic absorption spectrophotometry, inductively coupled plasma atomic emission spectrometry abbreviated ICP and the various nuclear methods, offer a high degree of sensitivity to various elements. The nuclear analytical technique, however, is generally considered superior to the other techniques with respect to accuracy and freedom from interferences.

Dag Brune, Bengt Forkman & Bertil Persson

Part A.
Fundamentals

1 The Nuclear Periodic System

1.1 Nucleons and nuclides

From atomic physics we know that a neutral atom consists
of a charged nucleus surrounded by a sufficient number of electrons to make it electrically neutral. The chemical properties
of the atom depend on the number and configuration of the electrons around the nucleus. The number of electrons and thus the
nuclear charge gives the atom its place in the atomic periodic
system.

We also know that the building-stones of the nucleus are
the single-charged proton and the uncharged neutron. These two
particles are closely related and are generally referred to as
nucleons. Their masses are known with high accuracy and are
slightly different

$$\text{mass of neutron } m_n = 1.008665012 \text{ u} \triangleq 939,5731 \text{ MeV} \qquad (1.1)$$
$$\text{mass of proton } m_p = 1.007276470 \text{ u} \triangleq 938,2796 \text{ MeV}$$
$$m_n - m_p = 0.001388542 \text{ u} \triangleq 1,2935 \text{ MeV}$$

where the masses are given both in the atomic mass scale with

$$1 \text{ u} = M(^{12}C)/12 = 1.6606 \cdot 10^{-27} \text{kg} \qquad (1.2)$$

and in the mass-energy scale where the mass-energy equivalence

$$mc^2 = E$$

is used,

$$1 \text{ u} \triangleq 931.503 \text{ MeV} = 149.244 \cdot 10^{-12} \text{Nm} \qquad (1.3)$$

(The sign $\triangleq$ denotes corresponds to).

The nuclear periodic system is related to the atomic periodic system by the charge number of the nucleus Z which also

gives the number of protons in the nucleus. This charge number
identifies the chemical element. The nuclear periodic system,
however, has a further degree of freedom, since it is necessary to specify the number of neutrons N in the nucleus, if
the nucleus is to be well-defined. The total number of nucleons
in the nucleus is the sum of the Z-number and the N-number and
is called the mass number A

$$A = N + Z$$

If the chemical symbol of an element is X, a particular
nucleus belonging to this element can be wholly described by
the notation

$$^A_Z X \quad or \quad ^A X$$

e.g., $^{208}_{82}Pb$ or ^{208}Pb

since we know that the charge number for lead is Z=82 and the
neutron number is given by N=A-Z.
    The proton is identical with the hydrogen nucleus ^1H. A
nucleus for which both the Z-number and the N-number is given
is called a <u>nuclide</u>. Nuclides with identical Z are called <u>iso-
topes</u>, a term which is often incorrectly used instead of "nuclide". For instance, ^{192}Hg is often called a "radioactive isotope".
It is a radioactive nuclide.Nuclides with idential A are called
<u>isobars</u>; those with identical N are called <u>isotones</u>. Finally,
an excited state of a given nuclide may be relatively long-lived, so that its decay time is directly observable. Such an
excited state is called an isomeric (or metastable) state, and
thus two nuclei of the same species but in different energy
states, of which at least one is metastable, are called <u>isomers</u>.

1.2 Stable nuclides

About 300 stable nuclides belonging to 81 different ele-

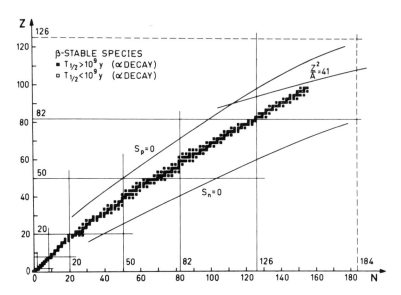

Fig. 1-1   The stable nuclides are plotted as a function of N
and Z. In the region below S_n = 0 and above S_p = 0
the nuclide in its ground state is neutron- or proton-
unstable respectively. In the region above $\frac{Z^2}{A}$ = 41
spontaneous fission occurs within seconds.
A. Bohr and B. Mottelson, Nuclear Structure I
(W.A. Benjamin 1969) p. 203.

ments are known. An element can thus often have several stable
isotopes which differ by the number of neutrons they contain.
It is interesting to see how the Z- and N-numbers for stable
nuclei are related (fig. 1-1).

The lightest nuclei have equal numbers of protons ·and
neutrons but in all others there is an excess of neutrons. The
stable nuclei can be divided in four groups according to the
numbers of protons and neutrons they possess (table 1-1).

The preference of stable even-even nuclei is caused by a
tendency for nucleons of the same kind to be grouped into stable pairs with spins opposed.

Table 1-1

Abundance of stable nuclei.

Number of protons Z	Number of neutrons N	Amount of stable nuclei (%)
even	even	63
odd	odd	1
even	odd	20
odd	even	16

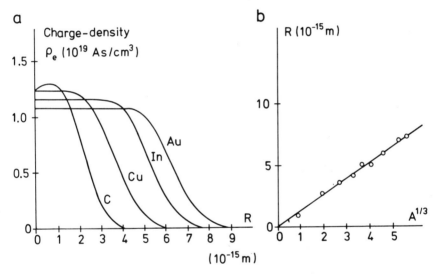

Fig. 1-2 a) Charge-density in different nuclei
          b) Nuclear radius as a function of A.

1.3 Nuclear size

To a reasonable approximation, the nucleus can be regarded as a sphere. The radii of nuclei have been determined by a number of different methods, all of which agree in showing that the nuclear density is approximately constant for all nuclei. The charge-density which is easy to measure is shown in fig.

1-2a. The constant density implies that the nuclear volume is proportional to the number of nucleons A or the radius proportional to $A^{1/3}$,

$$R = r_o A^{1/3} \qquad (1.4)$$

where r_o is a constant (fig. 1-2b).

As a good approximation over the whole mass region, we can use

$$r_o = 1.20 \cdot 10^{-15} m \qquad (1.5)$$

The observation that the nuclear density is constant is in agreement with the fact that the binding-energy per nucleon is almost constant. This implies that a nucleon in the nucleus is attracted only by its closest neighbours or, in other words that the nuclear force is short-ranged. This is the basic reason why the properties of nuclei are similar to those of droplets of water.

1.4 Nuclear mass

In atomic mass tables, the nuclear masses are normally not given separately, but rather only the masses of neutral atoms are listed. This means that to the mass of the nucleus are added the masses of the surrounding electrons minus the mass-equivalence of the electron binding energies. The last term is small and is in most cases neglected. The atomic mass of hydrogen $M(^1H)$ is

$$m_p = 1.007276 \ u$$
$$m_e = 0.000549 \ u$$
$$\overline{M(^1H) = 1.007825 \ u}$$

(atomic masses are symbolized by capital letters M and nuclear masses by letters m).

Analogously, the mass of a heavier nucleus is

$$m(Z,A) = Nm_n + Zm_p - \Delta m(N,Z) \qquad (1.6)$$

and the mass of the atom

$$M(Z,A) = Nm_n + Z(m_p + m_e) - \Delta m(N,Z) \qquad (1.7)$$

The term $\Delta m(N,Z)$ is the mass-equivalent of the binding energy (E_B) of the nucleons. This binding energy corresponds to the energy needed to decompose the nucleus into its constituent nucleons. Since energy has to be supplied for doing this, the masses of bound nucleons are smaller than these of the free ones.

The term $\Delta m(N,Z)$ in eq. 1.7 also contains the binding energy of the electrons to the atom but this energy is less than half percent of the total nucleon binding energy and is thus usually neglected.

The total nucleon binding energy divided with the mass number $E_B A^{-1}$ i.e. the average binding energy per nucleon is remarkably constant over the entire range of masses and is 7-9 MeV/nucleon. On an enlarged scale, however, a clear tendency can be seen (fig. 1-3). The average nucleon binding energy increases with the mass number for light nuclei, passes over a maximum in the Fe-Cu region and decreases for heavier nuclei. The curve is almost smooth but some wiggles which indicate mass regions of special stability are observed. Such stabilities are connected with the "magic numbers" (2,8,20,50,82,126).

Quantities related to the binding energy are the mass defect and the packing fraction. These are, in fact, more frequently tabulated than the binding energies. The mass defect Δ is the difference between the atomic mass M and the mass number A

$$\Delta = M-A \qquad (1.8)$$

with the norm $\Delta(^{12}C) = 0$

The packing fraction f is the mass defect divided by the mass number: $f = \Delta/A$.

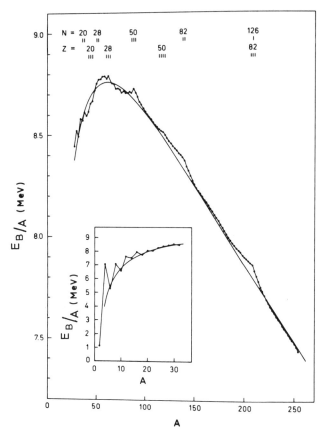

Fig. 1-3　The experimental binding energies. The smooth curve
represents the semi-empirical mass formula. A. Bohr
and B. Mottelson, Nuclear Structure I (W.A. Benjamin
1969) p. 168.

1.5 **Semiempirical mass equation**

It is possible to express the binding energy, E_B, of a
nucleus analytically to a good approximation. The expression
used is based on the analogy between a charged nucleus and
a charged droplet.

$$\Delta m(N,Z)c^2 = E_B = a_1 A - a_2 A^{2/3} - a_3 Z(Z-1)A^{-1/3} - a_4(N-Z)^2 A^{-1} \pm$$

$$\pm a_5 A^{-1/2} \tag{1.9}$$

24

This expression, which in combination with eq. 1.7 is called the semiempirical mass equation, fits the known nucleon binding energies over the whole mass region with parameters a_1 = 15.66 MeV, a_2 = 17.23 MeV, a_3 = 0.70 MeV, a_4 = 23.29 MeV and a_5 = 12 MeV. The smooth curve in fig. 1-3 represents this expression.

The first term in the expression gives the nucleon binding energy and is known as the volume energy. As in the theory of liquids, we have to correct this term by a surface energy term since the nucleons on the nuclear surface are more weakly bound than those inside. The second term which is proportional to the nuclear surface area makes this correction. The third term arises from the electrostatic repulsion between the protons in the nucleus. The fourth term is proportional to $(N-Z)^2 A^{-1}$ or $(A-2Z)^2 A^{-1}$ and arises from nucleon symmetry properties. It disappears when N and Z are equal showing that nuclei prefer the numbers of protons and neutrons to be equal. We know that this is the case in light nuclei (fig. 1-1). In heavier nuclei, the increasing electrostatic repulsion between the protons enhances the number of neutrons above that of protons. The fifth term takes into account the pairing tendency of nuclons already discussed, with the plus sign applying to even-even nuclei and thus minus sign to odd-odd nuclei. The term is zero for odd A nuclei.

The semiempirical mass-equation (eqs. 1.7, 1.9) is a quadratic function of Z. It follows that for constant A the masses of the isobars lie on parabolas. For odd A one parabola only is obtained, but for even A the pairing term in eq. 1.9 causes the masses to lie alternately on two distinct parabolas separated by an energy $2a_5 A^{-\frac{1}{2}}$ (fig. 1-4). From the semiempirical mass equation, it is possible to calculate the maximum stability curve of the nuclides as a function of A. The curve passes through the minima of the parabolas and is given by:

$$\frac{\partial m(Z,A)}{\partial Z} = 0$$

($\frac{\partial}{\partial Z}$ is the partial derivative, where the parameter Z is varying but the parameter A is kept constant) giving

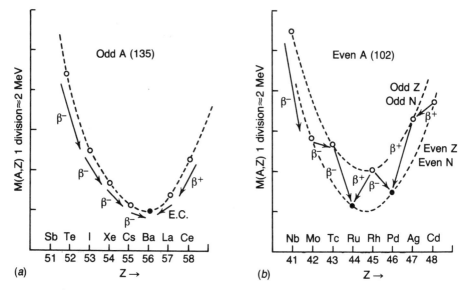

Fig. 1-4 Stability of isobars showing atomic mass M plotted
against atomic number Z. Open circles represent unstable and full circles stable nuclei.

a) Odd A (135) for which there is one stable nucleus
$^{135}_{56}$ Ba

b) Even A (102), for which there are two stable iso-
bars $^{102}_{46}$ Rn and $^{102}_{46}$ Pd, both of even type. Electron
capture (EC) is in principle also possible whenever
β^+-emission may take place W.E. Burcham, Elements
of Nuclear Physics (Longman 1979) p. 194.

$$m_p - m_n + a_3(2Z_s - 1)A^{-1/3} - a_4 4(A - 2Z_s)A^{-1} = 0$$

or

$$Z_s = \frac{m_n - m_p + a_3 A^{-1/3} + 4a_4}{2(a_3 A^{-1/3} + 4a_4 A^{-1})} \qquad (1.10)$$

(The most stable charge number Z_s given by this expression is not integral - a consequence of treating Z as a continous variable). It is worth noting that the volume energy $-a_1 A$

and the surface energy $a_2A^{2/3}$ do not contribute to the equation of the stability curve, since they depend only on A and not on how A is divided between protons and neutrons. In the limit of very small A, $Z_s \rightarrow A/2$, while in the limit of very large A, $Z_s \rightarrow$ $\rightarrow \dfrac{2a_4}{a_3} A^{1/3}$. The position of the most stable nucleus of given A is determined by the competing effects of Coulomb repulsion (reduced by reducing Z) and the symmetry energy (increased by reducing Z).

It is clear that for odd A there will only be one stable nucleus. For even A, there may be two or even three stable isobars. Transitions between isobaric nuclei take place in the direction of increasing stability by β^-- or β^+- emission or by electron capture (fig. 1-4). In this way, the abundances of the stable isotopes (table 1-1) can be understood.

1.6 Nuclear energy levels

We know from the shell model of the atom that the electrons move in practically independent orbits in an average field of the nucleus and other electrons. Strong evidence exists that the nucleons in the nucleus have sufficiently long mean free paths so that one is justified even in the nuclear case in assuming a fairly independent motion. For simplicity, we assume that the interaction between the nucleon, which is being studied, and all the other nucleons can be represented by a spherical symmetric potential. In such a potential a complete set of energy levels exists. The arrangement of the levels are dependent on the shape of the potential which is shown in fig. 1-5. The lowest energy state of a nucleus is called its ground state. When the nucleus is in this state all the nucleons are placed one by one from bottom of the potential well in the lowest unfilled available energy orbit consistent with the Pauli exclusion principle. This principle states that two electrons, protons or neutrons which all belong to a group of particles called fermions as they have half-integer intrinsic spin, never can occupy the same nondegenerate state.

27

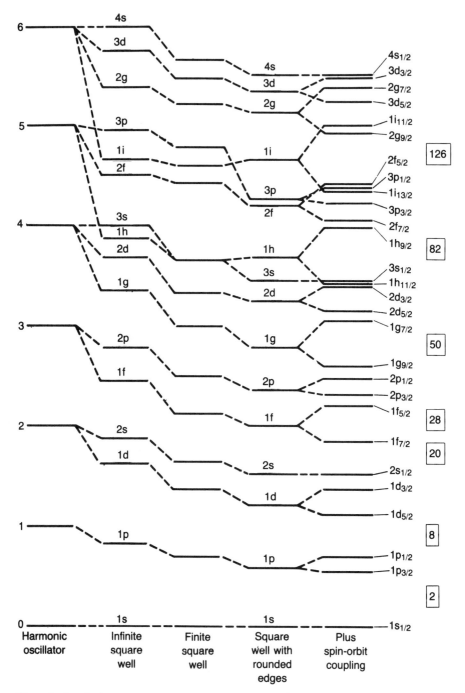

Fig. 1-5 Order of energy levels according to the independent-
          particle model with various assumptions for the shape
          of the nuclear potential and including the effects of
          spin-orbit coupling. B.T. Feld, Ann. Rev. Nucl. Sci. 2
          249 (1953).

The energy level distribution given in the column furthest to the right in fig. 1-5 agrees with that found in experiment. We recognize that energy gaps occur in the distribution at those places where the nucleon occupation numbers correspond to the magic numbers 2,8,20,28,50,82,126.

A nucleus can also be in definite excited states, where one or more nucleons have jumped to more energetic orbits in close analogy with excited atomic states, where electrons have jumped. The whole nucleus can also rotate and vibrate giving rotational and vibrational energy bands similar to those found in molecular spectra.

1.7 Nuclear spin and parity

Qualitatively we can picture a nucleus as consisting of nucleons moving around the centre of mass in certain orbits ("independent-particle model"). Protons and neutrons have the same intrinsic spin quantum number as the electron, $s = \frac{1}{2}$ and their orbital angular momenta have integral quantum numbers, just as in atomic physics. The same symbols are also used for the quantum numbers

$$\ell \begin{cases} 0,1,2,3,4,5,6 \\ s,p,d,f,g,h,i, \end{cases}$$

The resulting total intrinsic angular momentum I of the nucleus is the vector sum of the orbital angular momenta $\vec{\ell}$ and the intrinsic spin momenta $\vec{s}_i$ of every nucleon in the nucleus. The size of the intrinsic angular momentum (the vector length) is

$$|\vec{I}| = (I (I+1) \hbar^2)^{\frac{1}{2}}$$

where $\hbar = h/2\pi = 1.05459 \cdot 10^{-34}$ Js is the Planck's "barred" constant and I the intrinsic angular momentum quantum number (the nuclear spin quantum number).

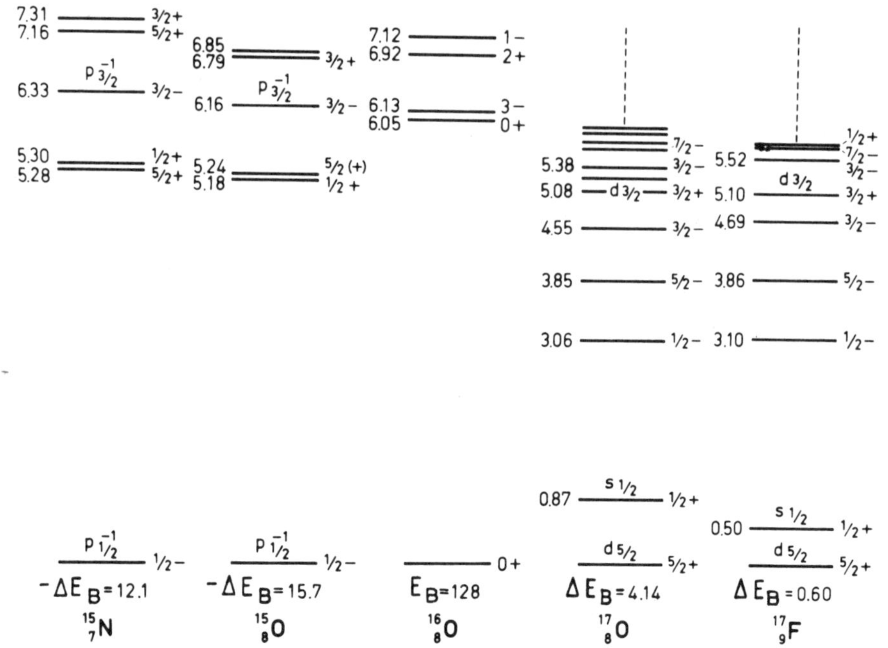

7.31 ——— 3/2+
7.16 ——— 5/2+
          6.85
     $p\frac{-1}{3/2}$ 6.79 ——— 3/2+   7.12 ——— 1−
6.33 ——— 3/2− $p\frac{-1}{3/2}$ 6.92 ——— 2+
               6.16 ——— 3/2−   6.13 ——— 3−
                                 6.05 ——— 0+

5.30 ——— 1/2+   5.24 ——— 5/2 (+)
5.28 ——— 5/2+   5.18 ——— 1/2 +
                                       7/2−   7/2−
5.38 ——— 3/2−   5.52 ——— 3/2−
5.08 —$d3/2$— 3/2+   5.10 —$d3/2$— 3/2+

4.55 ——— 3/2−   4.69 ——— 3/2−

3.85 ——— 5/2−   3.86 ——— 5/2−

3.06 ——— 1/2−   3.10 ——— 1/2−

                         0.87 — $s\,1/2$ — 1/2+
                                       0.50 — $s\,1/2$ — 1/2+
$p\frac{-1}{1/2}$ 1/2− $p\frac{-1}{1/2}$ 1/2−   ——— 0+ $d5/2$ 5/2+ $d5/2$ 5/2+
$-\Delta E_B = 12.1$ $-\Delta E_B = 15.7$ $E_B = 128$ $\Delta E_B = 4.14$ $\Delta E_B = 0.60$
$^{15}_{7}N$ $^{15}_{8}O$ $^{16}_{8}O$ $^{17}_{8}O$ $^{17}_{9}F$

Fig. 1-6 Level schemes for some light nuclei.
A. Bohr and B. Mottelson, Nuclear Structure I (W.A. Benjamin 1969) p. 321.

The quantum number I is therefore integral for nuclei with even A and half-integral for nuclei with odd A.

An energy level with the nuclear spin number I can be split into 2I+1 sublevels since an $\vec{I}$-vector can have 2I+1 different projection values on a fixed axis. In each sublevel only one neutron and one proton can be placed according to the Pauli exclusion principle. The occupation number of neutrons and protons respectively in a level with nuclear spin number I is thus (2I+1).

In its ground state, an even N, even Z nucleus has all its nucleons coupled in pairs. The total angular momentum of each pair is zero according to the coupling. Thus all even N, even Z nuclei have zero total intrinsic angular momentum in their ground states. In an odd A nucleus the single unpaired nucleon is the only one that contributes to the ground state momentum. When the nucleus is excited, nucleons move in new orbits and nuclear pairs may be broken. The vector rules valid for the ground state are hence not very useful for the excited states (fig. 1-6). As well as the I value, a plus or minus sign denoting the parity is usually given for nuclear states. These signs are of quantum-mechanical origin. The plus sign is associated with an even ℓ-number of nuclear angular momentum, the minus sign with an odd ℓ-number.

Since the nuclei are positively charged, a nucleus with intrinsic angular momentum must also have a nuclear magnetic moment, the vector of which ($\vec{\mu}_I$) is proportional to the intrinsic angular momentum vector ($\vec{I}$) and mostly parallel but in some cases (i.e. ^{15}N, ^{107}Ag) antiparallel to it. The nuclear magnetic moment may be expressed in nuclear magneton units (μ_N) and then be written

$$\vec{\mu}_I = g_I \mu_N \vec{I} \qquad (1.12)$$

$$\mu_N = \frac{e\hbar}{2m_p} = 5.0508 \cdot 10^{-27} \ JT^{-1} \ (joule \ tessla^{-1}) \qquad (1.13)$$

where m_p is the proton mass and the proportionality factor g_I is called the nuclear g-factor.

The magnetic moment can be measured with extreme accuracy using nuclear magnetic resonance methods (NMR). These methods are quite useful tools in studying the structure of molecules and the nature of chemical bonds since the environment surrounding the nucleus has a small but definitely measurable effect on the field sensed by the nucleus.

1.8 Nuclear and Coulomb potential

When a slow-moving neutron approaches a nucleus it does not feel any force, (as it is uncharged), until it reaches the edge of the nucleus. It then suddenly feels the strong nuclear force, is captured by the nucleus and falls down into the nuclear potential. Protons, α-particles and other charged particles on the other hand, experience difficulties in reaching the edges of nuclei since the nuclear electrostatic force (the Coulomb force) pushes them away. However if their kinetic energies are sufficiently large they can pass over the Coulomb

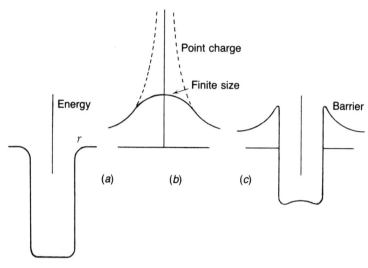

Fig. 1-7 Nuclear and Coulomb potentials
  a) Nuclear well, for neutrons
  b) Coulomb potential
  c) Addition of (a) and (b) for Z = 20.

potential and reach the nuclear edge, feel the nuclear force
and be captured in a similar way to neutrons.

The Coulomb potential is given by

$$V_c = \frac{1}{4\pi\varepsilon_o} \frac{zZe^2}{R} \qquad (1.14)$$

and has the value for single-charged particles

$$V_c = \frac{zZe^2}{4\pi\varepsilon_o r_o A^{1/3}} \simeq \frac{Z}{A^{1/3}} \text{ (MeV)} \qquad (1.15)$$

at the edge of the nucleus (fig. 1-7).

This value is called the Coulomb barrier height (see fig.
3-11). Even if the particle has not enough energy to pass over
the Coulomb barrier, some probability still exists for the particle to reach the nuclear edge by penetrating the barrier. This
penetration is a purely quantum mechanical effect and is called
tunnelling (see fig. 3-6).

In fig. 1-8 an actual nuclear potential is shown where the
energy level diagram from fig. 1-5 has been combined with the

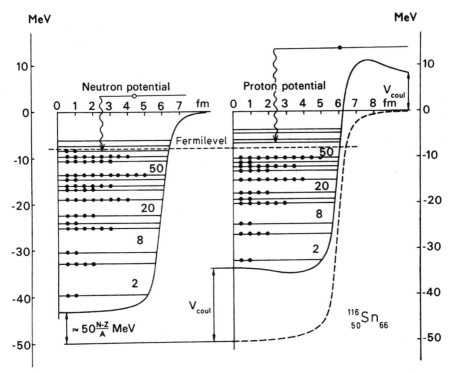

Fig. 1-8   Mean nuclear potential in which neutrons and protons
           move. The factor (N-Z)/A arises from the effects of
           the Pauli principle. An external nucleon is captured
           by the $^{116}_{50}$ Sn nucleus. S.G. Nilsson, private communi-
           cation.

potential outside the nucleus from fig. 1-7. The figure shows
how an external nucleon is captured and how it falls down to
the lowest unfilled energy level often passing over several intermediate levels.

1.9 Unstable nuclei

Unstable nuclei were originally found in nature, the heaviest nuclei being observed to be unstable to α-, β-, γ-emission or spontaneous fission (fig. 1-1). Even some light natural
nuclei, such as $^{40}K(β^-, EC)$, $^{115}In(β^-)$, $^{132}Gd(α)$, are unstable.

Later on, it was discovered that unstable nuclei could be produced artificially and now several hundred radioactive nuclei are known. In addition to the decay modes given above, a few nuclei can also disintegrate by delayed neutron emission.

1.10 Nuclear periodicity

Periodicity in atomic properties such as valency and ionization potential has been known for well over a century and forms the basis of the familiar periodic classification of the elements due to Mendeléev. It receives a natural interpretation in terms of the filling of the successive levels of a screened Coulomb potential by electrons whose number in a given sublevel is limited by the Pauli exclusion principle. A typical graph of an atomic property; the first ionization potential as a function of atomic number (Z), is given in fig. 1-9. In this chapter we have seen that a similar periodicity is to be expected for the nucleus since energy gaps occur in the energy level

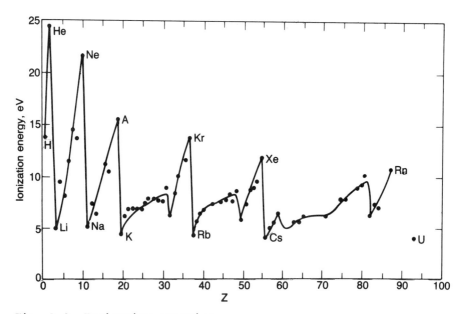

Fig. 1-9 Ionization energies.

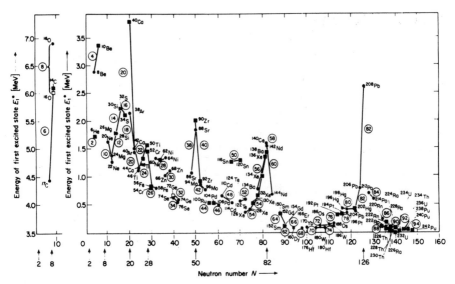

Fig. 1-10 Plots of the energies of first excited states of e-e
nuclei as a function of the neutron number N, showing
the distinct peaking of the magic numbers N = 2, 8,
20, 28, 50, 82 and 126. The points for different isotopes of the same element have been connected by
straight lines, the corresponding proton number Z
being indicated within a circle next to each line.
The data are represented by: $\square$: $I^\pi = 0^+$; $\bullet$: $I^\pi = 2^+$;
and $\blacksquare$: I^π = unknown or uncertain.
P. Marmier, E. Sheldon: Physics of Nuclei and Particles II (Academic Press 1970) p.1264.

distribution at the nuclear occupation numbers 2,8,20,28,50,82,
126.

Even if the shell structure of the nucleus is not of the
same importance in nuclear physics as the atomic shell structure in atomic physics, we end this chapter by showing a typical graph of nuclear periodicity (fig. 1-10).

1.11 References

1.11.1 INTRODUCTION TO NUCLEAR PHYSICS AND CHEMISTRY

1.1 W.E. Burcham, Elements of Nuclear Physics, Longman (1979).

1.2 R.D. Evans, The Atomic Nucleus, McGraw Hill (1955).

1.3 G.R. Choppin and J. Rydberg, Nuclear Chemistry, Theory and Applications, Pergamon Press (1980).

1.4 M. Haissinsky, Nuclear Chemistry and Its Applications, Addison-Wesley (1964).

1.5 B.G. Harvey, Introduction to Nuclear Physics and Chemistry, Prentice-Hall 2nd.ed. (1969).

1.6 P. Kruger, Principles of Activation Analysis, Wiley-Interscience (1971).

1.7 M. Lefort, Nuclear Chemistry, Van Nostrand (1968).

1.8 P. Marmier and E. Sheldon, Physics of Nuclei and Particles, Academic Press 2 vols. (1969).

1.9 D. De Soete, R. Gijbels, and J. Hoste, Neutron Activation Analysis, Wiley-Interscience (1972).

1.10 L. Yaffe (ed.), Nuclear Chemistry, Academic Press 2 vols. (1968).

1.11.2 FUNDAMENTAL PHYSICAL CONSTANTS

1.11 E.R. Cohen, The 1973 Table of the Fundamental Physical Constants, Atomic Data and Nuclear Data Tables 18, 581 (1976).

1.11.3 NUCLEAR MASSES

1.12 A.H. Wapstra and K. Bos, The 1977 Atomic Mass Evaluation, Atomic Data and Nuclear Data Tables 19, 175 (1977).

1.11.4 NUCLEAR MOMENTS AND RADII

1.13 V.S. Shirley, Table of Nuclear Moments in Hyperfine Structure and Nuclear Radiations, eds. E. Matthias and D.A. Shirley, North-Holland (1968).

1.14 G.H. Fuller and V.W. Cohen, Nuclear Spins and Moments, Nuclear Data 5A, 433 (1969).

1.15 L.R.B. Elton, R. Hofstadter, and H.R. Collard, Nuclear
     Radii, Landolt-Börnstein New Series vol 2, Springer (1967).

1.11.5 TABLE OF NUCLIDES

1.16 C.M. Lederer and V.S. Shirley (eds.), Table of Isotopes,
     Wiley-Interscience 7th ed. (1978).

1.17 D. Brune and J.J. Schmidt (eds.) Handbook on Nuclear
     Activation Cross Sectons, Techn. report series 156 IAEA
     (1974).

1.18 E. Bujdosó, I. Fehér and G. Kardos, Activation and Decay
     Tables of Radioisotopes, Elsevier (1973).

1.19 R.L. Heath, Gamma-ray Spectrum Catalogue for Ge(Li) and
     Si(Li) spectrometry TID-4500 (1974) (ANCR-1000-2).

2 Nuclear Decay

2.1 Different modes of decay

Radioactivity was discovered at the end of the last century and it was soon found that the radiation emitted consisted of three different types. These were named α-, β- and γ-rays for simplicity, and were found to be charged helium nuclei, fast electrons and electromagnetic radiation of very short wavelength similar to X-rays respectively. In natural radionuclides, α-processes and β-processes very often compete; that is, they may both be energetically possible. The γ-process is normally a very fast process but if the transition is complex competition between the γ-process and the other two processes may exist.

In α-emission, the mass number A changes by four units; in β-emission, the mass number A does not change, but Z and N change by one unit in opposite directions. In γ-emission, neither Z nor N changes. There is, then, what we may call β-decay correspondence between all the members of a chain of nuclides with the same nucleon number A (fig. 1-4), and α-decay correspondence between this chain and the chain in which the nucleon number is A-4. There are thus four families of natural radioactive nuclides, one with A=4n where n is an integer, and others with A=4n+1, 4n+2 and 4n+3. As an example fig. 2-1 shows the uranium-radium 4n+2 series marked out on a part of the chart of nuclides, with arrows indicating the directions of decay.

Fig. 2-1 shows only the branches of the chain which have been observed in natural radioactivity. Many radionuclides with A=4n+2 have been artificially produced on both sides of this chain, and they are found to be linked to it as side branches.

An artificially radioactive nuclide is produced in a nuclear reaction. Such a reaction is normally a fast process. If an excited nucleus with enough energy to emit a nucleon or a nuclear

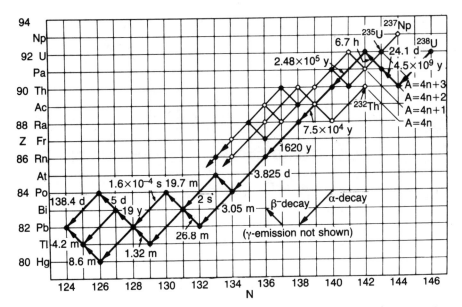

Fig. 2-1   The A = 4n + 2 radioactive series and the first links
of the other three series. Were branching occurs,
the half-life is written besides the main branch.

cluster is formed, particle emission occurs within about 10^{-15} s
after the primary interaction. If a fission channel is open,
this process will compete within the same time. After this fast
disintegration, the residual nucleus or the fission fragments
are usually still excited and decay to their ground states by
a cascade of γ-transitions. These transitions are also rather
fast ($t \lesssim 10^{-11}$ s) but, in some cases which will be discussed
later, the transitions are so complex that metastable states
with rather long lives are formed – the isomeric states.

When the nuclei have reached their ground states, they
are very often still unstable and decay by β-emission and
sometimes by α-emission through tunnelling, (fig.2-2). These
processes are both slow processes. Delayed neutron emission
may also occur but is a seldom process (fig. 2-17). To its type
it is a fast process but it is delayed since it follows a slow
β-process. Spontaneous fission which exists for the heaviest
nuclei is also a tunnelling process and hence slow. We will
now discuss the different modes of decay in more detail.

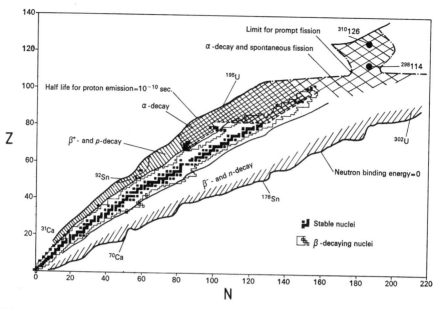

Fig. 2-2   The nuclear chart with different regions of decay.

2.2 Alpha decay

Nuclei are built up of nucleons and not of α-particles. However, short-lived combinations of two protons and two neutrons can be formed inside the nucleus producing α-like clusters. Such α-clusters have energies greater than the α-particle separation energy as kinetic energy is gained when two neutrons and two protons bind together. This is seen in fig. 2-3 where the α-particle energy lies above the zero energy line. In spite of this the α-particle cannot be emitted according to classical laws since it does not have energy enough to pass the Coulomb barrier. If the α-particle is treated quantum mechanically some probability exists for the α-particle to pass through (tunnel) the barrier even if it becomes negative in energy in the shell region between the nuclear surface at the distance r_1 out from centre and the potential surface at distance r_2, where the α-particle energy balances the Coulomb energy. Naturally the tunnel probability increases strongly with decreasing

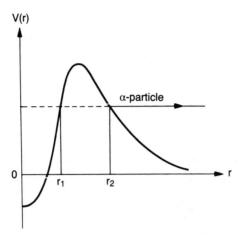

Fig. 2-3   The tunnelling effect. An α-cluster inside the
            nucleus with given energy reaches the Coulomb bar-
            rier at the radial distance r_1 and have to tunnel
            the distance r_2-r_1 through the barrier in order to
            be emitted.

barrier. Thus α-emitters with large disintegration energies have short half-lives and conversely. The variation is astonishing-
ly rapid. A factor of 2 in α-energy produces a factor of 10^{24}
in the half-life (fig. 2-4).

    The α-decay can be written

$$^A_Z X \rightarrow {}^{A-4}_{Z-2}Y + \alpha + Q_\alpha \qquad (2.1)$$

where $\alpha = {}^4_2He$

and Q_α is the total kinetic energy change of the reaction.

$$Q_\alpha = T_y + T_\alpha = (1 + \frac{m_\alpha}{m_y}) \, T_\alpha \qquad (2.2)$$

if nucleus X is initially at rest (see sec. 3.5).

    From the conservation of energy it follows that:

$$E = m_x c^2 = m_y c^2 + m_\alpha c^2 + T_y + T_\alpha \qquad (2.3)$$

and thus Q can also be written

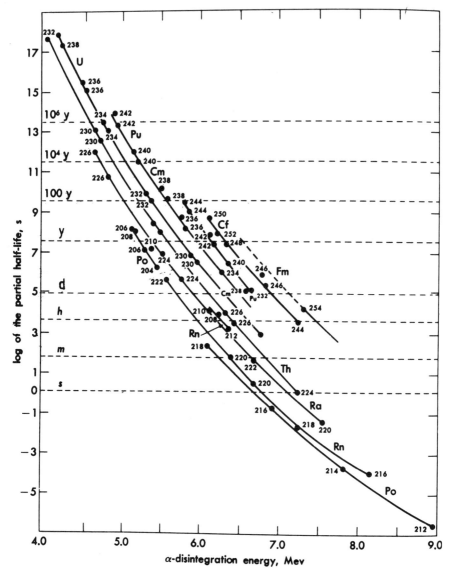

Fig. 2-4   The variation of half-life with α-disintegration for
           even Z.

$$Q = (m_x - m_y - m_\alpha)c^2 \qquad (2.4)$$

Branching in α-decay occurs when a nucleus decays to several levels of the daughter nucleus.

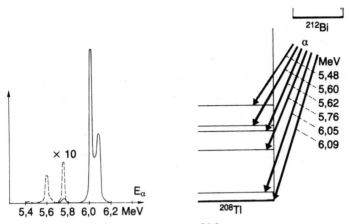

Fig. 2-5   An α-line spectrum from ^{212}Bi and the excitation
energies of ^{208}Tl.

In fig. 2-5 the α-particle spectrum from ^{212}Bi is shown.
When the α-particle is emitted with the highest possible energy,
$T_{\alpha o}$ = 6.09 MeV, the daughter nucleus is left in its lowest possible state, the ground state. When, however, the α-particle is
emitted with a lower energy, $T_{\alpha i}$, the daughter nucleus is left
with more energy, that is, in an excited state. The excitation
energy for a given level is simply:

$$E_x = (T_{\alpha o} - T_{\alpha i})(1 + \frac{m_\alpha}{m_y}) \hspace{3cm} (2.5)$$

2.3 Beta decay

In β-decay, three different kinds of disintegration are
distinguished; β⁻-decay, β⁺-decay, and electron capture. The
three processes are shown in fig. 1-4 where the stable nuclei
in a series of isobars represent the bottom of an energy valley.
Neutron-rich nuclides will decay by β⁻-emission along their isobar lines toward the bottom of the valley, and proton-rich nuclides will decay similarly with β⁺-emission or electron capture.
In the β-decay process, a negative (e⁻) or positive electron
(e⁺) is emitted from the nucleus, or an orbital electron is
captured by the nucleus. In all cases, an antineutrino or an

neutrino is simultaneously emitted. The electron and positron
are antiparticles, they can annihilate each other when brought
in close contact. In the same way, there are two kinds of neu-
trinos involved in β-decay, the neutrino (ν_e) emitted in β^+
(positron) decay and the antineutrino ($\bar{\nu}_e$) emitted in β^- (nega-
tron) decay. All four belong to a class of elementary particles
called leptons and have half-integer internal spin. The neutri-
nos are neutral and have negligible rest mass but, just like
photons, their energies can still be very large. In contrast
to α-spectra in which the α-particles show discrete energies,
β-spectra do not have this discrete character, but are instead
continuous. The existence of neutrino explains this continuous
character and ensures the conservation of angular momentum in
the decay.

2.3.1 β^--DECAY

In β^--decay, the nucleus $^A_Z X$ emits an electron (e^-) and an
antineutrino ($\bar{\nu}_e$). The residual nucleus $^A_{Z+1} Y$ has the same mass
number but the number of protons Z is increased and the number
of neutrons N decreased by one unit.

The decay can be written:

$$^A_Z X \rightarrow\; ^A_{Z+1} Y + e^- + \bar{\nu}_e + Q_{\beta^-} \tag{2.6}$$

The antineutrino has no mass and hence the Q-value for the
decay is (see sec. 3.5)

$$Q_{\beta^-} = m_x c^2 - m_y c^2 - m_e c^2 = T_y + T_{\beta^-} + T_{\bar{\nu}} \tag{2.7}$$

written with nuclear masses or kinetic energies.

It is assumed that the nucleus X is at rest. The equation
can also be written with atomic masses

$$Q_{\beta^-} = (M(X) - M(Y))c^2 \tag{2.8}$$

A typical energy diagram for a β^--decay is shown in fig. 2-6 a.
The recoil energy of the Y nucleus (T_y) can be neglected and
the available energy Q_{β^-} is shared between the electron and

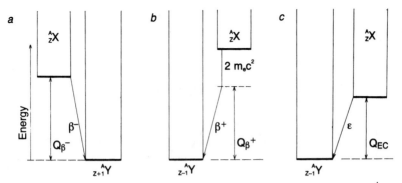

Fig. 2-6   Typical energy diagrams for a) β^--decay, b) β^+-decay
and c) electron capture.

the antineutrino. The distribution of the kinetic energy of the
electron is continuous with maximum energy

$$T_{\beta^-,o} = Q_{\beta^-} = (M(X) - M(Y))\ c^2 \tag{2.9}$$

reached when the antineutrino energy is zero (fig. 2-7).

2.3.2 β^+-DECAY

In β^+-decay, the nucleus $_Z^A X$ emits a positron (e^+) and a
neutrino (ν_e). The residual nucleus $_{Z-1}^A Y$ has the same mass number but the number of protons is decreased and the number of
neutrons increased by one unit:

$$_Z^A X \rightarrow\ _{Z-1}^A Y + e^+ + \nu_e \tag{2.10}$$

The Q-value for the decay is

$$Q_{\beta^+} = m_x c^2 - m_y c^2 - m_e c^2 \tag{2.11}$$

or written with atomic masses

$$Q_{\beta^+} = (M(X) - M(Y) - 2m_e)\ c^2 \tag{2.12}$$

since Y has one electron less than X. Since $m_e c^2$ = 0.511 MeV,
the maximum kinetic energy of the positron $T_{\beta^+,o}$ is

$$T_{\beta^+,o} = Q_{\beta^+} = (M(X) - M(Y))\ c^2 - 1.022\ \text{MeV} \tag{2.13}$$

45

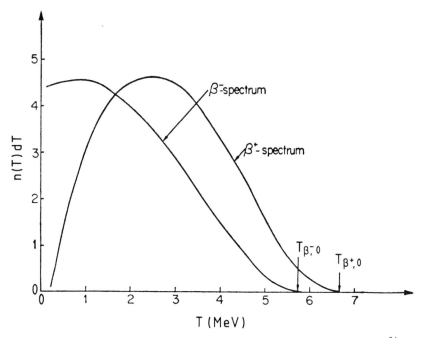

Fig. 2-7   Electron and positron spectra of the nuclide ^{64}Cu.

which occurs when the neutrino energy is zero (fig. 2-6b). The energy spectrum of the emitted positrons is continuous and resembles that of the electrons emitted in β^--decay. However, the Coulomb force is experienced in different ways by electrons and positrons and, for the latter, fewer low-energy particles are found and there is an upward shift in the mean energy. This is demonstrated in fig. 2-7, where the e^- and e^+-spectrum from ^{64}Cu-decay are shown. In this decay Q_{β^-} = 0.573 MeV and Q_{β^+} = 0.656 MeV.

Since positrons are antiparticles to electrons they can be annihilated by electrons. The possibility of annihilation is greatest when the positrons are at rest. Both the positron and the electron disappear and an energy of $2m_e c^2$ is released as electromagnetic radiation. This is the process inverse to pair production (sec. 5.5.4). Conservation of momentum requires that the annihilation energy appears as two oppositely-directed quanta each of energy $m_e c^2$ = 0.511 MeV, the so-called "annihilation radiation".

2.3.3 ELECTRON CAPTURE

In this process, the nucleus captures one of the electrons from the innermost electron shells and simultaneously emits a neutrino. In doing so, a proton changes to a neutron and thus the Z-number decreases and the N-number increases by one unit. Usually a K-electron is captured and the K-capture process can be written

$$_Z^A X + e^-_K \rightarrow \ _{Z-1}^A Y + \nu_e, \qquad (2.14)$$

$$Q_{EC} = (M(X) - M(Y))\ c^2 - E_{BE}^Y(K) = T_\nu + T_Y \simeq T_\nu \qquad (2.15)$$

where $E_{BE}^Y(K)$ is the binding energy of the K-electron in the Y-atom (fig. 2-6c).

The neutrino receives all the available energy in the decay, except for the negligible portion given to the recoiling nucleus. In heavy elements, the binding energy for a K-electron is over 100 keV, but the L- and M-electrons are bound by only about 20 and 5 keV, respectively. Thus, there is always a possibility for a proton-rich nuclide to decay to a stable one. If the excess energy is less than $2\ m_e c^2 = 1.022$ MeV, electron capture is the only mode of decay but at higher energies β^+-decay and electron capture compete with each other. In fig. 2-8, the branching ratio between K-capture and β^+-decay is given as a function of Q_{EC} or $Q_{\beta^+} + 2\ m_e c^2$.

Experimental study of the electron-capture process is not quite so convenient as β^- and β^+—spectroscopy. The neutrino is practically unobservable, so that the only way to detect electron capture in practice is to observe the atomic process following the disappearance of a K- (or an L-) electron. For K-capture, one can observe the K-lines in the X-ray spectrum of the nuclide produced, and for L-capture the L-lines. It is also usually possible to observe Auger electrons following K-capture. An Auger electron is an atomic electron which receives enough kinetic energy to be ejected, usually from the L-shell when another electron falls from the same shell to fill a vacancy in the K-shell. The process competes with photon (X-ray) emission and may be more convenient to observe experimentally.

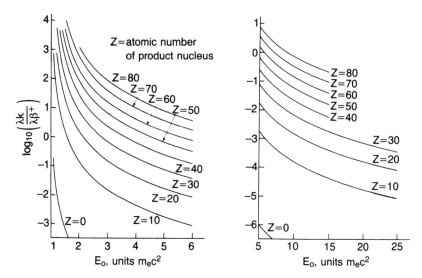

Fig. 2-8   Logarithmic plot of the ratio of decay constants
for allowed K-capture and positron emission versus
disintegration energy where E_0 is either Q_{EC} or
$Q_{\beta^+} + 2m_e c^2$. E. Feenberg and G. Trigg. Rev. Mod. Phys.
__22__, 539 (1950).

2.3.4 COMPARATIVE LIFE-TIME

The probability for a nucleus to decay by electron or positron emission is strongly dependent on the energy of the emitted
particle and is related to the state density of the electron-
antineutrino or positron-neutrino pair. The expression for this
state density is known. After correction for Coulomb interaction, the expression is integrated over all particle energies
and a factor, $f(Z, T_0)$, which contains the Z and T_0 dependence
of the probability is obtained. If the decay half-life $t_{1/2}(s)$,
which is proportional to the inverse of the decay probability
(eq. 2.22), is multiplied by the f-factor, a comparative life-
time is obtained. The value of $f \cdot t_{1/2}$ or $\log(f \cdot t_{1/2})$ tells us
if the β-transition is super-allowed (favoured), allowed or forbidden (hindered) (table 2-1). In the favoured case, the orbit
of the nucleon is practically unaltered in spite of the neutron ↔ proton transition.

Table 2-1

Log (f t$_{\frac{1}{2}}$)-values for different transitions.

Transition	Parity change	Change in nuclear spin	Approximate log (f t$_{\frac{1}{2}}$)
Super-allowed	no	0, ± 1	3-4
Allowed	no	0, ± 1	4-5
Non-allowed	yes or no	0, ± 1 ± 2 ...	> 6

2.4 Electromagnetic decay

In many α- and β-transitions, the daughter nuclide is formed in an excited state below the nucleon separation energy.
As the nuclide cannot deexcite by emitting a nucleon, it has
to decay by electromagnetic transitions or by internal conversion. Electromagnetic radiations are called γ-rays if they are
of nuclear origin. According to the particle-wave duality in
nature, electromagnetic radiation may appear as particle-like
quanta and these are then called photons. In a nuclear reaction,
the situation is often the same, viz., that a daughter nucleus
is formed with an excitation energy below the nucleon separation
energy. In the subsequent γ-decay, one or more photons are emitted in cascade (fig. 2-9).

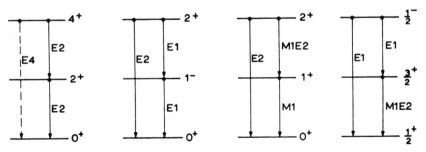

Fig. 2-9   γ-decay schemes with expected multipole transitions
noted.

In decay by internal conversion, the energy excess of the nuclide is transfered to an orbital electron which is emitted monoenergetically.

In the case of γ-ray emission, the γ-energy is given by

$$E_\gamma = E_i - E_f - T_r \qquad (2.16)$$

and $T_r = \dfrac{E_\gamma}{2m_r c^2}$ (from momentum conservation)

E_i = initial state energy

E_f = final state energy

T_r = recoil energy and

m_r = mass of recoil nuclide

The recoil energy of the nuclide T_r is normally negligible. In the case of internal conversion the kinetic energy of the emitted electron is

$$T_e = E_i - E_f - B_e - T_r \qquad (2.17)$$

and $T_r = \dfrac{m_e}{m_r} T_e$

where B_e is the binding energy of the emitted electron and m_e its mass. The recoil energy can still be neglected.

After emission of the conversion electron, the other orbital electrons are rearranged giving rise to X-rays which are characteristics of the decaying nuclide.

Here is a difference to the X-ray following an electron capture process: in that case, the emitted X-ray is characteristic of the nuclide produced.

The mass number and the differences in energy and spin between the initial and final states determine the speed of the transition and hence the life-time of the excited state (fig.2-10). Different types of γ-transitions can be defined according to the changes in the spins and parities of the nuclear states (table 2-2).

For low γ-energies and large changes in spin the life-times of the excited states are measurable. The states are then called isomeric states.

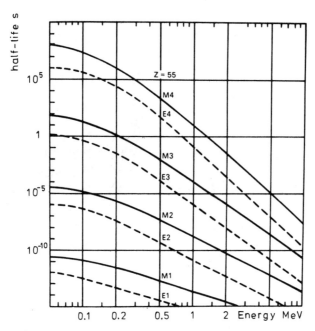

Fig. 2-10   Lifetime-energy relations for γ-radiation of different multipole types. The calculations are made for
Z = 55.

Table 2-2
Changes in spin and parity of nuclear states in E-M transitions.

Angular momentum change ΔL	1	2	3	4	5
Electric multipole	E1	E2	E3	E4	E5
Parity change	Yes	No	Yes	No	Yes
Magnetic multipole	M1	M2	M3	M4	M5
Parity change	No	Yes	No	Yes	No

Decay by internal conversion is favoured by increasing the
complexity of the γ-transition. The ratio between the probabilities of internal conversion and γ-transitions is called the
conversion coefficient α,

$$\alpha = \frac{\lambda_e}{\lambda_\gamma} \qquad (2.18)$$

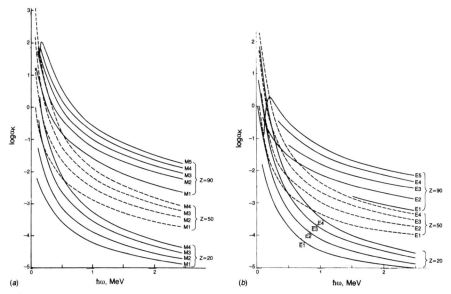

Fig. 2-11   K-shell conversion coefficient α_K for electric and
magnetic multipole transition with Z = 20, 50 and
90. M.A. Preston, Physics of the Nucleus (Addison-
Wesley) p. 307, 308 (1962).

This coefficient can be split into different terms

$$\alpha = \alpha_K + \alpha_{L_1} + \alpha_{L_2} + \alpha_{L_3} \tag{2.19}$$

depending on the position of the electron emitted prior to conversion. The branching ratio between internal conversion and
γ-decay is shown in fig. 2-11 for different multipolarities and
charge number.

In fig. 2-12 a β^--spectrum with an overlapping conversion
line from the 14 h isomeric state in ^{69}Zn is shown.

2.5 Fission

If a drop of water acquires electric charges, the electrostatic repulsion of the charges acts in opposition to the
surface force holding the drop together. If the acquired charge
is sufficiently large, the drop can split into two parts. In

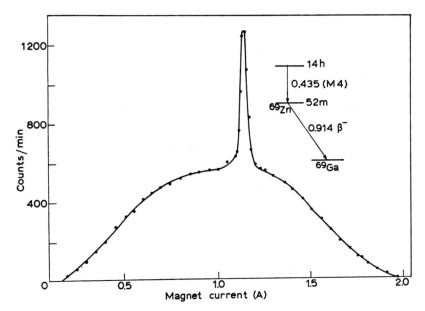

Fig. 2-12   A conversion electron line superimposed on a β-spect-
rum in the electron spectrum from ^{69}Zn decay.
K. Siegbahn (ed.) Alpha-, Beta- and Gamma-Ray Spectroscopy I (North-Holland)p. 487 (1965).

the same way, if the electric charge is large, a nucleus can
split. The phenomenon is called spontaneous fission and is observed for nuclei heavier than thorium. Thorium and uranium are
the heaviest nuclei found in nature and the half-lives for
spontaneous fission of these nuclei are extremely long, i.e.,
of the order of 10^{16} years or more.

In fig. 2-13, the potential energies of three different nuclei are illustrated as a functin of the nuclear deformation or
the fragment separation. In the case of a hypothetical very
heavy nucleus (A $\cong$ 300), the nucleus splits immediately after
its formation since no barrier for fission exists. For an uranium nucleus (A $\simeq$ 236), there exists a Coulomb barrier for fission with an height of about 6 MeV and spontanous fission can
occur as a tunnelling process through the barrier, similar to
the case of nuclear α-emission (sec. 2.2). For lighter nuclei,
the fission barrier increases and the nucleus must be excited
strongly if fission is to occur.

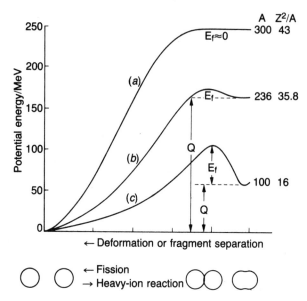

Fig. 2-13 Schematic representation of fission barriers of mass
number A ≈ 100, 236 and 300.
W.E. Burcham, Elements of Nuclear Physics (Longman)
p. 355 (1979).

When a thermal neutron is captured in ^{235}U, the ^{236}U nucleus formed is excited to about 6.5 MeV corresponding to the neutron binding energy. Fission can thus occur by passing above rather than through the fission barrier. If, on the other hand, a thermal neutron is captured in ^{238}U, the excitation energy of ^{239}U is not enough for this to happen. The neutron must bring with it a kinetic energy of almost 2 MeV in order to be able to split the ^{239}U nucleus formed.

The difference in excitation energy in these two cases is mainly due to the pairing tendency of nucleons. This effect is expressed in the fifth term of the mass formula (eq. 1.9). **The captured neutron is more strongly bound in the even-even nucleus** ^{236}U than in the odd-even nucleus ^{239}U.

The fission process may be represented schematically as in fig. 2-14.

When a thermal neutron has been captured in a ^{235}U nucleus, some time elapses before the nucleus takes the shape most suitable for passing over the fission barrier. This elapsed time

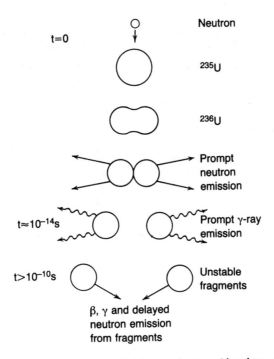

Fig. 2-14   Schematic representation of the fission process in
            uranium, showing emission of neutrons and γ-radiation
            and final decay of fission products. The time scale
            gives orders of magnitude only.
            W.E. Burcham, Elements of Nuclear Physics (Longman)
            p. 351 (1979).

is of the order of 10^{-15}s so that neutron and γ-emission are
competing processes of deexcitation. When the nucleus has passed
the barrier peak, it very quickly reaches the scission point
and splits into two fragments. These fragments have different
masses (fig. 2-15), which indicates that the nucleus is pear-
shaped when passing the barrier.

The primary fragments are excited and emit first prompt
neutrons and then prompt γ-rays. When the fragments have reached
their ground states, they are still unstable and decay further
by β- and γ-emission.

Nuclide yields in the fission process after the prompt
deexcitations but before the β-decays are shown in fig. 2-16
for the thermal neutron fission of ^{235}U. Two peaks are obser-

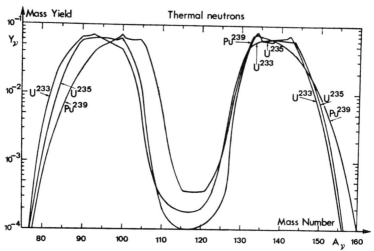

Fig. 2-15   Mass yield curves for the thermal neutron fission ^{233}U, ^{235}U, ^{239}Pu.

M. Barbier, Induced Radioactivity (North-Holland) p. 139 (1969).

ved in the region of maximum production with a steep descent of the yields on both sides, one towards the β-stability line, the other away from it. In the two highest regions the individual yields reach 3,6% of the total fission yield of ^{235}U.

The left parts of figs. 2-17a and b show how the fission fragment nuclides in an even A isobar chain from the lighter peak and in an odd A isobar chain from the heavier peak decay towards stability. (Compare this figure with fig. 1-4). In both cases a long-lived nuclide is formed, ^{90}Sr (28.8y) and ^{137}Cs (30.17y). Both are of importance for the radio-environment of mankind.

In some cases, the β-transition leaves the daughter nucleus in states above the neutron binding energy and a delayed neutron can be emitted. Examples are shown in both fig. 2-17a (^{90}Br) and 2-17b (^{137}Te, ^{137}I). Even if the neutron is emitted almost immediately after such a state is formed (10^{-15}s), the neutron time of production is determined by the life-time of the precursor. The nuclide ^{89}Kr is thus produced in three ways: directly as a fission fragment, by the β-decay of ^{89}Br (4.4 s) or by n-emission from an excited state of ^{90}Kr, which was formed by

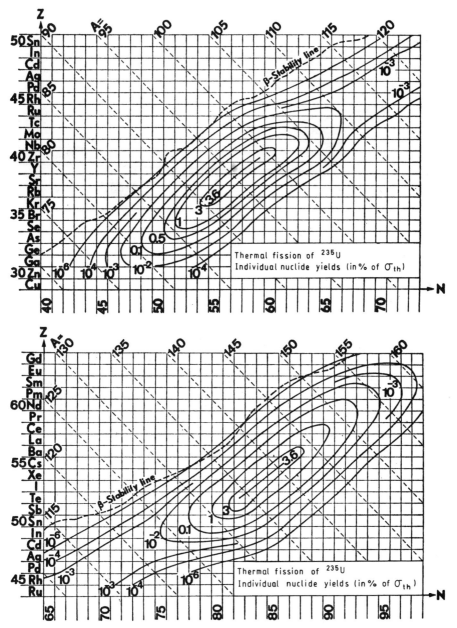

Fig. 2-16 Individual nuclide yields of fission products from
the thermal fission of ^{235}U, before β-decay of these fission products has set in, expressed in percent
of the thermal neutron fission cross section for
(a) Z = 30-50, (b) Z = 50-65.
M. Barbier, Induced Radioactivity (North-Holland)
p. 143 (1969).

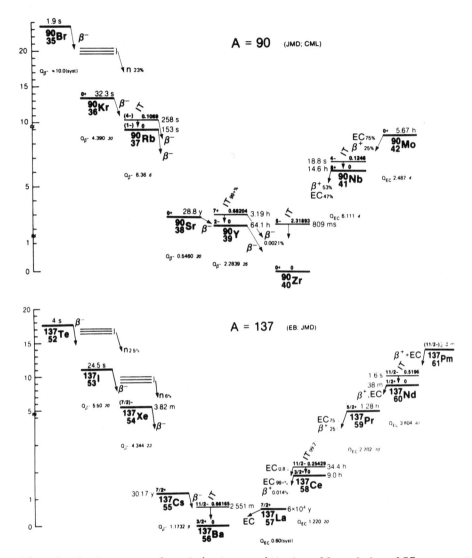

Fig. 2-17   Decay chain of isobars with A = 90 and A = 137.
M. Lederer and V. Shirley (eds.). Table of Isotopes
7th ed. (Wiley-Interscience 1978).

the β-decay of ^{90}Br (1.9 s). Delayed neutron emitters are of
great importance in nuclear reactor kinematics. They can also
be produced in nuclear reactions. Since delayed neutrons are
often easy to detect, such nuclides can be useful in nuclear
chemical analysis (sec.12.3.3).

Table 2-3

Energy released in the fission of ^{235}U (average values).

Kinetic energy of fission fragments	165 MeV
Prompt and delayed neutrons	5
Prompt γ-radiation	7
Delayed β-particles, associated antineutrinos and γ-radiation from radioactive decay	25
	202 MeV

A rather large amount of energy is released in the fission process, mainly as kinetic energy of the fission fragments but also to a smaller extent in the decay of the fragments. In table 2-3, the average energy released in the fission of ^{235}U is shown.

2.6 Decay schemes

Most of the characteristics of nuclides can be found in various tables of nuclides[2.1]. As an example, we will discuss the decay of isobars with mass number A = 34 (fig. 2-18). The ground state of ^{34}P has spin and parity 1$^+$. Its half-life is 12.4 s and it decays by β$^-$-emission to five or six levels in ^{34}S. The decay percentages in the five branches are quoted as 0.3%, 0.11%, 0.04%, 15% and 85%.

The maximum β$^-$-energy in the decay of ^{34}P to the ground state of ^{34}S is Q_β^- = 5.38 MeV. The small italic number after the Q-value gives the uncertainty in the last number. All the main β$^-$-decays are allowed, since the log ($f \cdot t_{1/2}$) values given adjacent to the percentage values are less than 6, which all fall in the allowed transition region. This is in agreement with the selection rule of ΔJ = 0, ± 1 and no parity shift (table 2-1).

The ground state of ^{34}S is 0$^+$ in agreement with the rule

59

A = 34 (EB; JMD)

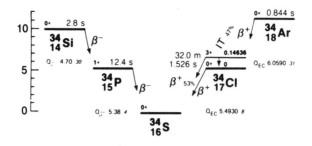

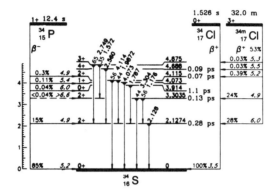

Fig. 2-18 Decay of isobars with A = 34.
M. Lederer and V. Shirley (eds.). Table of Isotopes
7th ed. (Wiley-Interscience 1978).

that in a nucleus with even N and Z numbers all the nucleons
are paired. The lower excited states in ^{34}S decay by E2 γ-emission, since there is a shift in nuclear spin of ΔJ = 2 and no
parity change (table 2-2). The transitions are fast as expected
(fig. 2-10) and no isomeric states are observed in ^{34}S. In ^{34}Cl,
both the ground state and the first excited state are long-lived.
The ground state decays by β⁺-emission to the ground state of
^{34}S, the transition being super-allowed. The maximum energy of
the β⁺-particle is Q_{EC} - 1.02 MeV = 4.47 MeV. The electron binding energy is negligible. Electron capture is too weak to have
been noted. The isomeric state at 0.146 MeV is more long-lived
than the ground state. It decays either by isomeric transition
IT, i.e., by γ-emission and internal conversion, or by β⁺-disin-
tegration. Incidentally, both types of transitions, have almost

the same probability. In the former case, the decaying γ-transition is of type M3 and the transition is slow, as is evident from the long half-life. The internal conversion constant α is not given in the scheme but has been measured to be 0.10. In the latter case, the main β^+-transitions are to the first two excited states of ^{34}S. No branching to the ^{34}S ground state is observed. The transition is strongly forbidden, since the change in spin is too great.

2.7 Radioactive-decay law

The decay probability of an unstable nucleus is at every moment constant which means that the nucleus has no memory of when it was formed. If N radioactive nuclei are present at time t and if no additional nuclei are introduced into the sample, the number of nuclei decaying in time inverval dt can be written.

$$dN = - \lambda N dt \qquad (2.20a)$$

$$d(\ln N) = - \lambda dt \qquad (2.20b)$$

where λ is the disintegration constant. Integrating this equation leads to

$$N(t) = N_0 e^{-\lambda t} \qquad (2.21)$$

where N_0 is the number of radioactive nuclei at t = 0.
The half-life $t_{1/2}$ is found by setting $N = \frac{1}{2} N_0$ at $t = t_{1/2}$

$$t_{1/2} = \frac{\ln 2}{\lambda} \qquad (2.22)$$

The disintegration rate is called the activity A

$$A = \lambda N \qquad (2.23)$$

It is generally expressed in units of bequerel (Bq)

1 Bq = 1 disintegration per second

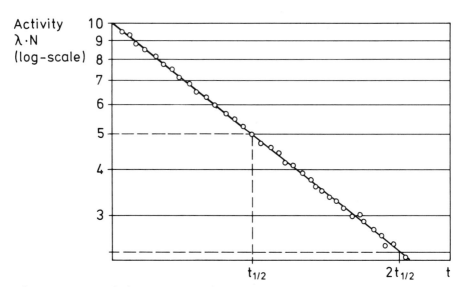

Fig. 2-19   Activity versus time.

An earlier unit often used is Curie (Ci)

$$1 \text{ Ci} = 3.7 \cdot 10^{10} \text{Bq} \qquad\qquad (2.24\text{b})$$

In the cases of decay discussed so far no new radioactive nuclei of the nuclide under study were produced in the sample. We can thus write the disintegration rate as the change of the number of decaying nuclei

$$A = \lambda N = - \frac{dN}{dt}$$

In the more general case new nuclei are produced and we define the growth $\frac{dN}{dt}$ as

$$\frac{dN}{dt} = \text{growth} = \text{production rate} - \text{disintegration rate} \qquad (2.25)$$

The activity or disintegration rate is always proportional to the counting rate in a detector recording the decay. If the activity is plotted versus time on a semilog scale (fig. 2-19), a straight line with a slope $-\lambda$ is obtained. The half-life can conveniently be determined from the plot. The average life ex-

pectancy of the radioactive nuclei is found by solving the mean
value equation

$$\tau = \frac{\int\limits_{t=0}^{t=\infty} t\,dN}{\int\limits_{t=0}^{t=\infty} dN} = \lambda \int\limits_{0}^{\infty} t\, e^{-\lambda t}\, dt = \frac{1}{\lambda} \qquad (2.26)$$

We see that the half-life is shorter than the average life by
the factor ln 2=0.693.

2.8 Counting rate

In practical work with radioactive materials, the number
of nuclei N is not directly evaluated and even the disintegration
rate is usually not measured absolutely. The usual procedure
is to determine a quantity proportional to λN which may be the
counting rate c in a given detector with an over all counting
efficiency ε where

$$c = - \varepsilon \frac{dN}{dt} = \varepsilon \lambda N \qquad (2.27a)$$

The efficiency ε can be estimated as a product of the following
parameters

$$\varepsilon = i\, g\, f_s\, f_a\, f_w\, b \qquad (2.27b)$$

where i = the detection probability for a particle entering
the detector to give a detectable response,

g = the geometrical factor i.e. the solid angle as a
fraction of 4π that the detector covers seen from
the source,

f_s = the probability for the particle to leave the source
without being absorbed or scattered in the source
material,

f_a = the probability for the particle to pass through the medium between the source and the detector without being absorbed or scattered,

f_w = the probability for the particle to pass through the detector window without being absorbeed or back-scattered and

b = the probability for a particle leaving the source in a direction outside the solid angle of detection but being scattered into it from the material that surrounds the source and the detector thus giving extra counts (the build-up factor).

2.9 Mixture of independent activities

It often occurs that two or more radioactive species are mixed together. The observed total counting rate is then the sum of the independent activities.

$$c = c_1 + c_2 = \varepsilon_1 \lambda_1 N_1 + \varepsilon_2 \lambda_2 N_2 \qquad (2.29)$$

The detection efficiences ε_1 and ε_2 are not necessarily the same and can be very different in magnitude. Consider, e.g., a mixture of ^{64}Cu with $t_{1/2}$ = 12.8 h and ^{61}Cu with $t_{1/2}$ = 3.3 h. The activity of such a sample versus time is plotted in fig. 2-20. Since ^{61}Cu decays more rapidly than ^{64}Cu, we expect that after a sufficiently long time the decay curve will asymptotically approach the 12.8 h line in a semilog plot.

We extrapolate this straight line backwards and take the difference between the curve and the line. We then plot this difference with the same scale and obtain the dashed straight line which represents the 3.3 h period. The intercepts of both straight lines on the vertical axis give the initial counting rate for each component. This method can be extended to mixtures with more than two periods.

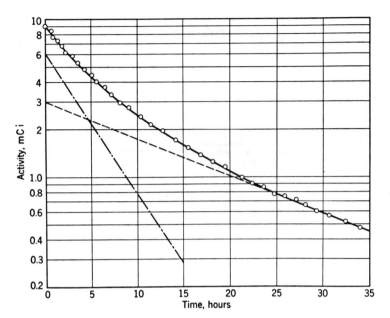

Fig. 2-20   Hypothetical decay curve for a sample containing ^{64}Cu (12.8 h) and ^{61}Cu (3.4 h).

2.10 Branching decay

We now consider a radionucleus which can decay in different ways, forming different nuclei. As an example we can choose the decay of ^{64}Cu (fig. 2-21).

The ^{64}Cu nuclide decays in three different ways;

β^- 40% to ^{64}Zn, β^+ 19% to ^{64}Ni, and EC 41% to ^{64}Ni.

Each decay is characterized by its partial disintegration constant λ_i and the total disintegration constant is given by

$$\lambda = \sum_i \lambda_i \qquad\qquad (2.30)$$

The disintegration rate can be written

$$\frac{dN}{dt} = \sum_i \frac{dN_i}{dt} = -\sum_i \lambda_i N,$$

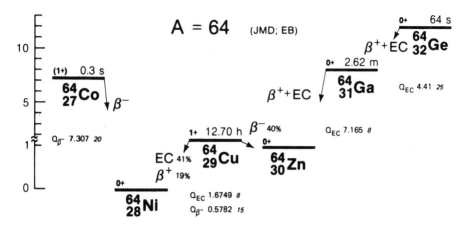

Fig. 2-21 Decay of isobars with A = 64.
          M. Lederer and V. Shirley (eds.). Table of Isotopes
          7th ed. (Wiley-Interscience 1978).

which gives

$$N(t) = N_o \, e^{-\sum\limits_{i} \lambda_i t} = N_o \, e^{-\lambda t} \tag{2.31}$$

For the ^{64}Cu nucleus, the disintegration constant λ is $1.52 \cdot 10^{-5} \, s^{-1}$, the partial disintegration constants being

$$\lambda_{\beta^-} = 0.61 \cdot 10^{-5} \, s^{-1}$$

$$\lambda_{\beta^+} = 0.29 \cdot 10^{-5} \, s^{-1}$$

$$\lambda_{EC} = 0.62 \cdot 10^{-5} \, s^{-1}$$

Independently of which kind of decay the detector records, we always get the same half-life $t_{1/2}$, since

$$\frac{dN_i}{dt} = -\lambda_i N = -\lambda_i N_o \, e^{-\lambda t} \tag{2.32}$$

Partial disintegration constants are of interest, however, as they can often be predicted theoretically.

2.11 Growth of radioactive daughters

Frequently a radionucleus A decays into a nucleus B which is also radioactive. Let us assume that B decays to a nucleus which is stable:

$$(A) \xrightarrow{\lambda_a} (B) \xrightarrow{\lambda_b} (C) \text{ (stable)}$$

The decay of A, the growth and decay of B and the growth of C are described by the following set of connected differential equations:

$$\frac{dN_a}{dt} = -\lambda_a N_a \tag{2.33}$$

(giving $N_a = N_{ao} e^{-\lambda_a t}$, see eq. 2.21)

$$\frac{dN_b}{dt} = \lambda_a N_a - \lambda_b N_b \tag{2.34}$$

(nucleus B is formed at the rate of decay of A but simultaneously decays at a rate $- \lambda_b N_b$) and

$$\frac{dN_c}{dt} = \lambda_b N_b \tag{2.35}$$

From eqs. 2.33 and 2.34, we obtain the following equation:

$$\frac{dN_b}{dt} + \lambda_b N_b = \lambda_a N_{ao} e^{-\lambda_a t} \tag{2.36}$$

We observe that the activity $A_b = \lambda_b N_b$ in this case is not equivalent to $-\frac{dN_b}{dt}$ (see eq. 2.25).

The eq. 2.36 can be solved. We multiply with $e^{\lambda_b t}$ on both sides of the equation and get

$$e^{\lambda_b t} \frac{dN_b}{dt} + e^{\lambda_b t} \lambda_b N_b = e^{\lambda_b t} \lambda_a N_{ao} e^{-\lambda_a t}$$

$$\frac{d}{dt} (e^{\lambda_b t} N_b) = \lambda_a N_{ao} e^{(\lambda_b - \lambda_a)t}$$

Hence

$$e^{\lambda_b t} N_b = \int^t \lambda_a N_{ao} e^{(\lambda_b - \lambda_a)t} \, dt + a$$

where $\underline{a}$ is an arbitrary constant.

We get

$$e^{\lambda_b t} N_b = \frac{\lambda_a N_{ao}}{\lambda_b - \lambda_a} e^{(\lambda_b - \lambda_a)t} + a$$

or

$$N_b = \frac{\lambda_a N_{ao}}{\lambda_b - \lambda_a} e^{-\lambda_a t} + a e^{-\lambda_b t}$$

We assume that $N_b = 0$ for $t = 0$ and the solution is then

$$N_b(t) = \frac{\lambda_a N_{ao}}{\lambda_b - \lambda_a} \left[e^{-\lambda_a t} - e^{-\lambda_b t} \right] \tag{2.37}$$

Finally, the growth of nucleus C is given by

$$\frac{dN_c}{dt} = \lambda_b N_b = \frac{\lambda_a \lambda_b N_{ao}}{\lambda_b - \lambda_a} \left[e^{-\lambda_a t} - e^{-\lambda_b t} \right] \tag{2.38}$$

and, if $N_c = 0$ for $t = 0$, we get

$$N_c(t) = \frac{\lambda_a \lambda_b}{\lambda_b - \lambda_a} N_{ao} \left[\frac{1 - e^{-\lambda_a t}}{\lambda_a} - \frac{1 - e^{-\lambda_b t}}{\lambda_b} \right] \tag{2.39}$$

2.11.2   TIME OF MAXIMUM ACTIVITY

The activity of species B($= \lambda_b N_b$) must pass through a maximum since N_b is assumed to be zero at $t = 0$ and must be zero at $t = \infty$. The time of maximum activity t_{max} occurs when $dN_b/dt = 0$. From eq. 2.36, we get

$$\frac{dN_b}{dt} = \lambda_a N_{ao} e^{-\lambda_a t_{max}} - \lambda_b N_b = 0 \qquad (2.40)$$

Using eq. 2.37, this can be written

$$e^{(\lambda_b - \lambda_a) t_{max}} = \frac{\lambda_b}{\lambda_a}$$

or

$$t_{max} = \frac{\ln(\frac{\lambda_b}{\lambda_a})}{\lambda_b - \lambda_a} \qquad (2.41)$$

2.11.3   SPECIAL CASES OF CHAIN DECAY

Two particular cases of chain decay can be distinguished in eq. 2.37 depending on whether the parent nucleus A or the daughter nucleus B has the shorter half-life.

In the first case, we have $t_{1/2}(A) < t_{1/2}(B)$ or $\lambda_A > \lambda_b$. Eq. 2.37 is then simplified when t becomes sufficiently large, i.e. for $t >> (\lambda_a - \lambda_b)^{-1}$

$$N_b(t) = \frac{\lambda_a}{\lambda_a - \lambda_b} N_{ao} e^{-\lambda_b t} \quad \text{(t large)} \qquad (2.42)$$

and the activity is

$$A(t) = \lambda_b N_b(t) = \frac{\lambda_a \lambda_b}{\lambda_a - \lambda_b} N_{ao} e^{-\lambda_b t} \quad \text{(t large)} \qquad (2.43)$$

It follows as anticipated that the daughter B decays with its own half-life (fig. 2-22). For smaller values of t, the general equation must be used.

If the straight portion of the total counting rate curve for large $\underline{t}$ in fig. 2-22 is extrapolated to t→0, one finds (eq.2.27a)

$$"c_b^o" = \varepsilon_b \frac{\lambda_a \lambda_b}{\lambda_a - \lambda_b} N_{ao}$$

If the first part of the total counting rate curve also is extrapolated to t→0 one gets

$$c_a^o = \varepsilon_a \lambda_a N_{ao}$$

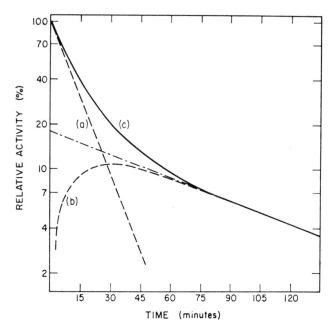

Fig. 2-22   Growth and decay curves for the radionuclides 8.7 min.
^{49}Ca → 57.5 min ^{49}Sc: (a) the decay curve for an ini-
tially pure source of ^{49}Ca; (b) the growth curve for
^{49}Sc in the source; (c) the observed total activity
of the source.
P. Kruger, Principles of Activation Analysis (Wiley-
Interscience)p. 99 (1971).

From these two equations, it follows that:

$$\frac{c_a^o}{"c_b^o"} = \frac{\varepsilon_a}{\varepsilon_b} \frac{\lambda_a - \lambda_b}{\lambda_b} \tag{2.44}$$

If $\lambda_a \gg \lambda_b$ this reduces to

$$\frac{c_a^o}{"c_b^o"} = \frac{\varepsilon_a \lambda_a}{\varepsilon_b \lambda_b} \tag{2.45}$$

If we now instead discuss the second case, we have
$t_{1/2}(A) > t_{1/2}(B)$ or $\lambda_a < \lambda_b$. When t becomes sufficiently large, i.e. for $t \gg (\lambda_a - \lambda_b)^{-1}$, eq. 2.37 now simplifies to

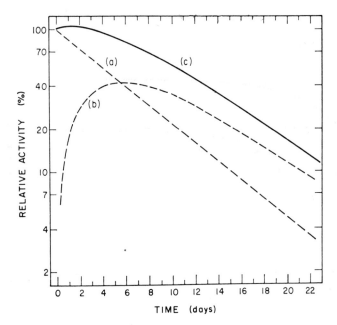

Fig. 2-23   Growth and decay curves for the radionuclides 4.5 d
^{47}Ca $\rightarrow$ 3.4 d^{47}Sc: (a) the decay curve for an initial-
ly pure source of ^{47}Ca, (b) the growth curve for
^{47}Sc in the source; (c) the observed total activity
of the source.
P. Kruger, Principles of Activation Analysis. (Wiley-
Interscience)p. 97 (1971).

$$N_b(t) = \frac{\lambda_a}{\lambda_b - \lambda_a} \ N_{ao} \ e^{-\lambda_a t} = \frac{\lambda_a}{\lambda_b - \lambda_a} \ N_a(t) \quad \text{(t large) (2.46)}$$

The daughter nucleus B thus decays with the half-life of
the parent nucleus A.   The ratio of the activities of B and
A is constant.

$$\frac{\lambda_b \ N_b}{\lambda_a \ N_a} = \frac{\lambda_b}{\lambda_b - \lambda_a} \qquad \text{(t large)} \qquad (2.47)$$

and the case is called <u>transient equilibrium</u> (fig. 2-23). If
$t_{1/2}(A) \gg t_{1/2}(B)$ or $\lambda_a \ll \lambda_b$, the two activities are the same for large values of t:

71

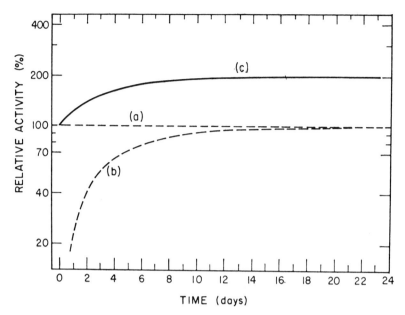

Fig. 2-24   Growth and decay curves for the radionuclides 27 y
^{90}Sr → 64 h^{90}Y: (a) the decay curve for an initially
pure source of ^{90}Sr; (b) the growth curve for ^{90}Y
in the source; (c) the observed total activity of
the source.
P. Kruger, Principles of Activation Analysis (Wiley-
Interscience) p. 98 (1971).

$$\frac{\lambda_b N_b}{\lambda_a N_a} = 1 \qquad \text{(t large)} \tag{2.48}$$

and the two species A and B are said to be in secular equilibrium (fig. 2-24).

2.12 Produced activity

Secular equilibrium also occurs when a radioactive species
is produced at a constant rate R (see sec. 3.6) from nuclear
reactions induced by an accelerator beam or in a neutron reactor.

2.12.1   GROWTH OF A RADIOACTIVE NUCLIDE

Consider first the simple case of a stable nuclide (A) being transformed at constant rate R by a nuclear reaction into the radioactive nuclide (B) which decays to the stable nuclide (C) with a decay constant λ_b

$$\text{(A) (stable)} \xrightarrow{R} \text{(B)} \xrightarrow{\lambda_b} \text{(C) (stable)}$$

The increase in B-nuclei is given by

$$\frac{dN_b}{dt} = R - \lambda_b N_b \tag{2.49}$$

since we can assume that the number of A-nuclei is almost unchanged. We assume that the number of B-nuclei is zero when the exposure starts and get from eq. 2.48

$$N_b(t) = \frac{R}{\lambda_b} (1-e^{-\lambda_b t}) \tag{2.50}$$

and hence for the activity when the exposure stops at time t_e

$$A_b(t_e) = \lambda_b N_b(t_e) = R(1-e^{-\lambda_b t_e}) \tag{2.51}$$

The factor $(1-e^{-\lambda_b t_e})$ is called the saturation factor S. For $t_e \gg \lambda_b^{-1}$ we get S=1 and the activity reaches a plateau.

For $t_e \ll \lambda_b^{-1}$ we get $S=\lambda_b t_e$ and the activity increases proportionally to t_e (fig. 2-25). The disintegration rate after an exposure time t_e and a waiting time t_o can easily be calculated from eqs. 2.20 and 2.49.

$$A_b(t_e \cdot t_o) = \lambda_b N_b(t_o) = R(1-e^{-\lambda_b t_e}) \, e^{-\lambda_b t_o} \tag{2.52}$$

2.12.2   GROWTH OF A RADIOACTIVE DAUGHTER

Another case of transformation which is also of practical interest is represented by the scheme

$$\text{(A) (stable)} \xrightarrow{R} \text{(B)} \xrightarrow{\lambda_b} \text{(C)} \xrightarrow{\lambda_c} \text{(D) (stable)}$$

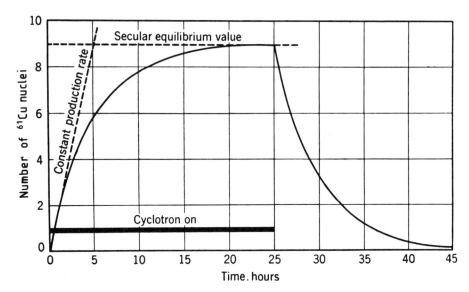

Fig. 2-25   A hypothetical plot of the number of radioactive
^{61}Cu nuclei present in a Ni target at various times
during and after bombardment with deuterons in a
cyclotron.

where nuclide (A) is transformed at a constant rate R to the
radioactive nuclide (B). The growth in the number of C- nuclei
is given by the equation (see eq. 2.34)

$$\frac{dN_c}{dt} = \lambda_b N_b - \lambda_c N_c \qquad (2.53a)$$

where $\lambda_b N_b$ during exposure is given by eq. 2.51 and we get

$$\frac{dN_c}{dt} = R(1-e^{-\lambda_b t}) - \lambda_c N_c \qquad (2.53b)$$

This equation can be solved in a way similar to that used
in solving eq. 2.36. For the disintegration rate of C-nuclei
at the end of the exposure as a function of exposure time, the
following equation can be written (fig. 2-26)

$$A_c(t_e) = \lambda_c N_c(t_e) = \frac{R}{\lambda_c - \lambda_b} \left[\lambda_c(1-e^{-\lambda_b t_e}) - \lambda_b(1-e^{-\lambda_c t_e}) \right]$$

(at exposure end) $\qquad (2.54a)$

74

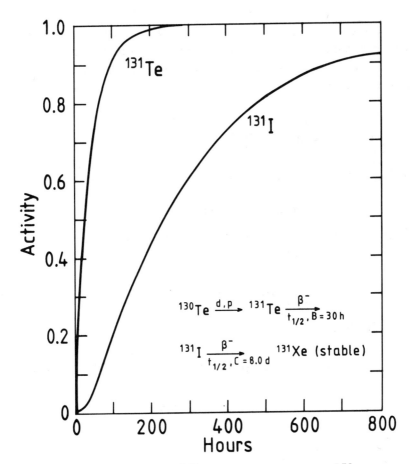

Fig. 2-26 Accumulation of ^{131}Te activity and of ^{131}I activity
in terms of the rate R of the reaction ^{130}Te(d,p)
^{131}Te taken as unity.
R. Evans, The Atomic Nucleus (Mc Graw-Hill) p 488
(1955).

or after a waiting time t_o

$$A_c(t_e t_o) = \frac{R}{\lambda_c - \lambda_b} \left[\lambda_c (1-e^{-\lambda_b t_e}) e^{-\lambda_b t_o} - \lambda_b (1-e^{-\lambda_c t_e}) e^{-\lambda_c t_o} \right]$$

(2.54b)

The determination of an element by measuring the disintegration rate of the daughter (C) is especially useful if (B) is

75

more shortlived than (C), i.e., $\lambda_b \gg \lambda_c$. Often (B) is expo-
sured to saturation, i.e., the saturation factor $S_b = 1$. If
the waiting time is long, so as to allow the B-nuclei to decay
completely into C-nuclei one can simplify the above equation
to:

$$A_c(t_e, t_o) \approx R(1 - e^{-\lambda_c t_e}) e^{-\lambda_c t_o} \qquad (2.54c)$$

(waiting time t_o, exposure time t_e, both times long)

2.12.3   BRANCHING TRANSFORMATION

Another frequently occurring case is represented by the
following transformation

$$
\text{(A) (stable)}
\begin{array}{c}
\overset{R_b}{\nearrow} \quad \text{(B) (isomeric state)} \\
\quad \Big\downarrow \lambda_b \\
\underset{R_c}{\searrow} \quad \text{(C) (ground state)} \overset{\lambda_c}{\longrightarrow} \text{(D) (stable)}
\end{array}
$$

The growth of the nuclides (B) and (C) is described by

$$\frac{dN_b}{dt} = R_b - \lambda_b N_b$$

or, from eq. 2.50

$$N_b = \frac{R_b}{\lambda_b} (1 - e^{-\lambda_b t})$$

and

$$\frac{dN_c}{dt} = R_c - \lambda_c N_c + \lambda_b N_b$$

or

$$\frac{dN_c}{dt} = R_c - \lambda_c N_c + R_b(1 - e^{-\lambda_b t}) \qquad (2.55)$$

76

Solving this equation in the same way as we solved eq. 2.36 we get at the end of the exposure

$$A_c(t_e) = \lambda_c N_c(t_e) = (R_b + R_c)\left[1 - e^{-\lambda_c t_e}\right] + \frac{R_b \lambda_c}{\lambda_b - \lambda_c} \cdot$$
$$\left[e^{-\lambda_b t_e} - e^{-\lambda_c t_e}\right] \quad \text{(at exposure end)} \quad (2.56a)$$

or after a waiting time t_o

$$A_c(t_e, t_o) = (R_b + R_c)\left[1 - e^{-\lambda_c t_e}\right]e^{-\lambda_c t_o} - \frac{R_b \lambda_c}{\lambda_b - \lambda_c} \cdot$$
$$\left[(1 - e^{-\lambda_b t_e})e^{-\lambda_b t_o} - (1 - e^{-\lambda_c t_e})e^{-\lambda_c t_o}\right] \quad (2.56b)$$

In practice, the isomeric state (B) is generally more short-lived than (C), i.e., $\lambda_b \gg \lambda_c$. Moreover (B) is usually exposed to saturation. Assuming that (B) has completely decayed into (C) after a sufficient waiting time one can simplify eq. 2.56b to:

$$A_c(t_e t_o) \simeq (R_b + R_c)\left[1 - e^{-\lambda_c t_e}\right]e^{-\lambda_c t_o} \quad (2.56c)$$

An example of this case is the transformation of ^{191}Ir by neutron irradiation:

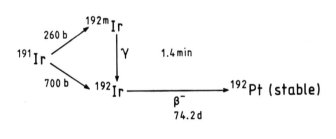

In other cases, $\lambda_b \ll \lambda_c$ and usually $t_e \gg \frac{1}{\lambda_b} \gg \frac{1}{\lambda_c}$

If both (B) and (C) are exposed to saturation we get the following simplification

$$A_c(t_e t_o) \simeq R_b e^{-\lambda_b t_o} + R_c e^{-\lambda_c t_o} \quad (2.56d)$$

An example is

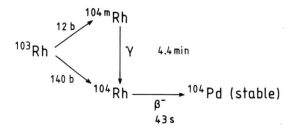

2.12.4 GROWTH OF A RADIOACTIVE DAUGHTER AFTER BRANCHING
TRANSFORMATIONS

We now study a more complex scheme where nuclide D in the
above section is also radioactive

$$
\begin{array}{c}
\text{(B) (isomeric state)} \\
R_b \nearrow \quad \downarrow \lambda_b \\
\text{(A) (stable)} \\
R_c \searrow \quad \text{(C) (ground state)} \xrightarrow{\lambda_c} \text{(D)} \xrightarrow{\lambda_d} \text{(E) (stable)}
\end{array}
$$

The growth of nuclide (D) is given by

$$\frac{dN_d}{dt} = \lambda_c N_c - \lambda_d N_d \tag{2.57}$$

where $\lambda_c N_c$ is given by eq. 2.56a or 2.56b.

This equation can be solved in the usual way. Here, we on-
ly give a simplified activation equation with $\lambda_b \gg \lambda_c$ and λ_b
$\gg \lambda_d$, which is the normal case in practice.

$$A_d(t_e t_o) \simeq \frac{R_b + R_c}{\lambda_d - \lambda_c} \left[\lambda_d (1 - e^{-\lambda_c t_e}) e^{-\lambda_c t_o} - \lambda_c (1 - e^{-\lambda_d t_e}) e^{-\lambda_d t_o} \right]$$

$$\tag{2.58}$$

2.13 References

2.13.1   GENERAL

2.1   M. Lederer and V. Shirley (eds.), Table of Isotopes with
       appendices 7th ed Wiley - Interscience (1978).

2.13.2   ALPHA DECAY, BETA DECAY AND ELECTRON CAPTURE

2.2   H. Behrens and J. Jänecke, Numerical Tables for Beta-
       Decay and Electron Capture, Landolt-Börnstein New Series
       Vol.4 (1969).

2.3   M.J. Berger et al., Response of Silicon Detectors to
       Mono-Energetic Electrons with Energies between 0.15 to
       5.0 MeV, Nucl. Instr. Meth. 69, 181 (1969).

2.4   H. Daniel, Shapes of Beta-Ray Spectra, Rev. Mod. Phys.
       40, 659 (1968).

2.5   G.P. Ford and D.C. Hoffman, Fermi-Function Integrals for
       Finding Relative Beta-group Intensities, Nuclear Data
       Tables A1, 411 (1966).

2.6   N.B. Gove and M.J. Martin, Log-f Tables for Beta Decay
       Nuclear Data Tables A10, 205 (1971).

2.7   H. Paul, Shapes of Beta Spectra, Nuclear Data Tables
       A2, 281 (1966).

2.8   A. Rytz, Catalogue of Recommended Alpha Energy and In-
       tensity Values, Atomic Data and Nuclear Data Tables 12,
       479 (1973).

2.13.3   GAMMA DECAY AND INTERNAL CONVERSION

2.9   W.W. Bowman and K.W. MacMurdo, Radioactive-decay Gammas
       Ordered by Energy and Nuclide, Atomic Data and Nuclear
       Data Tables 13, 90 (1974).

2.10   R.L. Heath, Gamma-Ray Spectrum Catalogue for Ge(Li) and
       Si(Li) Spectrometry. Tables and Graphs of Gamma-Energies
       and Intensities for over 300 Nuclides, ANCR-1000-2
       (TID 4500). AEC Report Series March (1974).

2.11   J.B. Marion, Gamma-Ray Calibration Energies, Nuclear
       Data Tables A4, 301 (1968).

2.12   F. Rösel, H.M. Fries, K. Alder, and H.C. Pauli, Internal
       Conversion Coefficients for All Atomic Shells, ICC Va-
       lues for (a) X = 30-67., (b) Z = 68-104, Atomic Data and
       Nuclear Data Tables 21, 110 (1978).

2.13   M.A. Wakat, Catalogue of γ-Rays Emitted by Radionucli-
       des, Nuclear Data Tables 8, 445 (1971).

2.13.4   DELAYED NUCLEON DATA

2.14   J.C. Hardy, Delayed Proton and Alpha Precursors, Nuclear
       Data Tables A11, 327 (1973).

2.15   L. Tomlinson, Delayed Neutron Precursors, Atomic Data
       and Nuclear Data Tables 12,179 (1973).

3 Nuclear Reactions

3.1 Different types of nuclear reactions

In this chapter we will discuss primarily neutron, proton and photon induced nuclear reactions, we will also mention reactions induced by light nuclei such as deuterons and α-particles. When a given target nucleus interacts with an incoming particle of sufficient kinetic energy, there are, in general, many ways for the reaction to proceed. The particle can be scattered elastically by the target nucleus or it can penetrate it and interact with its nucleons or nucleon clusters.

Three different types of reactions can be distinguished; elastic scattering, direct reactions and compound nucleus reactions. In fig. 3-1 is shown how the different reactions can proceed and their time-scales.

3.2 Elastic scattering and the nuclear force

If the particle is scattered elastically, a diffraction pattern which resembles that obtained when light is scattered by an opaque body is obtained (fig. 3-2).

Elastic scattering experiments have greatly enlarged our knowledge about nuclear sizes and densities and about the strong nuclear force between nucleons. Thus, it is found that the nuclear surface is diffuse but that inside the surface region the nucleon density is approximately constant at

$$\rho_o = 0.17 \cdot 10^{15} \text{ nucleon m}^{-3}$$

The nuclear force possesses the following characteristics:

a) Inside the nucleus, the average nuclear force is attractive and stronger than the repelling Coulomb force between the protons. Otherwise, no stable nuclei could exist.

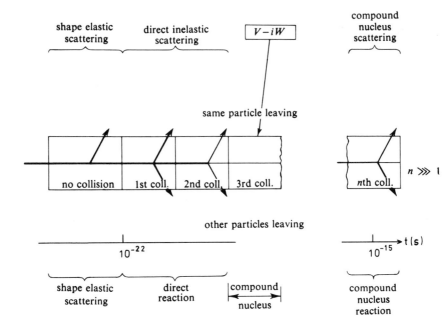

Fig. 3-1 Graphic representation of the course of a nuclear
reaction. As long as no collision takes place, only
shape-elastic scattering is possible. The first collision may produce direct reactions: later on, after
many secondary collisions, a compound nucleus is formed. E. Segré, Nuclei and Particles (Benjamin 1964)
p. 439.

b) The nuclear force is short-ranged, with a range of about
    1 to $2 \cdot 10^{-15}$m.

c) The nuclear force shows saturation properties, which means
    that a nucleon only binds a small number of surrounding
    nucleons.

d) Even if the nuclear force is strong, it barely binds a
    neutron-proton pair and then only if the spins of the nuc-
    leons are parallel. The break-up energy of deuterons, which
    corresponds to the parallel state, is 2.2 MeV. The anti-
    parallel state is not bound. The small binding energy ari-
    ses from the large relative kinetic energy between two
    nucleons which are close together.

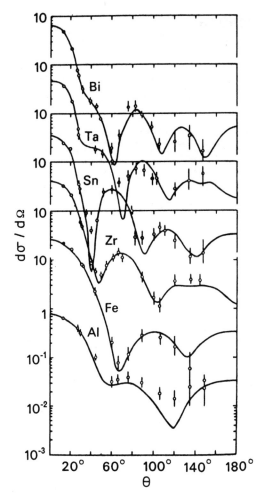

Fig. 3-2   Differential cross sections when 7 MeV neutrons are
scattered elastically from some target nuclei.

e)   The nuclear force is charge independent, which means that
the neutron and the proton have identical properties, except for the electric charge of the proton.

f)   The nucleon seems to have a hard core, which means that
two nucleons can never come closer than about $0.8 \cdot 10^{-15}$ m.

g)   The nuclear force shows exchange character, which means
that two nucleons interact with each other by exchanging
pi-mesons.

3.3 Direct reactions

If the incoming particle reaches the surface of the target nucleus, it interacts with some of the surface nucleons. It may then happen that the particle picks up the nucleon or knocks it out without disturbing the remainder of the target nucleus. If the incoming particle is a light nucleus, e.g., a deuteron, it may also happen that a nucleon is stripped from it and captured by the target nucleus. These types of interaction, in which also inelastic scattering is included, are called direct reactions. They are all very fast reactions, 10^{-22} s. The outgoing particles often have characteristic angular distributions determined by the angular momenta of the struck nucleons in the target nucleus or by the final momentum of the captured nucleon.

3.4 Compound nucleus reactions

Another type of reaction is compound nucleus formation, which lasts for a relatively long time, 10^{-16} - 10^{-15} s. Such a reaction is regarded as a two-step process, first a formation process and then a break-up one. In the formation process, the incoming particle penetrates into the target nucleus and is absorbed by it. The kinetic energy as well as the binding energy of the incoming particle is shared between many of the nucleons inside the target nucleus, none of them acquiring sufficient energy to escape immediately from the nucleus. All the energy supplied is kept by the compound nucleus as excitation energy. After a long time the compound nucleus will deexcite, when, by chance, enough energy is concentrated on one nucleon or a cluster of nucleons (e.g. an α-particle) for escape to occur or else by γ-emission. In both cases, a long time is needed, at least 10^{-16} s. This time is sufficiently long for no trace to be left identifying the particular process of formation. At the time of break-up, the compound nucleus has no memory of how it was formed. The second step, decay by particle or γ-emission is then simply the decay of a highly-excited state, very similar to the decay of natural radioactive nuclides. Pure compound nucleus reactions are most common at comparatively low energies of the projectile.

3.5 Nuclear kinematics

3.5.1 ELASTIC AND NON-ELASTIC COLLISIONS

Collisions between nuclear particles are governed by the laws of conservation of energy and momentum. These laws determine the relationship between the angles of scattering and recoil. They are independent of the detailed mechanism of the collision and, for our purposes, they can in most cases be treated non-relativistically

The collisions may be either <u>elastic</u> in which there is no change in the total kinetic energy or <u>non-elastic</u> in which either or both colliding particles absorb or emit energy, e.g., by the excitation of an internal vibration or oscillation. The particular case of nuclear transformation in which the particle system before collision is changed to another system after collision belongs to the class of non-elastic collisions.

3.5.2 LABORATORY SYSTEM

Consider a particle of mass m_1, e.g., from an accelerator incident with velocity v_1 on a particle m_2 at rest. We first assume that the particle m_1 collides elastically with m_2 and is deflected through an angle ψ_1, with respect to its initial direction. To calculate the velocity v_2 or the angle ψ_2 of recoil of m_2 we use the conservation laws for linear momentum and energy (fig. 3-3).

Conservation of linear momentum:

$$\vec{p}_1 = \vec{p}_1' + \vec{p}_2' \qquad (3.1a)$$

where $\vec{p}$ is the linear momentum vector $\vec{p} = m\vec{v}$. The primed symbols are related to the situation after collision.

The equation can also be written:

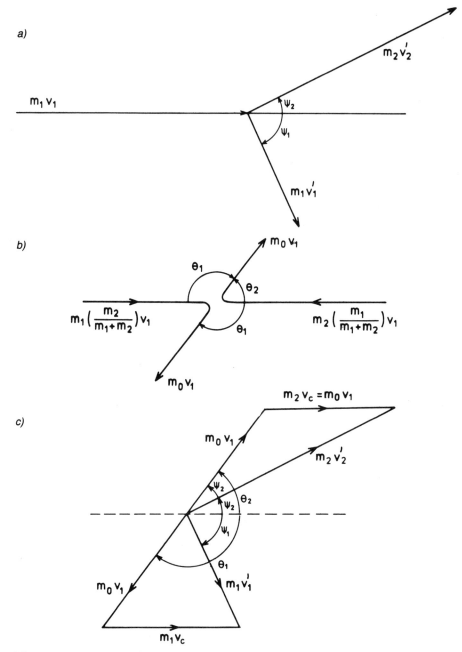

Fig. 3-3 Vector diagrams

a) elastic collision (laboratory system)

b) elastic collision (centre-of-mass system)

c) momentum diagram which relates laboratory to centre-of-mass angles.

86

$$\vec{p}_i = \vec{p}_f \qquad (3.1b)$$

where $\vec{p}_i$ and $\vec{p}_f$ are the total initial and final linear momentum. Conservation of energy for an elastic collision gives

$$T_1 = T_1' + T_2' \qquad (3.2a)$$

or

$$T_i = T_f \qquad (3.2b)$$

where T is the kinetic energy.

The relationship between momentum and kinetic energy is in the non-relativistic case, when all particle velocities are small compared to the velocity of light.

$$p^2 = 2 m T \qquad (3.3)$$

where p is the absolute magnitude of the vector $\vec{p}$.

We rearrange eq. 3.1a

$$\vec{p}_1' = \vec{p}_1 - \vec{p}_2'$$

and take the scalar product

$$(p_1')^2 = (p_1)^2 + (p_2')^2 - 2p_1 p_2' \cos \psi_2.$$

Converting to energies by using eq. 3.3

$$m_1 T_1' = m_1 T_1 + m_2 T_2' - 2(m_1 m_2 T_1 T_2')^{\frac{1}{2}} \cos\psi_2$$

and eliminating T_1' by using eq. 3.2a, we find

$$(m_1 + m_2) T_2'^{\frac{1}{2}} = 2(m_1 m_2 T_1)^{\frac{1}{2}} \cos\psi_2 \qquad (3.4a)$$

or

$$v_2' = 2 v_1 \frac{m_1}{m_1 + m_2} \cos \psi_2 \qquad (3.4b)$$

It is also possible to show that

$$\frac{m_2}{m_1} = \frac{\sin \psi_1}{\sin(2\psi_2 + \psi_1)} \qquad (3.4c)$$

87

Analogously we find for T_1'

$$T'^{\frac{1}{2}} = |m_1 \cos\psi_1 \pm (m_2^2 - m_1^2 \sin^2\psi_1)^{\frac{1}{2}}| \frac{T_1^{\frac{1}{2}}}{m_1 + m_2} \qquad (3.4d)$$

In non-elastic collisions the energy conservation equation still holds if a quantity Q is introduced into the energy conservation equation 3.2 to take care of the energy absorbed or released in the reaction. The masses after collision m_1' and m_2' must also be different from m_1 and m_2 since the particles, one or both of them, have gained or lost internal energy

$$T_f - T_i = Q \qquad (3.5)$$

where Q is defined as

Q = kinetic energy of the final system minus kinetic energy of the initial system.

For elastic collisions, Q = 0, while for non-elastic collisions Q may be either positive or negative. If Q > 0, the reaction is called exoergic and if Q < 0, it is called endoergic. The energy conservation law gives

$$E = m_i c^2 + T_i = m_f c^2 + T_f \qquad (3.6)$$

where m_i and m_f are the total masses in the initial and final systems.

The quantity Q can now be written as the mass-energy loss during the non-elastic collision

$$Q = T_f - T_i = (m_i - m_f) c^2 \qquad (3.7)$$

For a non-elastic collision, relation 3.4a becomes

$$(m_1 - m_1')T_1 + (m_1' + m_2')T_2' - 2(m_1 m_2' T_1 T_2')^{\frac{1}{2}} \cos\psi_2 = m_1' Q \qquad (3.8)$$

3.5.3 CENTRE-OF-MASS SYSTEM

Often and particularly in non-elastic processes, it is advisable to consider collision processes in the centre-of-mass system in which the centre-of-mass of the initial system (and thus also of the final system) is at rest rather than in the laboratory system in which the target nucleus is normally at rest. In the laboratory system, the centre-of-mass moves with velocity

$$\vec{v}_c = \frac{m_1 \vec{v}_1}{m_1 + m_2} \, , \tag{3.9}$$

while in the centre-of-mass system the two initial particles approach each other from opposite directions with equal and opposite momenta (fig. 3-3 b).

$$\vec{p}_1 = m_1(\vec{v}_1 - \vec{v}_c) = \frac{m_1 m_2}{m_1 + m_2} \vec{v}_1 = m_o \vec{v}_1 = -\vec{p}_2 \tag{3.10}$$

where $m_o = \frac{m_1 m_2}{m_1 + m_2}$ is the reduced mass of the initial pair.

In elastic scattering, the particles are each deflected through the same angle Θ_1, in the centre-of-mass system, since the total momentum must remain zero. The relation between Θ_1, which is needed in analysis and ψ_1, which is actually observed, is (fig. 3-3c)

$$\tan \psi_1 = \frac{\sin \Theta_1}{\cos \Theta_1 + m_1/m_2} \tag{3.11}$$

In all collision processes, an energy of $\frac{1}{2}(m_1 + m_2)v_c^2$ is associated with the centre-of-mass motion and is thus not avail‑able for producing internal effects such as nuclear excitations. The energy remaining from the limited kinetic energy is thus

$$\frac{1}{2} m_1 v_1^2 - \frac{1}{2}(m_1 + m_2) \frac{m_1^2 v_1^2}{(m_1 + m_2)^2} = \frac{1}{2} \frac{m_1 m_2}{m_1 + m_2} v_1^2 = \frac{1}{2} m_o v_1^2 \tag{3.12}$$

3.5.4 THRESHOLD ENERGY

When the bombarding particle converts the target nucleus to another one, a nuclear reaction is initiated. Such a reaction is written in a form similar to that of a chemical reaction

$$X + a \rightarrow Y + b + Q \tag{3.13}$$

where X is the target nucleus, <u>a</u> the incoming particle, Y the product nucleus and <u>b</u> the resulting particle. Usually, the reaction is written using an abbreviated form:

X (a,b) Y

The threshold energy for the reaction is the lowest energy particle <u>a</u> must have in order to be able to initiate the desired nuclear reaction. The target nucleus X is expected to be at rest before the collision. We know from eq. 3.12 that the energy available for the reaction is $\frac{1}{2}m_0 v_1^2$. Thus, for the reaction to occur, the following restriction holds:

$$\frac{1}{2} m_0 v_1^2 \geq Q \qquad\qquad\qquad (3.14)$$

(Q is negative for endoergic reactions)
giving

$$T_t = (\tfrac{1}{2} m_1 v_1^2)_{min} = - \frac{m_1 + m_2}{m_2} Q \qquad\qquad (3.15)$$

If a nucleus is bombarded with uncharged particles such as neutrons of energy T_t, this will be just sufficient to initiate the particular endoergic reaction. In the case of (n,p) and (n,α) reactions, the outgoing particle is positively charged and must have, according to classical theory sufficient energy to overcome the Coulomb barrier before a reaction can proceed (see fig. 2-3, 3-11). We know that quantum-mechanically a finite probability exists for a reaction to occur because of the tunnelling mechanism with outgoing particles having less energy. In practice, however, this probability drops rapidly (fig. 3-6) as the energy of the particle decreases below the barrier restriction.

This is important for activation analysis. Although T_t for the reaction ^{55}Mn(n,α)(T_t = 0.65 MeV) is lower than that for the reaction ^{52}Cr(n,p) (T_t = 3.27 MeV), the α Coulomb barrier is considerable higher than the proton Coulomb barrier for the two nuclides (V_α = 8,3 MeV, V_p = 4,7 MeV).

Hence it is possible to measure the amount of chromium by forming ^{52}V without interference from manganese if a neutron

energy of 7-8 MeV is chosen. Like outgoing charged particles,
incoming charged particles have difficulties in penetrating the
Coulomb barrier. Practically this means that the energy of the
particle has to be greater than the barrier energy for the desired
reaction to occur.

3.6 Cross section

The probability for a nuclear process to occur is generally expressed in terms of a cross section σ which has the di-
mension of an area. If one particle a traverses an area S in
which there is somewhere a single target nucleus A, then the
probability for the reaction A(a,b) B to occur can be written
σ S^{-1} which defines the cross section for the particular process.

The cross section is generally measured in barn:

$$1 \text{ barn} = 10^{-28} \text{ m}^2 \qquad (3.16)$$

which is of the order of the geometrical cross section of a
nucleus

$$\sigma_{geo} = \pi R^2 = \pi r_o^2 A^{2/3} = 4.5 \cdot 10^{-2} A^{2/3} \text{ barn} \qquad (3.17)$$

In a typical irradiation arrangement (fig. 3-4), in which
particles a can be scattered or absorbed, a thin homogeneous
foil of the element A of thickness dx is exposed to a parallel
beam of j incident a-particles per unit area and time (current
density). If σ_t is the total cross section for all processes
scattering or absorbing a-particles, the number of a-particles which disappears per unit time from a parallel beam with
unit area is

$$dj = -j \ \sigma_t N dx \qquad (3.18)$$

where N is the number of A-nuclei per unit volume.

If the target foil has thickness d and if the total cross

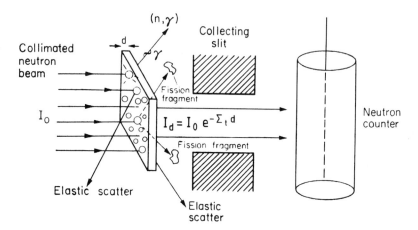

Fig. 3-4   Principle of transmission measurements of total
cross sections. Neutrons interacting with nuclei
in the sample do not reach the detector. A correction for small angle scattering has to be made.

section is not sensitive for the reduction in energy of the
particles when passing the foil, the number of a -particles in
the beam per unit area and time after traversing the foil is
obtained from the equation above by integrating:

$$j_d = j_o e^{-\sigma_t N d} = j_o e^{-\Sigma_t d} \qquad (3.19a)$$

or, introducing the beam current (intensity)

$$I_o = j_o S$$

we get

$$I_d = I_o e^{-\Sigma_t d} \qquad (3.19b)$$

In these equations we have introduced the total macroscopic
cross section Σ_t,

$$\Sigma_t = \sigma_t N = \sigma_t \frac{\rho N_A}{A}, \qquad /m^{-1}/ \qquad (3.20)$$

92

where ρ is the density and N_A = Avogadro's number. The reaction mean free path of particle $\underline{a}$ is given by

$$\lambda_t = \frac{1}{\Sigma_t} \qquad (3.21)$$

that means the average length in the medium the incoming particle runs, before the reaction occurs. The reaction rate for a parallel beam and a thick target is

$$R = (j_o - j_d)S = j_oS(1 - e^{-\sigma_t Nd}) \qquad (3.22a)$$

or

$$R = I_o(1 - e^{-\sigma_t Nd}) \qquad (3.22b)$$

For a target where $\sigma_t Nd \ll 1$, this expression can be approximated by

$$R = I_o\sigma_t Nd = j_o\sigma_t NSd = j_o\sigma_t NV \qquad (3.23)$$

where V is the volume irradiated and NV the number of nuclei irradiated. The reaction yield is defined as the reaction rate divided by the beam current giving for a thin target

$$Y = R/I_o = R/j_oS = \sigma_t Nd \qquad (3.24)$$

Sometimes the reaction is studied "in beam" i.e., a detector registers the outgoing particles $\underline{b}$ during the exposure.

The sensitive area of the detector falls within the solid angle $d\Omega$ and it is natural to define the differential cross section $d\sigma/d\Omega$ which is the probability for the reaction to emit a particle $\underline{b}$ into solid angle $d\Omega$. The reaction cross section is then given by

$$\sigma = \int_\Omega \frac{d\sigma}{d\Omega} \, d\Omega = 2\pi \int_0^\pi \frac{d\sigma}{d\Omega} \sin \psi d\psi, \qquad (3.25)$$

where ψ is the angle between the beam and the detector.

In another typical irradiation arrangement the incident

particles <u>a</u> on the target are multidirectional, e.g., as in
the case of neutrons in a reactor. The neutron flux φ = nv is
defined as the product of the neutron density times the neutron
velocity. The number of reactions of a particular kind between
the neutrons and the target nuclei per unit time will be given
by

$$R = n \frac{v}{\lambda} V = nv \, \Sigma V = nv \, \sigma N V = \varphi \sigma N V = \varphi \sigma \frac{\rho N_A}{A} V \qquad (3.26)$$

$\frac{v}{\lambda}$ = the distance moved by a neutron per unit time divided
by the mean free path for the particular reaction = probability
for the neutron to react in unit time in the particular way.

If the neutrons have a velocity distribution n(v) and the
cross section is dependent on velocity, we get the following
equations for a neutron exposure inside a reactor.

$$n = \int_0^\infty n(v) dv \qquad (3.27a)$$

$$\varphi = \int_0^\infty v \, n(v) dv \qquad (3.27b)$$

$$R = NV \int_0^\infty n(v) \, v\sigma(v) dv \qquad (3.27c)$$

Often many different types of reactions can occur when a
particle hits a target nucleus A. Each kind of reaction has
its own partial cross section. In addition to the various reaction cross sections, there also exists a probability for elastic scattering. Strictly speaking the elastic scattering cross
section σ_{sc} can be split up in two parts: the potential and
the resonant scattering cross sections.

$$\sigma_{sc} = \sigma_{ps} + \sigma_{rs} \qquad (3.28a)$$

The resonant scattering part belongs to the reaction cross
section since a compound state is formed as an intermediate
state, which then decays by re-emitting the incoming particle
and leaving the initial nucleus in its ground state.

The potential scattering is a pure elastic scattering process. The total cross section for reactions and scattering to occur is the sum of all the partial cross sections.

$$\sigma_t = \sigma_r + \sigma_{ps} \tag{3.28b}$$

where

$$\sigma_r = \sigma(n,\gamma) + \sigma(n,p) + \sigma(n,\alpha) + \sigma(n,2n) + \sigma(n,n') + \sigma_{rs} + \cdots \tag{3.28c}$$

Here $\sigma(n,n')$ is the inelastic scattering cross section.

Analogously, the total mean free path λ_t can be decomposed into partial mean free paths

$$\lambda_t^{-1} = \lambda_r^{-1} + \lambda_{ps}^{-1} \tag{3.29a}$$

where

$$\lambda_r^{-1} = \lambda_{n,\gamma}^{-1} + \lambda_{n,p}^{-1} + \lambda_{n,\alpha}^{-1} + \lambda_{n,2n}^{-1} + \lambda_{n,n'}^{-1} + \lambda_{rs}^{-1} \tag{3.29b}$$

A corresponding partial cross section is associated with each of the possible paths. It must be stressed that, for charged particles, the mean free path in matter is determined mainly by electromagnetic interactions and only secondarily by nuclear processes. This will be further discussed in chap. 5.

3.7 Nuclear resonances

Nuclear cross sections are often strongly energy dependent showing peaks corresponding to excited states or resonances in the compound nucleus formed. The energy dependence of the cross section in the vicinity of a maximum is accurately accounted for by the formula

$$Y(E) = \frac{C\ \Gamma}{(T-T_r)^2 + (\frac{\Gamma}{2})^2} \tag{3.30}$$

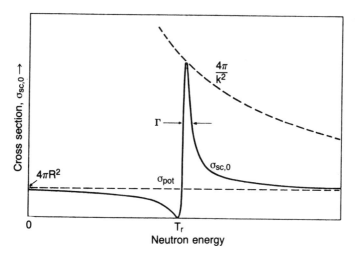

Fig. 3-5   Elastic scattering cross section for $\ell=0$ neutrons
near a resonance in the compound nucleus. The angular momentum of the target nucleus is $J=0$ and the
reaction cross section is neglected except for resonant
scattering. The maximum cross section is then
$\sigma_{sc,o} = 4\pi/k^2$. Resonant scattering interferes with
the potential scattering. At energies below the
resonance energy, the interference is destructive.
This coherence is characteristic of elastic scattering only.

where T is the particle kinetic energy in the centre-of-mass
system and T_r its value at resonance, Γ the resonance full-width
at half-maximum and C is a normalization constant (fig. 3-5).

Formulas similar to eq. 3.30 are obtained for a variety
of resonance phenomena. They express, for instance, the amplitude of the forced vibrations of a damped oscillator, in which
case T is a frequency and Γ a damping coefficient.

The interpretation of nuclear resonances is similar to
other resonance phenomena. We assume that the reaction occurs
in two steps:

$$
\left.
\begin{array}{ll}
X + a \rightarrow C^* & \text{formation of} \\
C^* \rightarrow Y + b & \text{decay of}
\end{array}
\right\} \text{the compound nucleus} \quad (3.31)
$$

96

where C* represents the excited compound state with excitation energy E_r corresponding to T_r and a mean life τ where

$$\tau = \frac{\hbar}{\Gamma} = \frac{6.6 \cdot 10^{-16}}{\Gamma \text{ (ev)}} \text{ s} \tag{3.32}$$

according to the Heisenberg uncertainty relation

$$\Delta E \Delta t \tilde{=} \hbar = h/2\pi$$

where h is the Planck constant.

With this assumption, the cross section for the reaction can be written as a product of two factors: the cross section for the formation of the compound state from the input channel a and the relative probability of its decay into the exit channel, b. Neglecting spins, the cross section for the reaction X(a,b) Y can be written

$$\sigma_{ba} = \sigma_{ca} \frac{\Gamma_b}{\Gamma} \tag{3.33}$$

where Γ_b is the partial width for exit channel b and

$$\Gamma = \sum_b \Gamma_b \tag{3.34}$$

is the total width of decay, which is the sum of all possible modes of decay; σ_{ca} then has the significance of the cross section for formation of the compound state through channel a. The cross section for the formation of a compound state σ_{ca} can in its turn be written as a product of three factors. The first one is the maximum cross section for absorption. From wavemechanical discussions and if only a-particles with $\ell=0$ are taken into account we can write

$$(\sigma_{abs})_{max} = \pi \bar{\lambda}_a^2 \tag{3.35}$$

where $\bar{\lambda}_a = \lambda_a/2\pi$ is the wave length of the particle and is related to the incoming particle linear momentum by

$$p_a = \frac{h}{\lambda_a} = \frac{\hbar}{\bar{\lambda}} = \hbar k_a \tag{3.36}$$

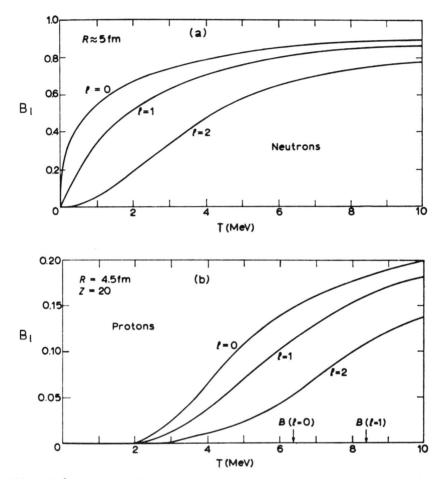

Fig. 3-6 Penetration factor for neutrons (a) and protons (b)
of various angular momenta and kinetic energies.
Barrier heights are indicated by arrows.

where k_a is the particle wave number. This cross section can
be much greater than the geometrical cross section of the
target nucleus which is

$$\sigma_{geo} = \pi R^2 = \pi r_o^2 A^{2/3} \tag{3.37}$$

The second factor is a penetration factor $B_{a\ell}$ (fig. 3-6).
This factor also depends on the angular momentum of the incoming particle and its charge since it has to penetrate a Cou-

lomb barrier but, if only neutrons with $\ell = 0$ are taken into account, which is approximately correct for slow neutrons, the factor can be written explicitly as:

$$B_{no} = \frac{4kK}{(k+K)^2} \simeq \frac{4k}{K} \qquad (3.38)$$

where $k = \lambda_{out}^{-1}$ (the wave number of the neutron outside the target) and $K = \lambda_{in}^{-1}$ (the wave number of the neutron inside the target nucleus).

The third factor is the energy dependence of the resonance which is given in eq. (3.30).

The cross section σ_{ca} can now be written

$$\sigma_{ca} = \lambda_a^2\, B_{a\ell}\, \frac{C\Gamma}{(T-T_r)^2 + (\frac{\Gamma}{2})^2} \qquad (3.39)$$

The product $B_{a\ell}\, C$ is often written

$$B_{a\ell} C = \Gamma_a \qquad (3.40)$$

which is the decay width through channel a corresponding to resonance scattering and we get

$$\sigma_{ba} = \pi\, \lambda_a^2\, \frac{\Gamma_a\, \Gamma_b}{(T-T_r)^2 + (\frac{\Gamma}{2})^2} \qquad (3.41)$$

(single-level Breit-Wigner formula for the case of a spinless particle with no angular momentum, see fig. 3-5. In the general case when the incoming particle a has the spin s, the target nucleus spin J_x and the compound nucleus formed spin J_c, the corresponding expression can be written:

$$\sigma_{ba} = \pi\, \lambda_a^2\, g\, \frac{\Gamma_a\, \Gamma_b}{(T-T_r)^2 + (\frac{\Gamma}{2})^2} \qquad (3.42)$$

(general single-level Breit-Wigner formula)

with

$$g = \frac{2J_c + 1}{(2s+1)(2J_x + 1)} \qquad (3.43)$$

99

As stated above T is the centre of mass energy and T_r its value at resonance. In practice, laboratory energies are frequently used in these equations. The same is valid for the wave number and the wave length.

3.8 Neutron-induced reactions

3.8.1 NEUTRON INTERACTION

Free neutrons are unstable particles, and decay to protons by β^--emission

$$n \rightarrow p + \beta^- + \bar{\nu}_e + 0,782 \text{ MeV} \tag{3.44}$$

The half-life is $t_{1/2}$ = 636 s.

Like the proton, the neutron has an intrinsic spin of 1/2 i.e., (eq. 1.11)

$$s = \{s(s+1)\}^{\frac{1}{2}}\hbar = (\frac{3}{4})^{\frac{1}{2}}\hbar \tag{3.45}$$

A slow neutron has no angular momentum ($\ell=0$) when it interacts with nuclei and it can interact easily as it is uncharged. If it is slow enough, in practice usually with a kinetic energy below 100 keV, only two types of reactions can occur, viz., elastic scattering and neutron capture (fig. 3-7).

$$n + X \rightarrow X + n \qquad \text{elastic scattering}$$

$$n + X \rightarrow Y + \gamma \qquad \text{neutron capture}$$

Exceptions to this behaviour are slow neutron reactions with Li and B which lead to particle emission, since the processes are exoergic (Q>0) and neutron-induced fission in some heavy elements.

The total neutron reaction cross section shows a large number of sharp resonances implying that the reaction passes through long-lived compound nuclear states. A representative

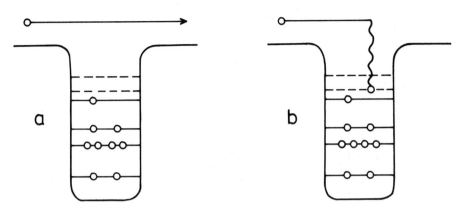

Fig. 3-7   Schematic diagram of (a) elastic scattering and (b)
           capture of neutrons.

cross section curve is exhibited by Gd and is shown in fig.
3-8.

The shape of the resonances follow the Breit-Wigner formula (eq. 3.42):

$$\sigma_r = \sigma_{rs} + \sigma_{n,\gamma} = \pi \, \lambdabar_n^2 \, g \, \frac{\Gamma_n(\Gamma_\gamma + \Gamma_n)}{(T_n - T_r)^2 + (\frac{\Gamma}{2})^2} \qquad (3.46)$$

with peak values

$$\hat{\sigma}_r = 4\pi \, \lambdabar_n^2 g \, \frac{\Gamma_n}{\Gamma} \quad \text{(for } T_n = T_r\text{)} \qquad (3.47)$$

The cross section integrated over a single and fairly narrow
$(\Gamma \ll T_r)$ resonance is

$$\int_0^\infty \sigma_r(T_n) \, dT_n \cong 2\pi^2 \, \lambdabar_n^2 \, g\Gamma_n = \frac{\pi}{2} \, \hat{\sigma}_r \, \Gamma \qquad (3.48)$$

At low energies, the γ-width Γ_γ is much greater than the
neutron width Γ_n. Representative values are $\Gamma_\gamma = 10^{-1}$ eV and
$\Gamma_n \simeq 10^{-3}$ eV for a medium-heavy nucleus. The Γ_γ-width is almost independent of the neutron energy T_n for slow neutrons,
but the Γ_n-width varies linearly with the neutron velocity as
can be shown from statistical considerations,

$$\Gamma_n \sim v_n \qquad (3.49)$$

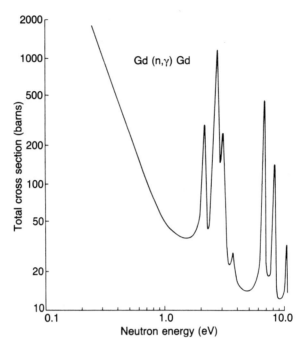

Fig. 3-8   Total cross section for the interaction of slow
            neutrons with gadolinium. D. Garber, R.R. Kinsey,
            Neutron Cross Sections (BNL 325) 3rd ed. 1976.

When a slow neutron is captured by a target nucleus, the
compound nucleus formed is excited to an energy somewhat above the neutron binding energy. If no resonance lies near zero
neutron energy, as in the case of Gd, then T_n may be neglected
in comparison with T_r in eq. (3.46) and we get

$$\sigma_{n\gamma} \sim \lambda_n^2 \, \Gamma_n \, \Gamma_\gamma \sim v_n \, \lambda_n^2 \sim \frac{1}{v_n} \qquad (3.50)$$

since $\Gamma_n \ll \Gamma_\gamma \ll T_r$ and $\lambda = \frac{1}{k} = \frac{\hbar}{p} \sim \frac{1}{v_n}$ $\qquad (3.51)$

This result is the familiar l/v law for neutron capture.
In a log-log cross section diagram, a straight line is obtained:

$$\log \sigma = -\frac{1}{2} \log T_n + \text{const.} \qquad (3.52)$$

For resonant scattering, on the other hand,

$$\sigma_{rs} \sim \lambda_n^2 \; \Gamma_n^2 \sim \text{const.} \tag{3.53}$$

As the neutron energy increases, the relative contribution of elastic scattering to the total cross section also increases and, in the 100-1000 keV range resonances are mainly due to scattering since now $\Gamma_n > \Gamma_\gamma$.

When the neutron energy increases still further, new reaction channels are opened. First, the inelastic scattering channels will be opened and then later channels for which multiple neutrons and charged particles are emitted. As the number of decay possibilities increases, the nonelastic cross section becomes larger and is in the MeV-region as large as the elastic cross section.

3.8.2   REACTION RATES AND EFFECTIVE CROSS SECTIONS

The neutron sources in practical use (see sec. 4.2) do not emit monoenergetic neutrons but their flux has a more or less broad energy distribution, a neutron spectrum. The reaction rate R of a sample with a nucleus density N in such a neutron flux is given by (eq. 3.27c)

$$R = NV \int_0^\infty n(v) v \sigma_t(v) \; dv \tag{3.54}$$

or if we define the reaction rate per nucleus (the microscopic reaction rate) r,

$$r = \int_0^\infty n(v) v \sigma_t(v) dv = \int_0^\infty \varphi(v) \sigma_t(v) dv \tag{3.55}$$

In order to evaluate this equation we have to know the shape of the neutron spectrum and the energy dependence of the cross section. Formulas exist which describe accurately the shapes of the true neutron spectra for all neutron sources used in practical applications. The neutron fluxes obtained from these formulas are called calculated fluxes φ_c;

$$\varphi \cong \varphi_c = \int n_c(v) v dv \tag{3.56}$$

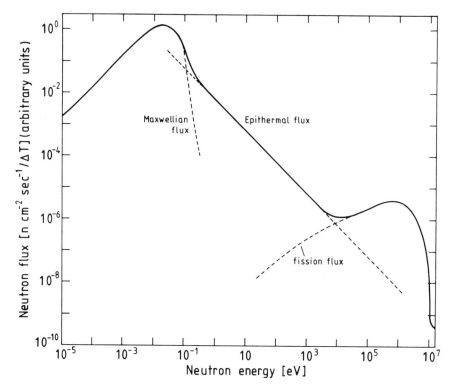

Fig. 3-9   Neutron flux of a nuclear fission reactor.

where n_c is the formula-given neutron spectrum.

If the calculated flux from a source is known; it is also possible to define an average or effective cross section from the experimental reaction rate

$$r = \varphi_c \sigma_{eff} \tag{3.57}$$

In this way σ_{eff} can be measured for a great number of reactions all of which of course are related to the specific neutron source.

3.8.3   THE REACTOR NEUTRON SPECTRUM

Reactor neutrons are in widespread use in nuclear chemical analysis since large fluxes of neutrons are obtained in fission reactors. The neutron spectrum consists of three com-

Table 3-1

Classification of neutrons.

Designation	Energy range
thermal	< 0.5 eV
epithermal	0.5 eV - 0.1 MeV
fast	> 0.1 MeV

ponents (fig. 3-9, table 3-1). Firstly we have the primary
neutrons from the fission process. They are fast, with kinetic
energies mainly above 10^5 eV. Below the fast neutron group
we find the second component with slowing down neutrons or
epithermal neutrons. In this group the kinetic energies are
between 0.5 eV and 10^5 eV. Finally we have the slow neutrons
or thermal neutrons. These neutrons have attained the same energy distribution as the molecules of the surrounding material
and have kinetic energies mainly below 0.5 eV.

These three components correspond to three regions in the
neutron reaction cross section in uranium, where the resonance
region separates the slow and fast reaction regions (fig. 3-10).
These three neutron components are present in all reactors,
but the ratios of the fast to the epithermal and to the thermal component vary from reactor to reactor and even within a
reactor from point to point (see figs. 4-15 and 4-16).

3.8.4   THERMAL NEUTRONS

The slow neutrons in a reactor are in thermal equilibrium
with the molecules of their surroundings. Their average energy
is related to the average energy of the molecules which is
itself dependent on the temperature of the moderator. The neutrons in the moderator behave similarly to the molecules of a
gas in a container with one essential difference, namely that
the neutrons may be captured and disappear. If there was no
removal, then the distribution of neutron energy would be the
same as the distribution in energy of the molecules of a gas in
a container. This distribution is the well known Maxwell-Boltz-

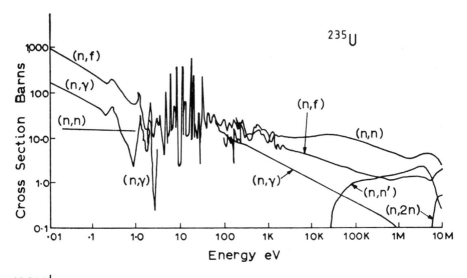

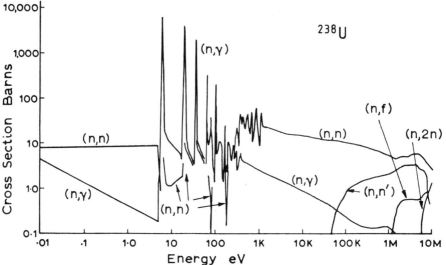

Fig. 3-10 Neutron cross-sections of (a) ^{235}U (b) ^{238}U. Data
taken from BNL and AWRE-0-28.

mann distribution. The deviations from a pure maxwellian distribution are, however, so small that usually all neutrons below about 0.5 eV in a reactor are called thermal neutrons.

The energy distribution of the neutron flux can thus be written

$$\varphi_c(T_n)dT_n = \frac{\varphi_{th}}{(k\theta)^2} \, T_n \, \exp\left(-\frac{T_n}{k\theta}\right) \, dT_n \qquad (3.58a)$$

where $\varphi_c(T_n)$ is the thermal neutron flux in an energy interval dT_n around T, θ is the absolute temperature of the surroundings and k is the Boltzmann konstant.

Analogously

$$n_c(T_n) = \frac{1}{v}\, \varphi_c\, (T_n) \tag{3.58b}$$

The total flux is

$$\int_0^\infty \varphi_c\, (T_n)dT_n = \varphi_{th} = n_{th}\, \frac{2}{\sqrt{\pi}}\, (\frac{2k\theta}{m_n})^{\frac{1}{2}} = n_{th}\, v_{th} \tag{3.59}$$

where $n_{th} = \int_0^\infty n_c(T_n)dT_n$

and $v_{th} = \dfrac{\int_0^\infty vn_c(T_n)dT_n}{\int_0^\infty n_c(T_n)dT_n} = \dfrac{\varphi_{th}}{n_{th}} = \dfrac{2}{\sqrt{\pi}}\, \dfrac{(2k\theta)^{\frac{1}{2}}}{m_n}$ (3.60)

which is the mean velocity of the thermal neutrons.

These results can be derived from the two definite integrals

$$\int_0^\infty x\, e^{-x}dx = 1$$

$$\int_0^\infty x^2\, e^{-x^2}\, dx = \frac{\sqrt{\pi}}{4}$$

The flux distribution $\varphi_c(T_n)$ peaks at

$$T_{peak} = k\theta \tag{3.61}$$

corresponding to

$$v_{peak} = (\frac{2k\theta}{m_n})^{\frac{1}{2}} = \frac{2}{\sqrt{\pi}}\, v_{th} \tag{3.62}$$

and has an energy average value

$$T_{av} = \frac{1}{\varphi_{th}} \int_0^\infty T_n\varphi_c(T_n)dT_n = 2k\theta \tag{3.63}$$

107

At low neutron energies, the cross section for neutron capture follows the 1/v law (eq. 3.51) which means that the reaction rate for producing a new nuclide is inversely proportional to the neutron velocity. If one knows the cross section σ_o for a certain velocity v_o, the cross section for any other velocity is given by

$$\sigma(v) = \frac{\sigma_o v_o}{v} \qquad (3.64)$$

It has been comparatively easy to measure cross sections for monoenergetic neutrons at low velocities. Such cross sections are usually tabulated for a velocity of 2200 m/s which is the peak velocity in the maxwellian flux distribution for 20^oC and corresponds to an energy of 0.0253 eV.

It is often convenient to define a conventional thermal flux φ_o by measuring a specific reaction rate from the thermal neutron source and using the 2200 m/s cross section σ_o as an effective cross section (eq. 3.57).

$$\varphi_o = \frac{r}{\sigma_o} = \frac{1}{\sigma_o} \int v n_c(v)\sigma(v)dv = v_o \int n_c(v)dv = v_o n_{th} \quad (3.65)$$

The conventional thermal flux is not the maxwellian flux which is given by eq. 3.59.

$$\varphi_{th} = v_{th} n_{th}$$

The connection between the two fluxes are

$$\varphi_o = \frac{v_o}{v_{th}} \varphi_{th} = (\frac{\pi \theta_o}{4\theta})^{\frac{1}{2}} \varphi_{th} \qquad (3.66a)$$

or if measurements are made at 20^oC

$$\varphi_o = (\frac{\pi}{4})^{\frac{1}{2}} \varphi_{th} \qquad (3.66b)$$

The advantage of the φ_o-definition is that, for all reactions whose σ_o is known, a reaction rate can be calculated if the conventional thermal flux has been measured according to the definition. This convention requires no assumptions as to the shape of the neutron spectrum and is valid in so far as the reaction cross section obeys to the 1/v law.

The thermal neutron flux is often measured by using an 1/v-detector. Even here of course the calibration constant is dependent on temperature.

3.8.5 EPITHERMAL NEUTRONS

Above energies of about 0.5 eV reaction cross sections no longer follow the 1/v-law and the conditions found for thermal neutrons cannot be applied. In this region the reaction cross section is characterized by resonance peaks. Even if these resonances can be described by Breit-Wigner formulas (eq. 3.42), in many cases there is a lack of knowledge of the parameters required which makes calculation of the effective cross sections complicated. However, it can be shown that the slowing down of fast neutrons by elastic collisions leads to a neutron flux distribution which can be described by an $1/T_n$ curve.

$$\varphi_c(T_n)dT_n = vn_c(T_n)dT_n = \varphi_{epi}\frac{dT_n}{T_n} = \varphi_{epi}d(\ln T_n) \quad (3.67)$$

where φ_{epi} is the flux per unit logarithmic energy interval.

This assumption is good if the slowing down medium is weakly absorbing in the range from fission to thermal energies and if the scattering cross section is constant.

Some reservations must thus be made about the $1/T_n$ variation in hydrogenous media, such as water, since the scattering cross section Σ_s is energy dependent for hydrogen. The flux distribution thus has to be written

$$\varphi(T_n)dT_n = \frac{const}{T_n\Sigma_s(T_n)} dT_n \quad (3.68)$$

The total flux is the sum of the thermal flux and the epithermal flux but at energies above 0.5 eV the contribution of the maxwellian thermal flux is negligible compared to the epithermal $1/T_n$ flux. This energy is called the cadmium threshold, because cadmium foil of 0.5 - 1 mm thickness is a filter which only neutrons of energies above 0.5 eV can pass through.

The correct cut-off energy depends somewhat on the geometrical conditions and the foil thickness. To calculate reaction rates for the $1/T_n$-flux region

$$r_{epi} = \int_{0.5\ eV}^{0.1\ MeV} n(T_n)v\sigma(T_n)dT_n = \varphi_{epi} \int_{0.5\ eV}^{0.1\ MeV} \frac{\sigma(T_n)}{T_n} dT_n \approx$$

$$\varphi_{epi} \int_{0.5eV}^{\infty} \frac{\sigma(T_n)}{T_n} dT_n \qquad\qquad (3.69)$$

an effective cross section I, usually called the resonance integral, has been defined

$$I = \int_{0.5\ eV}^{\infty} \frac{\sigma(T_n)}{T_n} dT_n \qquad\qquad (3.70)$$

so that

$$r_{epi} = \varphi_{epi} \cdot I \qquad\qquad (3.71)$$

In some cases, I-values can be determined directly. Such a case is the reaction $^{197}Au(n,\gamma)^{198}Au$ where only one resonance peak is observed. This resonance has been thoroughly investigated and its resonance integral is known to be 1551±20 barn. Gold is therefore often used to measure epithermal fluxes and as a calibration standard to measure other resonance integrals. For all other reactions the uncertainty in I is much greater.

In some publications the resonance integrals I as defined by eq. 3.70 are not given but instead the reduced resonance integrals I_r. These are correlated to I by

$$I = I_r + I_{1/v} \qquad\qquad (3.72)$$

It is clear that the resonance peaks are imposed on the $1/v$ cross section curves, and the $1/v$ curves continue above the cadmium threshold. I_r is given because for some purposes in neutron physics it is necessary to know the contribution of pure resonances separately, while on the other hand, it is

110

possible in some cases to calculate the integrals for well defined resonance peaks. The recalculation of I from I_r is possible in every case since

$$
I_{1/v} = \int_{T_{cd}}^{\infty} \frac{\sigma(T_n)\, dT_n}{T_n} = \sigma_0 v_0 \int_{T_{cd}}^{\infty} \frac{dT_n}{vT_n} = 2\sigma_0 v_0 \int_{v_{\sigma d}}^{\infty} \frac{dv}{v^2} = \frac{2\sigma_0 v_0}{v_{cd}}
$$

$$
= 2\sigma_0 \left(\frac{T_0}{T_{cd}}\right)^{\frac{1}{2}} = 0.45\, \sigma_0 \tag{3.73}
$$

for a cut-off energy of T_{cd} = 0.5 eV.

Calculation of I_r is possible if the resonance parameters
(Γ_n, Γ_γ, Γ, T_r) occuring in the Breit-Wigner formula (eq. 3.42)
are known.

In the case of a single resonance peak the following equation can be used

$$
I_r = 4.10 \cdot 10^6 g \Gamma_n \Gamma_\gamma (T_r^2 \Gamma)^{-1} \text{ barns} \tag{3.74}
$$

where resonance energy and level widths are expressed in eV and
g is the statistical factor (eq. 3.43).

3.8.6   FAST REACTOR NEUTRONS

The distribution of the fast neutron flux is approximately described by Watt's equation

$$
\varphi_f(T_n) = 0.484 \exp(-T_n) \sinh(2T_n)^{\frac{1}{2}} \tag{3.75a}
$$

where T_n is expressed in MeV. This function agrees with the
experimental data from 0.075 to 15 MeV and is shown in fig.
3-9. Other approximations have been proposed by Cranbey

$$
\varphi_f(T_n) = 0.4527 \exp(-1036T_n)\sinh(2.29T_n)^{\frac{1}{2}} \tag{3.75b}
$$

and by Leachman

$$
\varphi_f(T_n) = 0.7725 \sqrt{T_n} \exp(-0.775\, T_n) \tag{3.75c}
$$

The three representations differ only slightly, the largest deviations occuring at high energies.

Cross sections for (n,γ)-reactions at energies of 100 keV and above, which are considered here, are very small and mainly (n,p)-, (n,α)-, (n,2n)-, and (n,n')-reactions have to be taken into account. All of these are threshold reactions which means that up to a certain neutron energy T_t, the cross section is zero. Above this threshold, the cross section is energy dependent. In the case of emission of a charged particle (p or α), the cross section is determined mainly by the penetrability of the Coulomb barrier (fig. 3-6).

Some of the cross sections for ^{235}U and ^{238}U are shown in fig. 3-10.

The reaction rate r_f for fast neutrons is given by

$$r_f = \int_{T_t}^{\infty} \sigma(T_n)\varphi_f(T_n)dT_n \qquad (3.76)$$

which is called the response integral.

The average fast neutron cross section

$$\sigma_{av} = \frac{\int_{T_t}^{\infty} \sigma(T_n)\varphi_f(T_n)dT_n}{\int_{T_t}^{\infty} \varphi_f(T_n)dT_n} \qquad (3.77)$$

can be determined by measuring the reaction rate and evaluate the integrated flux from one of the distributions given in equations (3.75). Obviously σ_{av} can also be calculated in principle if the cross section curve $\sigma(T_n)$ of the reaction is known.

3.8.7 CALCULATION OF A REACTION RATE FOR REACTOR IRRADIATION

If a nuclide is irradiated in a reactor neutron spectrum, thermal, epithermal and fast neutron activation are possible. Consequently the microscopic reaction rate has to be written

$$r_{tot} = \int_0^{\infty} n(v)v\sigma(v)dv = \int_0^{v_1} n(v)v\sigma(v)dv + \int_{v_1}^{v_2} n(v)v\sigma(v)dv +$$

$$+ \int_{v_2}^{\infty} n(v)v\sigma(v)dv \qquad (3.78)$$

where v_1 corresponds to the cadmium cut-off energy $T_{cd} = 0.5$ eV and v_2 corresponds to an energy of 0.1 MeV where the fast neutron region starts.

Since we study (n,γ) reactions we can almost neglect the fast reaction rate contribution since the contribution to the total (n,γ) reaction due to fission neutrons is usually very small.

We thus write

$$r_{tot} = n_{th} v_o \sigma_o + \varphi_{epi} I = \varphi_o \sigma_o + \varphi_{epi} I \qquad (3.79)$$

which follows from eqs. 3.65 and 3.71. Eq. 3.79 allows the calculation of the reaction rate if φ_o and φ_{epi} are known. The conventional thermal flux φ_o and the epithermal flux φ_{epi} are determined by absolute counting of the activity, which during a given irradiation time is induced in $1/v$ detectors, whose activation cross section are accurately known:

$${}^{59}Co(n,\gamma){}^{60}Co \qquad \sigma_o = 37.2 \text{ b} \qquad \lambda = 4.168 \cdot 10^{-9} \text{ s}^{-1}$$

$${}^{197}Au(n,\gamma){}^{198}Au \qquad \sigma_o = 98.8 \text{ b} \qquad \lambda = 2.98 \cdot 10^{-6} \text{ s}^{-1}$$

To avoid self-shielding (see sec. 10.1.6), these elements are irradiated as dilute Al-alloys eg. 0,1-1 weight per cent Co or Au in Al. Due to resonance peaks in the $\sigma(T_n)$ curve, they are irradiated both without and with cadmium cover.

The induced activity without cadmium cover is given by eqs. 2.52 and 3.54.

$$A(t_e, t_o) = NVr (1-e^{-\lambda t_e})e^{-\lambda t_o} \qquad (3.80)$$

With cadmium cover we get

$$A_{epi} = NV\varphi_{epi} I (1-e^{-\lambda t_e})e^{-\lambda t_o} \qquad (3.81)$$

In the equations NV is the number of target nuclei, t_e is the irradiation time and t_o the waiting time before measurement. The net thermal activation is thus given by

$$A_{th} = A-A_{epi} = NV\varphi_o \sigma_o (1-e^{-\lambda t_e})e^{-\lambda t_o} \qquad (3.82)$$

In absolute counting of the activities the fluxes φ_o and φ_{epi} are determined as σ_o, NV, λ, t_e and t_o are known.

As will be explained in (sec. 10.2.2), the knowledge of the absolute thermal neutron flux is not required in activation analysis, as a comparator method can be used in the assessment of the analysis.

3.8.8   FAST GENERATOR NEUTRONS

Common neutron generators use the reaction $^{3}H(d,n)^{4}He$ to produce neutrons. The neutrons have energies of about 14 MeV and are nearly monoenergetic. However, the energy varies with the emission angle of the neutrons from the tritium target. For instance, its range is 13.4-14.7 MeV for a 150 keV accelerating potential. For 14 MeV neutrons the total reaction and scattering cross section σ_t can be written

$$\sigma_t = 2\pi R^2 \tag{3.83}$$

Cross sections for monoenergetic neutrons of about 14 MeV can be determined by measuring the reaction rates, the error in the energy determination being usually about ± 200 keV. The flux is defined by the "Texas Convention"

$$\varphi_{14} = \frac{r_{cu}}{\sigma_{cu}} \tag{3.84}$$

Here cu refers to the (n, 2n)-reaction on ^{63}Cu, for which a cross section of 500 mb is adopted.

3.9 Proton-induced reactions

All the types of nuclear reactions initiated by neutrons, which we so far have discussed can also be initiated by protons. Elastic and inelastic scattering, proton capture and reactions where neutrons and different charged particles are emitted, are all well-known. The competition between the different reaction modes is however somewhat different, since protons are charged. The height of the Coulomb barrier (eq. 1.15) for stable nuclides

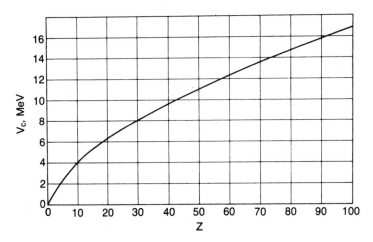

Fig. 3-11 Height of the Coulomb barrier for protons, deuterons
and tritons as a function of Z.

is given in fig. 3-11 and the proton barrier penetration factor for a nuclide with Z = 20 is shown in fig. 3-6b. We see,
that the barrier-penetration factor is substantially less than
unity even when the kinetic energy is a few MeV above the Coulomb barrier.

The effect of the charge is two-fold. Firstly, it is difficult for the proton to reach the edge of the nucleus and
experience the nuclear force. Secondly, the protons are scattered by electrons and lose energy when they passes through matter.

Let us consider a proton beam bombarding a target of thickness d. The stopping power will be discussed in sec. 5.1 and
is given by eq. 5.1. Here we will write the equation for the
stopping power as:

$$- \frac{dT_p}{dx} = \epsilon_p N \qquad (3.85)$$

where N is the atomic density and ϵ_p the stopping cross section
per atom (eq.5.2).

The energy loss in passing through the target, if it is
thin, is then

$$\eta_p = \epsilon_p N d \qquad (3.86)$$

The yield of a reaction with an energy dependent cross section $\sigma(T_p)$ is obtained from eq. 3.24 and is

$$Y = \int_0^d \sigma(T_p)N\,dx = \int_{T_p-\eta_p}^{T_p} \frac{1}{\varepsilon_p}\, \sigma(T_p)\,dT_p \qquad (3.87a)$$

For a very thin target with a weakly energy-dependent cross section this equation can be written

$$Y = \frac{\eta_p}{\varepsilon_p}\, \sigma(T_p) \qquad (3.87b)$$

In the vicinity of an isolated resonance, we use the Breit-Wigner expression for the cross section (eq. 3.42) and obtain from eq. 3.87a:

$$Y = \int_{T_p-\eta_p}^{T_p} \frac{1}{\varepsilon_p}\, \pi\lambda_a^2 g\, \frac{\Gamma_a\Gamma_b}{(T_p-T_r)^2+(\frac{\Gamma}{2})^2}\, dT_p \qquad (3.88)$$

or, if we use the peak cross section $\hat{\sigma}_r = \sigma_{ba}^{max}$

$$\hat{\sigma}_r = \pi\, \lambda_a^2 g\, \frac{\Gamma_a\,\Gamma_b}{(\frac{\Gamma}{2})^2} \qquad (3.89)$$

$$Y = \int_{T_p-\eta_p}^{T_p} \frac{1}{\varepsilon_p}\, \hat{\sigma}_r\, \frac{\Gamma^2/4}{(T_p-T_r)^2+(\frac{\Gamma}{2})^2}\, dT_p =$$

$$= \frac{\hat{\sigma}_r\Gamma}{2\varepsilon_p}\left[\tan^{-1}\frac{(T_p-T_r)}{\frac{1}{2}\Gamma} -\tan^{-1}\frac{(T_p-\eta_p-T_r)}{\frac{1}{2}\Gamma} \right] \qquad (3.90)$$

The maximum yield for a target of thickness d is obtained when the protons have their resonance energy in the middle of the target, i.e., $T_p - \frac{1}{2}\eta_p = T_r$

$$Y_{max}(T_r,\eta_p) = \frac{\hat{\sigma}_r}{\varepsilon_p}\,\Gamma\, \tan^{-1}\frac{\eta_p}{\Gamma} \qquad (3.91)$$

which, for a thick target, gives

$$Y_\infty(T_r) = \frac{1}{2}\pi\, \frac{\hat{\sigma}_r\Gamma}{\varepsilon_p} \qquad (3.92)$$

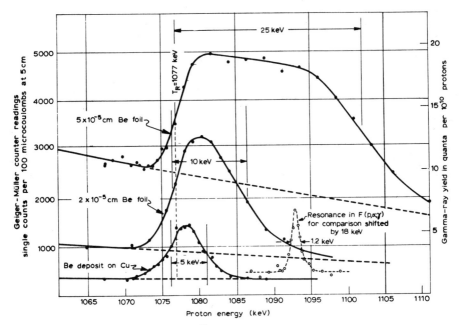

Fig. 3-12 The yield of the ^{9}Be(p,γ) reaction for different
target thicknesses near the resonance at T_r=1077 keV.
E.B. Paul, Nuclear and Particle Physics (North-Holland
1969) p. 214

The ratio

$$\frac{Y_{max}(T_r,\eta_p)}{Y_\infty(T_r)} = \frac{2}{\pi} \tan^{-1} \frac{\eta_p}{\Gamma} \qquad (3.93)$$

thus enables us to find the ratio of the target thickness to
resonance width.

The area under a yield-energy curve for a very thin target depends on the target thickness and is (eq. 3.87)

$$G(T_r,\eta_p) = \int_0^\infty Y dT_p = \frac{\eta_p}{\varepsilon_p} \int_0^\infty \sigma(T_p) dT_p \simeq \frac{\eta_p}{\varepsilon_p} \hat{\sigma}_r \frac{\Gamma}{2} \int_{-\infty}^{+\infty} \frac{d\left(\frac{2T_p}{\Gamma}\right)}{1+\left(\frac{2T_p}{\Gamma}\right)^2}$$

$$= \frac{\pi}{2} \frac{\eta_p}{\varepsilon_p} \hat{\sigma}_r \Gamma = \eta_p Y_\infty \qquad (3.94)$$

A measurement of Y_∞ and $G(T_r,\eta_p)$ enables us to determine
the thickness of the thin target and, combining eqs. 3.93 and
3.94, we also obtain the resonance width (fig. 3-12).

3.10 Photon-induced reactions

3.10.1 PHOTON INTERACTION

The interaction of photons with nuclei is of electromagnetic character. This type of interaction is two orders of magnitude smaller than the strong interaction between incident nucleons and the target nucleus. This implies that the target nucleus is almost transparent to photons. The photonuclear absorption cross section is thus small and has a maximum value of only a few millibarns per nucleon.

In the energy region of interest for nuclear chemical analysis (10-40 MeV) the dominant feature of the photonuclear absorption cross section is the giant resonance which occurs in all elements. Light nuclei have a resonance peak at energies of around 22 MeV, which value decreases to about 13 MeV for heavy nuclei, approximately following the expression

$$E_r = c_1 \, A^{-1/c_2} \tag{3.95}$$

where c_1 = 47.9 MeV and c_2 = 4.27.

The resonance width is about 4 MeV for spherical nuclei and increases to about 8 MeV for strongly deformed ones (fig. 3-13). The energy dependence of the giant resonance absorption cross section for medium and heavy nuclei has often been approximated by a Lorentz-shaped resonance line, with width Γ

$$\sigma(E) = \hat{\sigma}_r \, \frac{E^2 \, \Gamma^2}{(E^2 - E_r^2)^2 + E^2 \Gamma^2} \tag{3.96}$$

where E_r, $\hat{\sigma}_r$ and Γ are the resonance energy, peak cross section and full width at half-maximum, respectively.

For deformed nuclei, the giant resonance splits into two main peaks (fig. 3-14). For light nuclei, the giant resonance shows considerable fine structure related mainly to the properties of individual levels, but also to collective surface vibrations.

We have already discussed (sec. 2.4) the fact that the electromagnetic interaction is divided into different electric

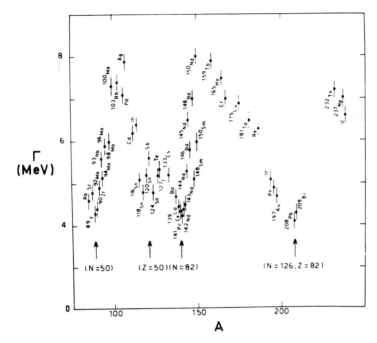

Fig. 3-13 Giant resonance width as a function of A. P. Carlos
et al., Nuclear Physics A219 (1974) 61.

and magnetic multipole interactions. The strongest of these
is the E1 interaction mode. The giant resonance results primarily from E1 absorption and can be explained as a vibration of
the groups of neutrons and protons confined in a nucleus with
a rigid surface. The cross section integrated over the resonance is about two times the classical value of the dipole sum

$$\int_{0}^{E_{\pi}} \sigma(E)dE \simeq 2 \cdot 0.06 \frac{NZ}{A} \text{ (MeV barns)} \qquad (3.97)$$

The energy E_{π} is the meson production threshold. Consequently, the integrated cross section is approximately proportional to the mass number. In the energy region above the giant
resonance, the photon interacts mainly with n-p clusters (quasi-
deuterons) inside the nucleus at medium energies, but above
the photomesom threshold at about 150 MeV, new resonances appear
in the cross section curve. The average photonuclear cross sec-

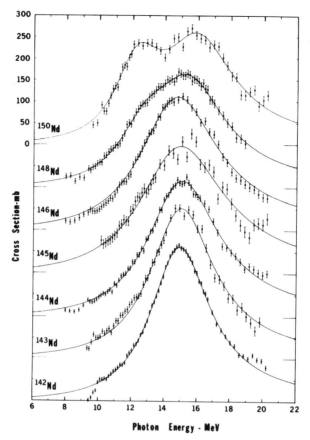

Fig. 3-14 Total photoneutron cross sections for the neodymium
isotopes showing the "evolution" of the giant resonance as one makes the transition from spherical to
statically deformed nuclei. P. Carlos et al, Nuclear
Physics <u>A219</u> (1974) 61.

tion is about 0.30 mb per nucleon for photon energies between
300 MeV and 1000 MeV.

In the giant resonance region, the most probable result
of photonuclear absorption is the emission of a single neutron,
but other processes, such as the emission of γ-rays, the emission of more than one neutron and, particularly for light nuclei, the emission of charged particles, must also be considered.

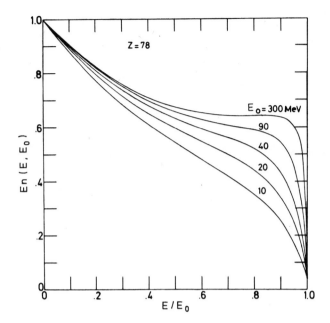

Fig. 3-15 Dependence of the bremsstrahlung spectrum shape on
the electron kinetic energy for a platinum target
normalized to unity at zero energy. H.W. Koch, J.W. Motz,
Rev. Mod. Phys. 31 (1959) 920.

3.10.2   BREMSSTRAHLUNG REACTION YIELD

Since there exist no intense monochromatic photon sources,
the bremsstrahlung beam obtained when energetic electrons from
electron accelerators hit a target is used as the photon source
in almost all photoactivation studies. The energy spectrum of
the photons in a bremsstrahlung beam from a thin target is well
known and is shown in fig. 3-15.

The spectra shown are calculated from the spectrum formulas
given by Schiff [3.29]. For relativistic electrons the average
angle of emission of the photons, $\bar{\theta}$, is

$$\bar{\theta} = \frac{m_e c^2}{E_o} \qquad (3.98)$$

with E_o being the maximum photon energy.

The bremsstrahlung beam has a continous photon energy spectrum. Let $n(E,E_o)dE$ be the number of photons with energies bet-

ween E and E + dE per unit radiation thickness per second in a bremsstrahlung beam, where E_0 is the maximum photon energy.

The energy content of the beam in the sample is then

$$U(E_0) = \int_0^{E_0} En(E,E_0)f(E)dE \qquad (3.99)$$

where $f(E)$ is a correction factor which accounts for the distortion of the bremsstrahlung spectrum by the effect of photon absorption in the machine target, in the walls of the accelerator chamber and in the sample.

We define the number of equivalent quanta Q by

$$Q = \frac{1}{E_0} U(E_0) \qquad (3.100)$$

i.e., Q is the number of quanta with energy E_0 which have the same energy content as the bremsstrahlung beam and we define the cross section per equivalent quantum, σ_q, by

$$\sigma_q(E_0) = \frac{\int_0^{E_0} \sigma(E)n(E,E_0)f(E)dE}{Q} \qquad (3.101)$$

The bremsstrahlung reaction yield measured in monitor response units can be written

$$y(E_0) = \frac{\int_0^{E_0} \sigma(E)n(E,E_0)f(E)dE}{U(E_0)R(E_0)} = \frac{\sigma_q}{E_0 R(E_0)} \qquad (3.102)$$

The monitor measures only some part of the energy content of the beam and thus $R(E_0)$ gives the sensitivity of the monitor for a E_0-bremsstrahlung beam.

While $f(E)$ and $R(E_0)$ are quantities specific for a particular laboratory, $n(E,E_0)$ has been tabulated for bremsstrahlung spectra.

In table 3-2 and table 3-3, the energy content and the energy spectrum of an undistorted bremsstrahlung beam for dif-

122

Table 3-2

Energy content of an undistorted bremsstrahlung beam. (A.S. Penfold and J.E. Leiss, Analysis of Photo-Cross Sections, Physical Research Laboratory, University of Illinois 1958).

E_0	$\int_0^{E_0} En(E,E_0)dE$
MeV	MeV
10	03.55488
13	04.78641
16	06.04665
19	07.32788
22	08.62522
27	10.81470
32	13.02999
36	14.81454
44	18.41618
52	22.04727

ferent end point energies in the giant resonance region are given. It is easy to obtain the photon spectrum from the second table by dividing the energy spectrum value with the photon energy E.

Table 3-3

Energy spectrum of an undistorted bremsstrahlung beam
$\Phi = En(E,E_o)$. (A.S. Penfold and J.E. Leiss, Analysis of Photo-Cross Sections, Physical Research Laboratory, University of Illinois (1958). The table values are 10^5 times too high.

E MeV	Φ 10MeV	E MeV	Φ 13MeV	E MeV	Φ 16MeV	E MeV	Φ 19MeV	E MeV	Φ 22MeV
10·	02933	13	02814	16	.02739	19	02687	22	02649
09	16426	12	16239	15	16140	18	16080	21	16040
08	22660	11	22271	14	22105	17	22024	20	21981
07	27063	10	26222	13	25883	16	25731	19	25661
06	30926	09	29321	12	28676	15	28386	18	28250
05	34821	08	32093	11	30985	14	30476	17	30229
04	39085	07	34829	10	33081	13	32262	16	31852
03	43959	06	37720	09	35136	12	33907	15	33275
02	49629	05	40900	08	37269	11	35519	14	34601
01	56252	04	44469	07	39564	10	37176	13	35905
00	64010	03	48504	06	42083	09	38935	12	37239
		02	53068	05	44875	08	40838	11	38643
		01	58215	04	47975	07	42918	10	40146
		00	64010	03	51414	06	45198	09	41772
				02	55217	05	47700	08	43539
				01	59406	04	50438	07	45461
				00	64010	03	53426	06	47548
						02	56676	05	49812
						01	60200	04	52258
						00	64010	03	54895
								02	57728
								01	60764
								00	64010

Table 3-3
Continued

E MeV	Φ 27MeV	E MeV	Φ 32MeV	E MeV	Φ 36MeV	E MeV	Φ 44MeV	E MeV	Φ 52MeV
27	02604	32	02573	36	02554	44	02527	52	02508
26	15997	31	15970	34	21935	42	21938	50	21943
25	21949	30	21938	32	28178	40	28224	48	28273
24	25620	29	25619	30	31479	38	31534	46	31613
23	28170	28	28163	28	33550	36	33553	44	33631
22	30068	27	30037	26	35057	34	34931	42	34967
21	31563	26	31483	24	36335	32	35990	40	35936
20	32805	25	32647	22	37568	30	36909	38	36713
19	33891	24	33625	20	38869	28	37797	36	37402
18	34891	23	34483	18	40308	26	38720	34	38070
17	35855	22	35270	16	41931	24	39725	32	38761
16	36820	21	36020	14	43771	22	40841	30	39504
15	37813	20	36759	12	45849	20	42090	28	40322
14	38855	19	37508	10	48182	18	43488	26	41229
13	39964	18	38283	8	50782	16	45046	24	42236
12	41151	17	39096	6	53657	14	46772	22	43352
11	42428	16	39958	4	56816	12	48674	20	44583
10	43803	15	40875	2	60265	10	50757	18	45936
09	45283	14	41854	0	64010	8	53025	16	47412
08	46873	13	42902			6	55482	14	49017
07	48577	12	44021			4	58129	12	50753
06	50401	11	45216			2	60971	10	52621
05	52347	10	46490			0	64010	8	54623
04	54418	09	47846					6	56762
03	56618	08	49285					4	59039
02	58948	07	50810					2	61454
01	61411	06	52423					0	64010
00	64010	05	54124						
		04	55915						
		03	57799						
		02	59774						
		01	61844						
		00	64010						

3.11 References

3.11.1 GENERAL NUCLEAR REACTIONS

3.1   F. Ajzenberg-Selove (ed.), Nuclear Spectroscopy 2 vols.
      Academic Press (1960).

3.2   J. Cerney (ed.), Nuclear Spectroscopy and Reactions 4 vols.
      Academic Press (1974).

3.3   P.M. Endt and M. Demeur (eds.), Nuclear Reactions I,
      North-Holland (1959).

3.4   P.M. Endt and P.B. Smith (eds.), Nuclear Reactions II,
      North-Holland (1962).

3.5   J.B. Marion and F.C. Young, Nuclear Reaction Analysis
      Graphs and Tables North-Holland (1968).

3.11.2   COMPILATION OF CHARGED PARTICLE-INDUCED REACTION DATA

3.6   F.K. McGowan and W.T. Milner, Charged-particle Reaction
      List 1948-1971 Atomic and Nuclear Data Reprints (ed. K.Way)
      Academic Press (1973).
      1971-1972 Nuclear Data Tables All, 1 (1972).
      1972-1973 Atomic Data and Nuclear Data Tables 12, 499 (1973).
      1973-1974 ibd. 15, 189 (1975).
      1974-1976 ibd. 18,1 (1976)

3.7   K.A. Keller et al. Q-values and Excitation Functions of
      Nuclear Reactions, Landolt-Börnstein New Series vol. 5
      Springer (1974).
      a) Q-values.
      b) Excitation functions for charged particle induced
         nuclear reactions.
      c) Estimation of unknown excitation functions and thick
         target yields for p, d, ^{3}He and α-reactions.

3.8   J. Lorenzen and D. Brune, Excitation Functions for Char-
      ged-particle-induced Nuclear Reactions in Light Elements
      at Low Projectile Energies. Handbook on Nuclear Activa-
      tion Cross-Sections, Techn. report series 156 IAEA (1974).

3.11.3 COMPILATION OF NEUTRON-INDUCED REACTION DATA

3.9 W.E. Alley and R.M. Lessler, Neutron Activation Cross Sections Nuclear Data Tables All, 621 (1973).

3.10 G.A. Bartholomew, L.V. Groshev et al, Compendium of Thermal-Neutron-Capture γ-Ray Measurements.
Part I: $Z \leq 46$ Nuclear Data Tables A3, 367 (1967).
Part II: Z = 47 to Z = 67 (ag to Ho) ibd. A5,1 (1968).
Part III: Z = 68 to Z = 94 (Er to Pu) ibd. A5, 243 (1969).

3.11 J.R. Bird et al, Compilation of keV-Neutron-Capture Gamma-Ray Spectra, Nuclear Data Tables All, 433 (1973).
BNL-325, Third edition, National Neutron Cross-Section Center, Brookhaven National Laboratory.
Vol. 1: Thermal cross sections, resonance properties, resonance parameters and bibliography (dec. 1973).
Vol. 2: Curves and bibliography (jan. 1976).

3.12 BNL-400 Third edition, Angular Distributions in Neutron-induced Reactions, National Neutron Cross-Section Center, Brookhaven National Laboratory.
Vol. 1: Z = 1-20 (jan. 1970).
Vol. 2: Z = 21-94 (june 1970).

3.13 D. Brune and J.J. Schmidt (eds.), Handbook on Nuclear Activation Cross Sections, Techn. report series 156 IAEA (1974).

3.14 E. Bujdosó, I. Fehér and G. Kardos, Activation and Decay Tables of Radioisotopes, Elsevier (1973).

3.15 CINDA A (1935-1976). An Index to the Literature on Microscopic Neutron Data.
Vol. 1: $Z \leq 50$,
Vol. 2: $Z \geq 51$, IAEA (1979).

3.16 G. Erdtmann, Neutron Activation Tables, Verlag Chemie (1976).

3.17 R.B. Schwartz et al., MeV Total Neutron Cross-sections, Graphical Plots of the Results of the Neutron Time-of-flight Measurements Performed at the US NBS for 15 Elements resp. Isotopes. NBS Monograph 138 (jan. 1974). (U.S. Government Printing Office Washington, D.C. 1974).

3.11.4   COMPILATION OF PHOTON-INDUCED REACTION DATA

3.18 B.L. Berman, Atlas of Photoneutron Cross-sections Obtai-
ned with Monoenergetic Photons, Atomic Data and Nuclear
Data Tables 15, 319 (1975). Revised Edition UCRL-78482
Lawrence Livermore Laboratory (dec. 1976).

3.19 B. Bülow and B. Forkman, Photonuclear Cross-sections, Handbook on Nuclear Activation Cross-sections, Techn. report
series 156, IAEA (1974).

3.20 M.G. Davydov et al., The Caracteristics of Photo-activa-
tion of Light Elements, INDC (CCP)-104/LN 1977 Translated
from Atomnaya Energiya.

3.21 E.G. Fuller et al., Photonuclear Reaction Data, 1973 NBS
Special Publications 380 (U.S. Government Printing Office
Washington D.C. 1973) and Supplement to NBS-380, (1976).

3.11.5   GAMMA LINES OF RADIONUCLIDES

3.22 R.W. Carr and J.E.E. Baglin, Table of Angular Distribution
Coefficients for (Gamma-Particle) and (Particle-Gamma)
Reactions, Nuclear Data Tables 10, 143 (1972).

3.23 G. Erdtmann and W. Soyka, Gamma-lines of Radionuclides, a
Tabulation of Gamma-lines for Radionuclides Produced by
Neutron-induced or Any Other Reactions.
Part 1   Z = 2-57, J. Radioanal. Chem. 26, 375 (1975).
Part 2   Z = 58-100, ibd. 27, 135 (1975).

3.24 S.A. Lis et al., Gamma-ray Tables for Thermal Neutron,
Fast Neutron (14 MeV) and Photon Activation Analysis,
J. Radioanal. Chem. 24, 125 (1975).

3.25 R.J. de Meijer, A.G. Drentje, Tables for Reaction Gamma-
Ray Spectroscopy
Part 1   A = 6-20 Atomic Data and Nuclear Data Tables 13,
         1 (1974).
Part 2   A = 21-32 ibd. 15, 391 (1975).
Part 3   A = 33-44 ibd. 17, 211 (1976).

3.26 M.E. Toms, A Compilation of Photonuclear Reaction Products
and Associated Gamma-energies, J. Radioanal. Chem. 20, 17
(1974).

3.11.6 ENERGY LEVELS OF LIGHT NUCLEI

3.27 For light nuclei, there exist some very informative and
illustrative schemes of nuclear levels and cross-section
data.
A = 3, 4 S. Fiarman, S. Hanna, W.E. Meyerhof, Nucl. Phys.
A251, 1 (1975), A206, 1 (1973).
A = 5-10, 11-12, 13,15, 16-17, 18-20
F. Ajzenberg-Selove, Nucl. Phys. A320, 1 (1979).
A336, 1 (1980) A268, 1 (1976), A281, 1 (1977), A300, 1
(1978).
A = 21-44 P. Endt, C. Van der Leun, Nucl. Phys. A310, 1
(1978).

3.28 For heavier nuclei compilations of the energy levels appear
in Nuclear Data Sheets, Academic Press. The level schemes
present the energy level diagrams for the entire A-chain
showing the levels and the known transitions between them.
The following information is given; nuclear level properties,
energy, half-life, spin and parity, in deformed region
the Nilsson model quantum numbers, letter code indications
which nuclear reactions are known to excite the level,
transition properties, α-, β-, γ-energies and intensities,
multipolarities of γ-rays, log ft's of β-transition, mass
differences, total β-disintegration energies and neutron
and proton separation energies.

3.11.7 BREMSSTRAHLUNG SPECTRUM

3.29 L.I. Schiff, Phys. Rev. 83, 252 (1951).

4 Nuclear Radiation Sources

4.1 Introduction

As in all chemical analysis the main problem in nuclear chemical analysis is to decide which method is the most accurate and convenient one to solve a given task. Using different radiation sources, different activation products are produced from the target nuclide and the choice of activation product thus determines the irradiation source required.

In many cases, it is more convenient to use standard methods applicable to a great number of targets even if the accuracy is impaired, while in others it is necessary to make a careful choice of radiation source to solve the problem in a satisfactory way. Very often financial aspects and the local equipment are decisive for the choice of source. In this chapter, we will briefly discuss the main sources of neutrons, protons, deuterons and α-particles, electrons and photons.

4.2 Neutron-producing devices

4.2.1 RADIONUCLIDE SOURCES

Radionuclide neutron sources are based on the (α,n) reaction with beryllium as the standard target, on the photoneutron i.e. (γ,n) reaction with beryllium or deuterium (D_2O) as target, or on spontaneous fission, with ^{252}Cf being the prime example:

$$^{9}_{4}Be + \alpha \rightarrow {}^{12}_{6}C* + n + 5.71 \text{ MeV} \tag{4.1a}$$

where C* denotes that the carbon nucleus mostly is left excited,

$$^{9}_{4}Be + \gamma \rightarrow {}^{8}_{4}Be + n - 1.67 \text{ MeV} \tag{4.1b}$$

where the ^{8}Be nucleus immediately decays. The half-life is $7 \cdot 10^{-17}$ s.

$$^8_4 Be \rightarrow 2\alpha + 0.09 \text{ MeV} \tag{4.1c}$$

$$^2_1 H + \gamma \rightarrow {}^1_1 H + n - 2.23 \text{ MeV} \tag{4.1d}$$

$$^{252}_{98} Cf \rightarrow 2f + 3.8n + 200 \text{ MeV} \tag{4.1e}$$

where the spontaneous fission half-life is 2.65y and the average number of released fission neutrons 3.8.

4.2.1.1 Sources of type (α, n)

In (α, n) sources the 5.3 MeV α-particles from 138-d ^{210}Po have until recently been used. Other naturally occurring α-particle emitters used include 1600-y ^{226}Ra (in equilibrium with its daughters) and 21.8-y ^{227}Ac. Modern sources include the transuranium elements plutonium, americium and curium. Light elements other than beryllium can be used as the target, but since beryllium yields the largest output of neutrons per α-particle, it is almost exclusively used as the target element in large sources. Characteristics of available sources are summarized in table 4-1.

Theoretically, the neutron spectrum could be calculated from the known α-particle energy, the Q value, the kinematics of the (α, n) reaction and the nuclear properties of ^{9}Be. The neutrons are not monoenergetic, because the α-particles lose energy in the material and all angles of incidence are possible. Furthermore, in some 80% of the reactions, the ^{12}C nucleus is left in a 4.43 MeV excited state, which then decays with emission of 4.43 MeV γ-rays. Another complication is the degradation of neutron energy by collisions in the source itself. In practice, then, the spectrum is broad and has to be measured by a suitable neutron spectrometer. Neutron spectra from some typical (α, n) sources are shown in fig. 4-1.

It is of interest to know the γ-ray dose rates from different radionuclides since this facilitates handling and shielding and some experiments are improved by low dose rates. Such information is given in table 4-1 and it can also be seen that α-particle heating is usually no problem.

Table 4-1

Characteristics of radionuclide neutron sources. Radiation sources for laboratory and industrial use. The Radiochemical Centre Amsterdam 1974/75.

Radionuclide	$t_{1/2}$	neutron yield n/s per Ci	γ-radiation from an encapsulated source Exposure rate mR/h at 1 m per Ci
$^{227}Ac(\alpha)$	21.8y	$1.5 \cdot 10^7$	120
$^{241}Am(\alpha)$	433y	$2.2 \cdot 10^6$	< 2.5
$^{252}Cf(f)$	2.65y	$4.3 \cdot 10^9$	300 (unshielded)
$^{242}Cm(\alpha)$	163d	$2.5 \cdot 10^6$	2.5
$^{238}Pu(\alpha)$	87.8y	$2.2 \cdot 10^6$	< 1
$^{226}Ra(\alpha)$	1600y	$1.3 \cdot 10^7$	780
$^{124}Sb(\gamma)$	60d	$5.2 \cdot 10^6$	980
$^{228}Th(\alpha)$	1.91y	$2 \cdot 10^7$	600

$1Ci = 3.7 \cdot 10^{10} Bq$
$1R = 2.58 \cdot 10^{-4}$ coulomb/kg.

4.2.1.2 S o u r c e s   o f   t y p e (γ,n)

In (γ,n) sources, the γ-ray emitter must have an energy greater than the (-Q)-value, 1.67 MeV for $^9Be(\gamma,n)$ or 2.23 MeV for $^2D(\gamma,n)$. The energy of the neutron is given by

$$T_n \simeq \frac{A-1}{A} (E_\gamma + Q) - \frac{E_\gamma^2}{2Auc^2} + \delta \cos \psi_n \qquad (4.2)$$

with the correction term

$$\delta^2 = \frac{2(A-1)}{A^3 uc^2} E_\gamma^2 (E_\gamma + Q) \qquad (4.3)$$

where ψ_n is the angle between the incident photon and the emitted neutron, u is the atomic mass unit and uc^2 corresponds

132

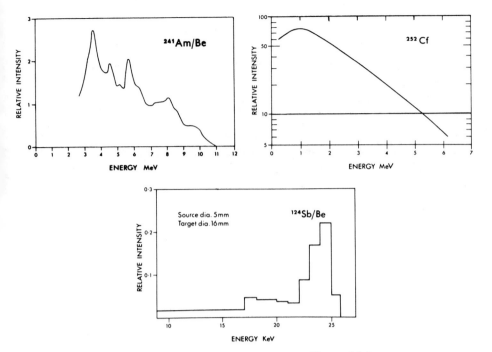

Fig. 4-1   Neutron spectra from some radionuclide sources.

to 931 MeV. Observe that Q is negative since the reactions are endoergic.

The spread in neutron energy is given by

$$\Delta T_n = 2\delta \tag{4.4}$$

but since the emitted neutron scatters in the source material, the actual spread is considerably larger (fig. 4-1 c). The most commonly-used photoneutron source is ^{124}Sb + Be with T_n = 26 ± 2 keV. ^{124}Sb is produced by the irradiation of normal antimony by the ^{123}Sb(n,γ) reaction. The ^{124}Sb has a half-life of 60.2 days and emits several γ-rays, including a 1.692 MeV γ-ray in 49% of the disintegrations.

The possibility of producing nearly monoenergetic neutrons is interesting but, as is seen in table 4-1, the neutron yield is low and, with the further disadvantages of shorter half-lives and added hazard due to penetrating γ-radiation, photoneutron sources are not widely used for neutron-activation purposes.

4.2.1.3 S p o n t a n e o u s   f i s s i o n   s o u r c e s

Californium -252 is now being produced in milligram to gram
quantities by irradiating plutonium or transplutonic nuclides
in high flux reactors. About 3.2% of its decays are spontaneous
fission, releasing 3.76 neutrons per fission. From eqs. 2.20 a
and 2.22, the neutron emission rate is calculated to be
$2.34 \cdot 10^{12}$ $ns^{-1}g^{-1}$. This is quite high and results in very compact sources. The spectrum (fig. 4-1 b) is close to that of
neutron-induced fission (fig. 3-7) which opens new possibilities
for reactor simulating experiments and for instrument calibration.

4.2.2 NEUTRON GENERATORS

Fast neutrons can be produced in accelerators of various
types, accelerating protons, deuterons, α -particles and even
electrons. In neutron activation analysis, we are especially
concerned with accelerators giving neutrons with energies up
to some 15 MeV. These can be provided by (p,n), (d,n), and (α,n)
reactions where the charged particle has been accelerated at
energies of 0.1 to 10 MeV and by (γ,n) reactions where the photons are derived from electrons accelerated to 10-100 MeV. Different types of p-, d-, α- and e-accelerators will be discussed
later in this chapter. Here, we will concentrate on the neutron
generator which is a small accelerator in which deuterons are
accelerated to energies of some hundred keV. Table 4-2 shows
the characteristics of some nuclear reactions for producing
fast neutrons. The neutron generator generally operates with
the T(d,n) ^{4}He reaction

$$^{3}H + {}^{2}H \rightarrow {}^{4}He + n + 17.6 \text{ MeV} \qquad (4.5)$$

At zero bombarding energy and in forward direction, the
neutrons have an energy of 14.1 MeV. At low bombarding energies,
the emission is nearly isotropic. The cross section for the T(d,n)
reaction peaks at 105 keV, and this would be the optimum bombarding energy for a thin target. However, we are generally
more concerned with copious neutron production than with strict
monochromaticity and hence thick targets are used (see sec. 3.9).

Table 4-2

Neutron-producing reactions for ion accelerators.

	Endoergic		Exoergic	
Reaction	T(p,n)	^{7}Li(p,n)	D(d,n)	T(d,n)
Q (MeV)	-0.764	-1.646	3.266	17.586
Threshold (MeV)	1.019	1.882		
Min. E_n at 0° (MeV)	0.0639	0.0294	2.448	14.05

The D(d,n) reaction is less frequently used in activation analysis since the neutrons are generally too low in energy to initiate the required nuclear reaction.

4.2.2.1 Pumped neutron generators

Two systems are used for neutron production, the pumped neutron generator and the sealed-tube pulsed neutron generator, which will be discussed in the next section.

The pumped neutron generator is composed of the following basic components:

i) source of deuteron ions, ii) acceleration system, iii) power supply, iv) target and v) vacuum pump system.

The ion source generates a relatively copious supply of charged particles, but at energies of only a few eV. The ion beam is extracted, focused and accelerated by an electromagnetic field, finally striking a target, where the nuclear reactions occur. Since particles can be scattered by gas and thus lost to the beam, the apparatus is generally evacuated to about 10^{-5} torr or less. A good vacuum is also required to avoid arcing. The pumped neutron generator is either equipped with a high voltage supply of the Cockroft-Walton type (e.g. in the Kaman neutron generator) with a normal voltage range of 150-500 kV (fig. 4-2) or works as a single stage Van de Graaff accelerator with a normal accelerating voltage of 90-600 kV(eg., the french-produced Sames generator). The principle of the VdG generator is outlined in sec. 4.3.

135

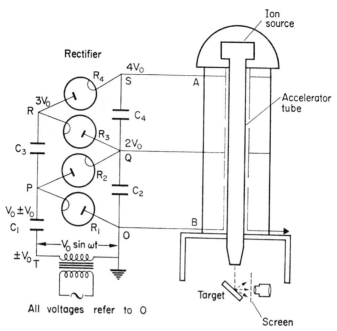

Fig. 4-2   Cockroft-Walton accelerator. Two condenser banks
C_1, C_3 and C_2, C_4 are combined with rectifiers R_1,
R_2, R_3, R_4 to quadruplicate the transformer peak
potential V_o.

The ion source in the CW and VdG machines is usually of
the radiofrequency (RF) or Penning type, although duoplasmatrons
are also coming into use. An RF source is shown in fig. 4-3.
It consists of a cylindrical glass bottle with an electrode
("probe") at one end and a canal, or hole, for extraction of
ions at the other end. Hydrogen, deuterium, or helium gas
(depending on the ions to be generated) is introduced from a
bottle of compressed gas at about 0.1 torr through either a
needle-valve or a heated palladium "leak". A gas discharge is
produced by the application of a radiofrequency field (typically ~ 100 MHz) of some 100 watts input power to the ring electrodes. The resulting low-density plasma (~ 0.01% of neutral gas
density) is concentrated by axial magnetic field of a solenoid
magnet of about 0.1 tesla. The ions are extracted from the plasma boundary at the canal by application of an electrostatic
field. In practice, the maximum ion current is about 1 mA.

136

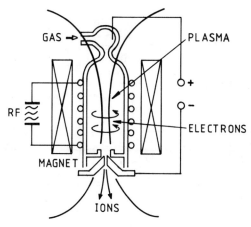

Fig. 4-3   The principle of the RF ion source (ref. 4.8).

The duoplasmatron employs a hot-cathode gas discharge
with a concentrating magnetic field at the canal. Currents
of over 10 mA may be obtained. However, the power and cooling
requirements are greater than for the RF ion source.
Various types of ion sources have been described in detail
by Wilson and Brewer[4.9].
The ions are focused and accelerated by the gap lens and
the accelerating column. The ions are focused on the target by
the lens and a focal spot on target of 3-10 mm diameter is achi-
even in CW and VdG accelerators. The beam can be pulsed by
applying an RF or square-wave voltage to plane electrodes bracke-
ting the beam and deflecting it into water-cooled stops. The
basic principles of the pumped neutron generator system are shown
in fig. 4-4 (Kaman type).

4.2.2.2 S e a l e d - t u b e   n e u t r o n   g e n e r a t o r s

Sealed-tube pulsed neutron generators operate with the
ion source and accelerating gap at the same pressure, thus
obviating the need for a vacuum pumping system. The maximum
potential drop which can be maintained without sparking under
vacuum or at pressures suitable for ion source operation is
of the order of 100 kVcm^{-1}. With an accelerating gap of 1 to
2 cm, a potential drop of 120 to 150 kV is obtained. This is
quite suitable for neutron generation by the $T(d,n)^4He$ reaction.
Negative high voltage pulses are applied to the target from

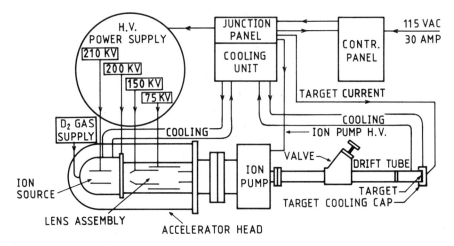

Fig. 4-4   Principles of the pumped neutron generator system.

a capacitor network discharging through a thyratron into a step-
up transformer and deuteron currents of several hundred mA may
be obtained. With such large beam currents, the peak neutron
generation rate may be over 10^{13} ns^{-1}. However, the permissible pulse width and duty cycle are relatively small, and the average effective neutron production rate may be only $\sim 10^{8}$ns^{-1}.
A typical tube is operated at 2.5 μs pulse width with up to
10 pulses/s. In the sealed tube generator, the tritium in the
target is continuously replenished. This is a great advantage
since, during running, the tritium in the target normally becomes depleted. The replenisher is loaded with a mixture of
deuterium and tritium which recirculates. Tritons as well as
deuterons are then accelerated from the ion source via the
extractor into the target so that the level of tritium remains
roughly constant during operation. A diagram of a pulsed neutron generator is given in fig. 4-5. For mechanical protections,
the glass tube is encapulated in a steel housing.

4.2.2.3 T a r g e t s

Various types of targets are used for neutron production.
The conventional tritium target usually consists of a titanium
(or zirconium) foil, with a copper backing for cooling purposes. Tritium gas is diffused into this foil. The life-time of

138

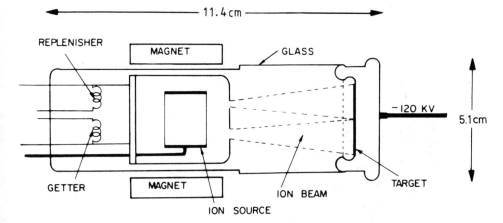

Fig. 4-5   Diagram of a pulsed neutron generator.

such a target usually amounts to less than a few hours at full
neutron output. However, by rotating the target, the life-time
can be considerably prolonged. Rotating target devices are available on the commercial market. In the $T(d,n)^4$He reaction, deuterium exchanges with the tritium on the target at a rate proportional to the beam current. This reduces the target life-time
and the neutron output. Also, tritium is released from the target due to heating effects produced by deuteron stopping in the
target.

In the sealed neutron tube, tritium is replenished and
the useful life of the tube may be 10^5 to $5 \cdot 10^5$ pulses, after
which the tube must be returned to the factory for rebuilding.

At the Lawrence Radiation Laboratory, Livermore, Cali-
fornia, a high intense neutron source has been developed for
studies of radiation effects in materials [4.10]. The system
contains basically the ion source of the duoplasmatron type,
a deuteron accelerator (400 kV) and a rotating tritium target.

The target rotates at 1100 rpm while the deuteron beam
is held fixed. In extended runs, an average flux of $1.4 \cdot 10^{12}$ n
$cm^{-2}s^{-1}$ has been achieved. Rotating target assemblies for intense 14 MeV neutron sources have been described in detail by e.g.
Booth and Barshall [14.11].

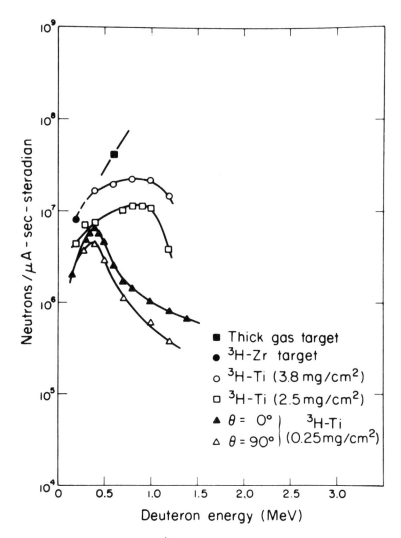

Fig. 4-6 Yield of T(d,n)^{4}He raction (ref. 4.12).

4.2.2.4 F a s t   n e u t r o n   o u t p u t   a n d   d i s t r i-
         b u t i o n

        The full neutron output of a high capacity neutron genera-
tor amounts to about $2 \cdot 10^{11} ns^{-1}$ (4π geometry).

        The neutron output is a function of various parameters,
mainly the following:

- The energy of the bombarding particles (see fig. 4-6).

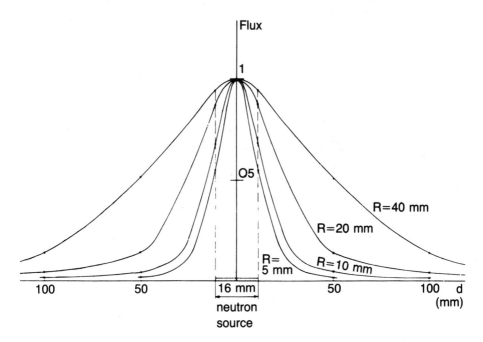

Fig. 4-7 Lateral flux distribution around a target (ref. 4.13).

- The beam current, which determines the number of incident
   particles on the target.
- The amount of tritium in the target.

     Beam currents from high capacity generators are usually
of the order of 1-2 mA. The loading may be expressed in $Ci \cdot cm^{-2}$
or in the thickness of the tritide $mg \cdot cm^{-2}$. A typical amount
of tritium on the target is 1 $Ci \cdot cm^{-2}$ or $3.7 \cdot 10^{10}$ $Bq \cdot cm^{-2}$.

     In the generator, not only single atoms of deuterium
are ionized and accelerated but D_2^+ ions are also accelerated.
D_2^+ ions are accelerated with velocities lower than those of
the D^+ ions, and will consequently give rise to a lower neutron
output.

     The fast neutron distribution around the target is non-
uniform. At distances exceeding 10 cm from the target, the fast
flux decreases roughly with distance (d) as $1/d^2$. Close to the
target, the flux decrease can be described approximately by
the 1/d relationship. The lateral flux distribution, i.e., the

flux in planes parallel to a disc source, has been calculated by Op de Beeck[4.13]. The results are given in fig. 4-7.

The fast neutrons have an energy range of about 13.5 -
15 MeV, at backward angles to forward angles taking into account the neutron angular distribution as well as the deuteron
energy[4.14]. In analysis, it is convenient to place the samples to be irradiated as close as possible to the target. Usually a standard is also irradiated simultaneously. Furthermore,
the sample and standard are generally rotated in front of the
target in order to obtain equal neutron doses, due to the non-
uniform neutron distribution.

4.2.3   OTHER NEUTRON-PRODUCING REACTIONS

4.2.3.1   P r o t o n (d e u t e r o n) a c c e l e r a t o r s

Proton (deuteron) machines can supply essentially monochromatic well-defined neutrons from about 1 keV to 20 MeV,
using thin solid or gas targets of tritium, deuterium or lithium.
The reaction data are given in table 4-2. The neutron energy
varies with the angle to the ion beam according to the following relationship (eq. 3.8):

$$(m_3+m_n)T_n+(m_1-m_3)T_1-2(m_1m_nT_1T_n)^{\frac{1}{2}}\cos\psi_n = m_3Q \qquad (4.6)$$

where m_1, m_n and m_3 are the masses of the projectile, the neutron and the residual nucleus respectively, T_1 and T_n the
kinetic energies of the projectile and the outgoing neutron
and Q the Q-value of the process. In fig. 4-8 the relationship
between T_n and T_1 is shown at the observation angles of $\psi=0^\circ$
and 150° for the most interesting reactions. The $^7Li(p,n)^7$ Be
reaction is not strictly monoenergetic; there is a second group
($\sim$ 10% of the main group) observed at bombarding energies above
2.38 MeV, because of excitation of the 435 keV level in ^{7}Be.
The $T(p,n)^3$He reaction has also a second low intensity group
at proton energies below 1.15 MeV, because of centre-of-mass
motion.

$D(d,n)^3$He neutrons are no longer monoenergetic above
a deuteron energy of 4.45 MeV because of a tertiary reaction
$D(d,np)D$.

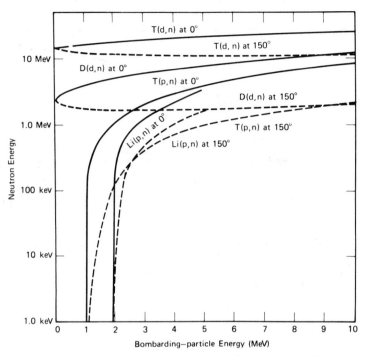

Fig. 4-8 Neutron energies from $T(p,n)^4$He, $D(d,n)^3$He, $T(p,n)^3$He $T(p,n)^3$He and ^{7}Li$(p,n)^7$Be reactions.

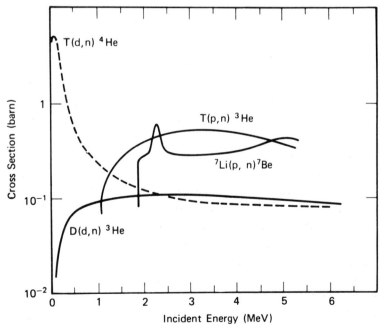

Fig. 4-9 Cross sections for some reactions producing neutrons.

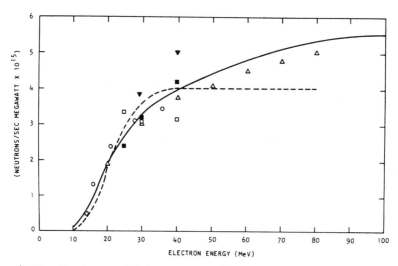

Fig. 4-10   Neutron yield from a thick target of depleted maximum as a function of incident electron energy (ref.
4.15).

All the reactions mentioned here are of considerable
interest in fusion physics. Their cross sections are shown in
fig. 4-9.

4.2.3.2 E l e c t r o n   a c c e l e r a t o r s

Electron accelerators may be used conveniently as neutron
sources. Usually the electrons are converted to photons by
slowing down in a heavy element such as tantalum or tungsten,
which yields bremsstrahlung beams of photons. These photons
then interact with suitable targets such as Be, D or heavy
metals such as W, Pb or U to produce neutrons by the (γ,n)
reaction.

The neutrons can be used either directly as fast neutrons
or be moderated to thermal energies by suitable water or paraffin
assemblies placed around the accelerator target.

At electron energies above 20 MeV or above the giant
resonance in uranium (eq. 3.95), greater yields can be achieved
with a depleted uranium target. At 25 MeV, the yield is about
0.01 neutrons per electron and increases with increasing energy. Measured and calculated[4.15] source strengths for a thick
target expressed in terms of neutrons /s per MW of beam power

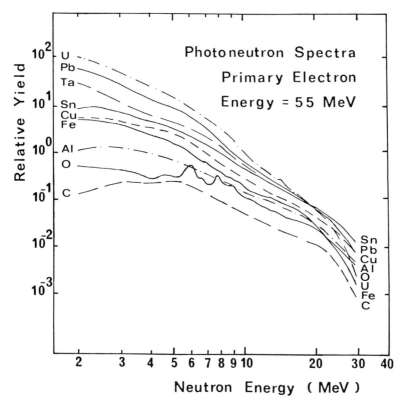

Fig. 4-11 Neutron spectra from different targets (ref. 4.16).

are shown in fig. 4-10. At 25 MeV, the emission rate is about
$3 \cdot 10^{15}$ ns^{-1} per MW. An average emission rate of $4 \cdot 10^{13} ns^{-1}$ and
a peak rate of $5 \cdot 10^{16} ns^{-1}$ or $2 \cdot 10^{19}$ n/pulse, may be achieved
in a powerful linac (linear accelerator).

    Neutron spectra from different targets are shown in fig.
4-11[4.16]. The spectra are of the "evaporation" type more or
less resembling the neutron spectrum from fission. The emergent
spectra are modified by neutron interactions (particularly
inelastic scattering) within the targets themselves.

    The greatest thermal neutron flux density is achieved
with the target immersed in a large, e.g., 30 to 50 cm radius,
moderator with a high density of hydrogen nuclei, for example,
water or dense polyethylene or paraffin $(CH_2)_n$. The flux density depends on the source energy and on the geometrical and
absorbing properties of the target assembly.

Table 4-3

Neutron production and fluxes available from electron irradiation of beryllium targets. Adapted from data collected by Malcolm H. MacGregor, Applied Radiation Corporation News, (July 1958) for the neutron yields from the electron irradiation of a beryllium target located behind a thin, high Z converter.

Electron Energy (MeV)	Total Neutron Production (n/s-kW)	Thermal Neutron Flux in a Water-Moderated Source (n/cm^2-s-kW)
2	$5.0 \cdot 10^8$	$1.2 \cdot 10^7$
3	$6.7 \cdot 10^9$	$1.0 \cdot 10^8$
4	$1.8 \cdot 10^{10}$	$2.1 \cdot 10^8$
6	$3.0 \cdot 10^{10}$	$3.0 \cdot 10^8$
20	$7.0 \cdot 10^{11}$	$5.0 \cdot 10^8$

Table 4-3 lists the neutron production and water-moderated thermal neutron fluxes available from a 20 MeV, 12-kW electron linac operated at full power. The thermal neutron flux of $6 \cdot 10^{10}$n cm^{-2}s^{-1} at full power is about equivalent to the thermal neutron flux available in a 2 kW nuclear reactor and is sufficient for the activation of microgram quantities of many elements.

4.2.4 NUCLEAR REACTORS

4.2.4.1 Chain reaction

The operation of a nuclear reactor is based on the fission process which was described in sec. 2.5. Nuclear fission reactions occur spontaneously with the very heaviest transuranium nuclides, for example, ^{252}Cf (sec. 4.2.1.3), and by irradiation of all heavy elements with sufficiently energetic charged particles, neutrons and photons. In general, these reactions all require considerably more energy to occur than they produce.

So far the only reactions of practical importance are the neutron fission of uranium and plutonium. The three nuclides fissionable with thermal neutrons are ^{233}U, ^{235}U and ^{239}Pu, of which only ^{235}U (0.71% of natural uranium) occurs in nature. The other two can be produced from the fertile nuclides ^{232}Th and ^{238}U by the neutron reactions

$$^{232}Th\ (n,\gamma)\ ^{233}Th \xrightarrow{\beta^-} {}^{233}Pa \xrightarrow{\beta^-} {}^{233}U \qquad\qquad (4.7a)$$

$$^{238}U\ (n,\gamma)\ ^{239}U \xrightarrow{\beta^-} {}^{239}Np \xrightarrow{\beta^-} {}^{239}Pu \qquad\qquad (4.7b)$$

From fig. 3-10 where elastic and inelastic scattering cross sections and capture and fission cross sections of the isotopes ^{235}U and ^{238}U are represented, it apperars that the fission probability for ^{235}U is highest for thermal neutrons, whereas fission with ^{238}U only occurs with neutrons having energies above 1 MeV. At neutron energies above 5 MeV, fission cross sections of ^{235}U and ^{238}U are of the same magnitude. In order to optimize the fission yield, the fuel has to be enriched in ^{235}U and the neutrons have to be slowed down by a moderator to thermal energies.

Both nuclear power reactors and research reactors are normally operated using thermal neutrons. Here the fuel elements are surrounded by a moderator of light nuclei as H, D, Be, C which allows the neutrons to reach thermal energies before they induce a new fission. The average distance the neutron has to travel in the moderator before being thermalized is listed in table 5-1 as the scattering length L_s and is of the order of 10 cm.

Fast reactors also exist. They are operated with fast neutrons and are used to convert the fertile materials into fissile materials. We will somewhat below briefly discuss the conditions needed for a fast converter and breeder reactor.

In thermal reactors ^{235}U and ^{239}Pu are the usual fuels. Moderating materials are present to slow down the fast neutrons The fission reaction begins with a neutron capture reaction in ^{235}U which creates a compound ^{236}U nucleus. This excited nucleus can fission into two heavy nuclei with the release of new

(ν) neutrons. The ^{236}U compound nucleus can also deexcite by γ-emission.

$$^{235}U + n \longrightarrow ^{236}U^* \left\{ \begin{array}{l} \nearrow\; ^{A_1}Z_1 + {}^{A_2}Z_2 + \bar{V} + Q_f \;\;(fission) \\ \searrow\; ^{236}U + \gamma + Q_c \;\;\;\;\;\;\;\; (capture) \end{array} \right. \tag{4.8}$$

where Q_f is ~ 200 MeV and $\bar{\nu} \simeq 2.5$ neutrons per fission for ^{235}U, $\bar{\nu} \simeq 2.9$ for ^{239}Pu and $\bar{\nu} = 3.8$ in the spontaneous fission of ^{252}Cf.

The conservation of nucleons requires

$$Z_1 + Z_2 = 92$$
$$A_1 + A_2 + \nu = 236 \tag{4.9}$$

for the primary fission fragments.

Of greatest interest is the number of fast neutrons released per neutron absorbed by the fissionable material. This is of course smaller than ν, since some of the compound nuclei decay to the ground state by γ-emission, rather than by fission. We define a factor η, which is the number of fast neutrons released per neutron absorbed. This is given by

$$\eta = \frac{\nu \sigma_f}{\sigma_f + \sigma_\gamma} \tag{4.10}$$

where σ_f and σ_γ are the cross sections for fission and γ-emission at the incident neutron energy considered (see fig. 3-10). The variation of η with neutron energy for different nuclear fuels are given in fig. 4-12. If the η-factor is greater than 2 then one of these neutrons can be used to maintain the fission chain reaction in the reactor and the other to form a fissionable nucleus from a fertile one, thus replacing the nucleus fissioned which produced the neutrons with a new fissionable one. If more fissionable nuclei are produced than are consumed then the reactor works as a breeder. Otherwise it is a converter. From fig. 4-12, we see that the reactor has to be operated with ^{233}U as fuel if thermal neutrons are used for breeding but in fast breeder reactors ^{239}Pu can also be used as fuel.

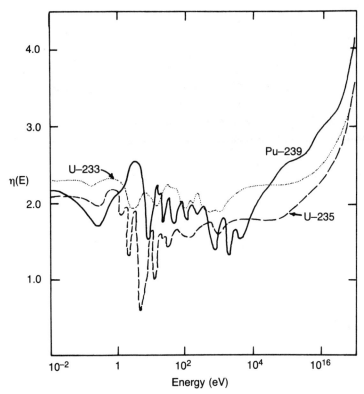

Fig. 4-12 Variation of η with neutron energy for nuclear fuels.

4.2.4.2 R e s e a r c h   r e a c t o r s

Nuclear reactors designed specifically to serve as sources of neutrons for experimental purposes, and with powers ranging from about 10 kW to a few megawatts, are known as research
reactors. Practically all the research reactors available today
are thermal. They are loaded with enriched uranium up to about
90% ^{235}U because a higher flux density per watt is available
than with normal uranium. The moderator and coolant are usual
water or heavy water. Normal uranium reactors require graphite
or heavy water moderation, and are large and expensive. The
highest thermal fluxes exist at the centre of the pile, but
the thermal neutrons are accomplished by neutrons of all other
energies in the fission spectrum. This may not be a disadvantage if methods of monochromatization are used. If a pure ther-

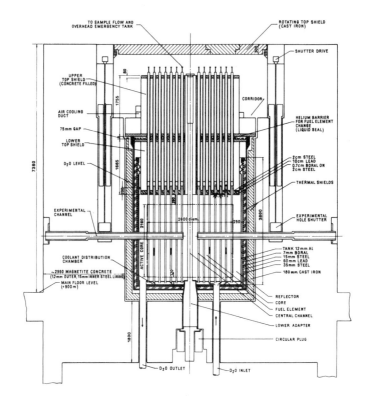

TO SAMPLE FLOW AND
OVERHEAD EMERGENCY TANK

ROTATING TOP SHIELD
(CAST IRON)

SHUTTER DRIVE

UPPER
TOP SHIELD
(CONCRETE FILLED)

CORRIDOR

HELIUM BARRIER
FOR FUEL ELEMENT
CHANGE
(LIQUID SEAL)

AIR COOLING
DUCT

75mm GAP

LOWER
TOP SHIELD

D₂0 LEVEL

2cm STEEL
10cm LEAD
0.7cm BORAL ON
2cm STEEL

THERMAL SHIELDS

EXPERIMENTAL
CHANNEL

2600 diam

EXPERIMENTAL
HOLE SHUTTER

ACTIVE CORE

COOLANT DISTRIBUTION
CHAMBER

TANK 12mm AL
7mm BORAL
15mm STEEL
60mm LEAD
35mm STEEL

180mm CAST IRON

~2550 MAGNETITE CONCRETE
(12mm OUTER,15mm INNER STEEL LINING)

MAIN FLOOR LEVEL
(+900 m)

REFLECTOR

CORE

FUEL ELEMENT

CENTRAL CHANNEL

LOWER ADAPTER

CIRCULAR PLUG

D₂0 OUTLET

D₂0 INLET

VERTICAL SECTION REACTOR FR-2   all dimensions in mm

Fig. 4-13a.   Vertical section of the research reactor FR2 at
Karlsruhe, Germany, Directory of Nuclear Reactors Vol. II p. 308/309 IAEA 1959.

mal flux (of lower intensity) is required, a thermal column,
which is essentially a graphite reflector extension passing
through the shield wall, may be used. Research reactors are
designed to make neutron irradiations easily available. They
have many irradiation positions near the core, in port holes,
in the thermal column, etc., apparatus for transfer of samples
quickly into and out of the reactor, temperature controls,
and other features.

In fig. 4-13 a and b the vertical and the horizontal sections of the research reactor FR2 at the reactor station Karlsruhe Germany are shown. This research reactor is fueled by natural uranium and moderated and cooled by heavy water. In such

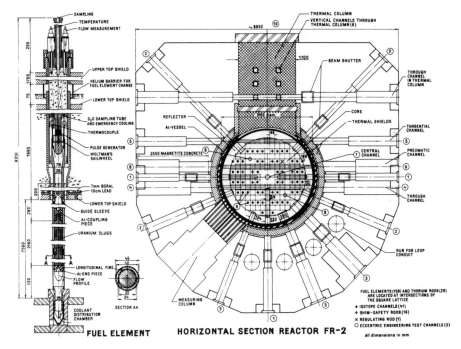

FUEL ELEMENT   **HORIZONTAL SECTION REACTOR FR-2**

Fig. 4-13b.   Horizontal section of FR2

a reactor, one obtains high flux density of thermal neutrons over a large volume. In FR2 the average flux is $\varphi_0 = 1.4 \cdot 10^{13}$ n cm^{-2}s^{-1}.

4.2.4.3 Neutron spectra and spatial distributions

The spectra characteristic of enriched uranium reactors moderated by water and heavy water are illustrated schematically in fig. 4-14. A detailed discussion of the reactor neutron spectrum is already given in sec. 3.8 but from this figure we observe that the ratio of thermal flux to epithermal flux is larger in D_2O than in H_2O reactors, because of the smaller absorption cross section (i.e., the longer thermal neutron lifetime, see table 5-4). This may be worth considering if epithermal neutrons are wanted.

Typical distributions of the thermal flux with radius are illustrated in fig. 4-15, for cylindrical H_2O- and D_2O-moderated and reflected reactors. The distributions in the axial

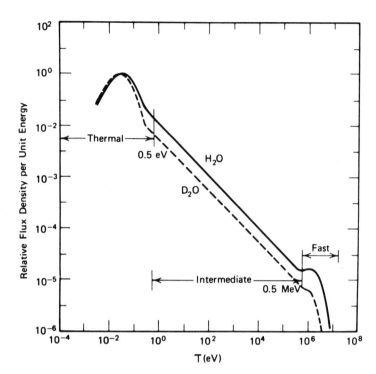

Fig. 4-14 Typical thermal reactor neutron fluxes. E. Profio,
Experimental Reactor Physics (Wiley-Interscience
1976) p. 206.

direction are similar. Fine structure, as would be evident in
a real heterogenous reactor, is not shown. However, we can see
some important general features.

In the water reactor, the flux at first decreases with
radius, then increases to reach a second maximum in the reflector and finally drops off rapidly. The hump in the reflector
occurs because fast neutrons leaking from the core are thermalized in the reflector, and the absorption cross section is less
than in the core. Water-moderated reactors are often fitted
with graphite reflectors because this increases the neutron
multiplication, and the variation of flux with radius is much
smaller.

The enriched-uranium fueled, D_2O-moderated and reflected
reactor is characterized by relatively small absorption in its
core and reflector and thus by a large volume with a fairly

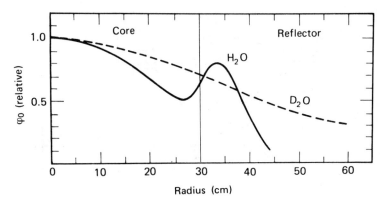

Fig. 4-15 Schematic radial flux distributions in research
reactors moderated by H_2O and D_2O, E. Profio, Experimental Reactor Physics. (Wiley-Interscience 1976) p.
209.

large flux. However, inserting an absorbing sample tends to
depress the flux more severely than in H_2O-moderated reactors.
It will be noted that the maximum flux is usually available
only at the centre of the core. The maximum-to-average ratio,
considering both axial and radial distributions, is generally
around 2. The flux available in the reflector, where beam holes
and the larger irradiation facilities are located, may be considerably less than the core average.

4.3 Ion accelerating devices

4.3.1 ELECTROSTATIC GENERATORS

In principle, the simplest method for accelerating ions
is to inject them into a region over which a constant electrostatic potential is maintained. This constant electrostatic
potential can be produced by rectification of a voltage (Cockroft-Walton accelerator fig. 4-2) or by the mechanical transport of electricity from ground potential to the high-tension
terminal (e.g. in Van de Graaff accelerators).

In a Van de Graaff generator, also called an electrostatic
generator the high voltage is generated by mechanically trans-
ferring charge on a motor-driven belt or chain from ground to

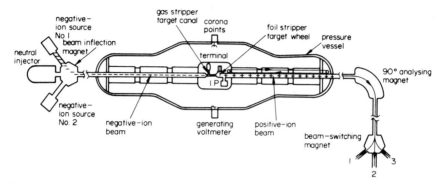

Fig. 4-16   Schematic of a two-stage (tandem) Van de Graaff accelerator.

the inside (field-free region) of an insulated hollow terminal
Transfer of charge to the belt or chain is made at the low-
potential end by a corona discharge from spray points. At the
terminal end, the transfer process is reversed and, by allowing the terminal pulley to reach a higher potential then the
terminal itself, negative charge may be conveyed to the down-
going belt. Electrostatic generators are normally enclosed in
a pressure vessel filled with an insulating gas to permit the
terminal voltage to be raised to a level corresponding to the
overall axial voltage accelerating tube. In early electrostatic accelerators, ions were injected into the accelerating
tube from a gaseous discharge tube housed in the terminal.
However, in the tandem accelerators, as shown in fig. 4-16, the
ion source is at earth potential and produces negative ions
(e.g., H^-, He^-, $^{16}O^-$). These are accelerated to the terminal
through an extension of the accelerating tube mentioned above.
In the terminal, H^- ions, for instance, move with a velocity
v given by the equation

$$T_p = \frac{m_p v^2}{2} = eU \qquad (4.11)$$

where U is the terminal potential and passage through a thin
stripper, e.g., a carbon foil of thickness about 50 µg cm^{-2} or
a tube containing gas at low pressure, which removes both the
extra electron and a further one so that positive ions of H^+ is
available for acceleration to earth potential from the terminal,

yielding a duplication of energy (2eU). Negative ions of heavier elements may be stripped to a positive charge state in
the terminal and the final kinetic energy in this case corresponds to the charge change times the potential (ΔqU).

Energy sharpness to about 0.5 keV is possible in a tandem
generator with a spread about the mean energy of not more than
1-15 keV even in the largest accelerators of, say, 20 MV terminal voltage supplying proton currents of, say, 10 µA.

4.3.2 CYCLOTRONS

The principle of operation of the cyclotron resonance
accelerator is based on the fact that the frequency of circulation of ions in the magnetic field in the nonrelativistic
range is independent of the energy of the ion. We assume that
ions of charge q and mass m are moving in a plane perpendicu-
lar to the magnetic field B. The circular orbit is then described by the equation

$$qvB = \frac{mv^2}{r} \tag{4.12}$$

where r is the orbit radius and v is the velocity of the ion.
We solve eq. 4.12 for the angular frequency $\omega = v/r$ and obtain

$$\omega = \frac{qB}{m} \tag{4.13}$$

To the extent that the mass m can be regarded as a constant
when an ion is accelerated in a cyclotron, eq. 4.13 shows that
the frequency of circulation is constant. Fig. 4-17 shows the
basic principle of a cyclotron.

Ions are produced at the centre of the cyclotron and are
then deflected in a magnetic field within two hollow conductors
known as 'dees', inside a closed vessel containing hydrogen
gas at low pressure. The potential between D_1 and D_2 must
change sign just as the ions cross the gap; they are then
accelerated. The conventional flat-field cyclotron has been
very much used for charged-particle induced nuclear reactions
but the machine is limited to a maximum energy at which the
relativistic increase in mass of the accelerated particles

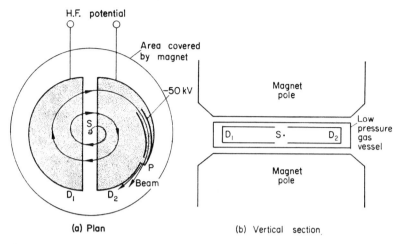

(a) Plan   (b) Vertical section

Fig. 4-17   Simplified diagram of cyclotron showing position of
dees.

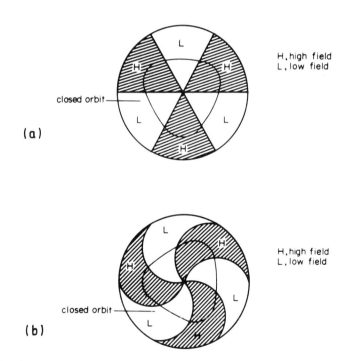

Fig. 4-18   Schematic representation of an N=3 (a) radial ridge
pole profile (b) spiral ridge pole profile.

causes them to lose phase with the radiofrequency accelerating voltage.

The AVF cyclotron is a modern version of the conventional cyclotron. It has a periodic azimuthal variation field due to ridges of radial or spiral type as shown in fig. 4-18a and 4-18b. Such fields cause the particles to remain exactly in phase with the r.f. voltage throughout the acceleration process. The modern AVF cyclotron is now approaching the performance of the tandem Van de Graaff accelerator, both in energy resolution and in beam emittance. Energy variation of an AVF cyclotron is still considerably more difficult to carry out than the corresponding change on the tandem Van de Graaff. However, for many purposes it can be of more use than the tandem accelerator, especially where high-intensity pulsed beams are concerned. The very narrow pulse structure which can be obtained in the beam is well suited to time-of-flight techniques.

4.4 Electron accelerating devices

4.4.1 LINEAR ACCELERATORS

The electron linear accelerator uses an oscillating electric field to accelerate electrons in a straight line at relativistic speeds. This condition implies that the phase velocity of the accelerating field must be constant, equal to the velocity of light. For an operating frequency of the order of 3000 MHz, typical of electron linacs, the spatial period of the accelerating field must be the light velocity divided by the frequency or about 10 cm.

In familiar types of waveguide the phase velocity of the travelling wave is always greater than the velocity of light. It is, however, possible to reduce the phase velocity by placing metallic devices in the form of irises at intervals along the cylindrical waveguide. Electrons, travelling at the velocity of light are injected into the waveguide in bursts at just the proper time to ride along the crest of the electrodynamic

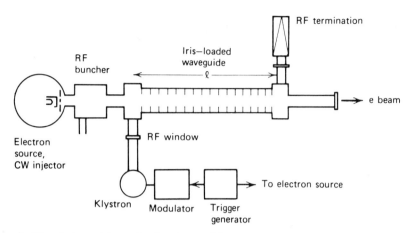

Fig. 4-19 Schematic of electron linac.

wave and reach the maximum energy. The principal features of the electron linac are shown in fig. 4-19.

It is also possible to design the electron linac to operate as a standing-wave accelerating structure. In such a structure, the RF power is coupled into the centre cell and by multiple reflections a standing wave field in this structure results from the superposition of two waves travelling in opposite directions of which only the wave travelling in the same direction as the electrons contributes to the energy gain.

Electron linacs giving energies up to some twenty or thirty MeV are now available commercially from several manufactures. They are able to achieve peak pulse electron currents of the order of amperes and beam power around 20 kW.

4.4.2 MICROTRONS

Microtrons sometimes called electron cyclotrons are constructed by mounting a resonant accelerating cavity near the periphery of uniform magnetic field. The cavity, is driven by an external fixed frequency oscillator in the GHz region. Every time the electrons pass through the cavity they gain energy and in the microtron it is so arranged that the energy gained per turn is constant. Successive orbits of the electrons are thus circles of larger and larger radius with a common tangent at the point where they thread the cavity, see fig. 4-20.

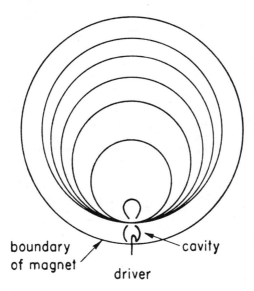

boundary
of magnet

cavity

driver

Fig. 4-20   Orbits in a microtron.

The fundamental cyclotron relation, (eq. 4.13) expressed in terms of the rotation period τ and the total energy E, is

$$\omega = 2\pi\nu = \frac{2\pi}{\tau} = \frac{eB}{m} = \frac{eBc^2}{E} \qquad (4.14a)$$

or

$$\tau = \frac{2\pi E}{eBc^2} \qquad (4.14b)$$

If the magnetic field is constant in space and time, then

$$\Delta\tau = \frac{2\pi}{eBc^2} \, \Delta E \qquad (4.15)$$

I.e., the change in period is directly proportional to the change in energy, quite independent of the actual value of the total energy. This is the basic fact on which the microtron operates. Microtrons are compact, stable and reliable in running as they use a constant magnetic field. They are available commercially and are cheaper than linacs but their intensity is somewhat lower.

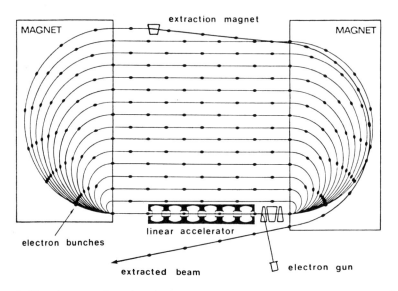

Fig. 4-21 A race-track microtron to produce electron energies
up to 50 MeV.

A very interesting construction is the race-track microtron, a hybrid between linac and microtron. Such an accelerator
has improved intensity, resolution and duty cycle. The duty cycle is defined as the fraction of operating time that an acceleration beam is actually produced. An average current of 20-
200 μA is available with an energy resolution of 0.1-1% and a
duty cycle from 0.01% in the pulsed version up to 100% when
the linac is continuosly driven. In fig. 4-21 a schematic picture of a racetrack microtron is given. These are now being
constructed for energies up to about 500 MeV and they are relatively inexpensive.

4.5 Photon sources

4.5.1 BREMSSTRAHLUNG

In nuclear chemical analysis, photons from a bremsstrahlung beam obtained when electrons from an electron accelerator
hit a target are often used. The characteristics of the bremsstrahlung beam are discussed in sec. 3.10.2.

The relationship between the number of photons per MeV per

incident electron per g cm^{-2} of a radiator (Z,A) can be written

$$n(E,E_o) = \frac{N_A}{A} \, \sigma \, (E,E_o)_{brems}$$

$$= 32 \, \frac{N_A}{A} \, \frac{Z^2 r_e^2 \alpha}{E} \, \phi(Z,E,E_o) (\text{MeV}^{-1} \, g^{-1} \, cm^2) \quad (4.16)$$

where
E is the photon energy, E_o the electron energy, r_e is the classical electron radius, N_A the Avogadros number, α the fine structure constant (1/137) and σ_{brems} the bremsstrahlung cross section.

$\phi(Z,E,E_o)$ is a slow-varying energy-dependent function and is tabulated in table 3-3.

From the fomulas it is seen that the number of photons increases almost as $\frac{Z^2}{A}$ when the radiator mass increases. We also see that the number of photons per MeV is almost proportional to the inverse energy of the photons.

4.5.2   CAPTURE GAMMA RAYS

Capture γ-ray photons from the reaction $^3H(p,\gamma)$ reaction[4.17] have proven useful as a photon source in the energy just above 20 MeV, where 150 keV resolution has been achieved. Even γ-rays from resonance capture of reactor neutrons have been used to study properties of single nuclear levels just below the particle emission thresholds. This technique brings to bear high intensities and very high energy resolution, but suffers from the fact that the energy of the monoenergetic photons is not easily varied; the method depends on the chance overlap of the sharp-line source with the nuclear level under study. Experiments using this technique have been summerized by Arad and Ben-David[4.18].

4.6 References

4.6.1   GENERAL

4.1   G. Bell and S. Glasstone, Nuclear Reactor Theory, Van Nostrand (1970).

4.2   J. Cerny (ed.), Accelerators in Low and Intermediate Energy
      Nuclear Physics, Nuclear Spectroscopy and Reactions Part
      A, Chap. 1, Academic Press (1974).

4.3   J.B.A. England, Accelerators, Technique in Nuclear Struc-
      ture Physics Part 1, Chap. 3, MacMillan (1974).

4.4   D. Jakeman, Physics of Nuclear Reactors, The English Uni-
      versities Press (1966).

4.5   J.J. Livingood, Principles of Cyclic Particle Accelerators,
      Van Nostrand (1961).

4.6   E. Profio, Experimental Reactor Physics, Wiley-Interscience
      (1976).

4.7   S.J. Skorka, Tandem Accelerators, Nucl. Instr. Meth. 146,
      67, (1977).

4.6.2 SPECIAL

4.8   C.J. Cook, Rev. Sci. Instr. 33, 649 (1962).

4.9   R.G. Wilson and G.R. Brewer, Ion Beams with Applications to
      Ion Implantation, J. Wiley & Sons, 64 (1973).

4.10 R.A. Van Konynenburg, UCRL-76857 (1975).

4.11 R. Booth and H.H. Barschall, Nucl. Instr. Meth. 99, 1 (1972).

4.12 E.A. Burill, Neutron Production and Protection, High
      Voltage Eng. Corp. Burlington, Mass., USA (1963).

4.13 J. Op de Beeck, J. Radioanal. Chem. 1, 313 (1968).

4.14 J.L. Fowler and J.E. Brolley Jr, Rev. Mod. Phys. 28, 103
      (1956).

4.15 D.E. Groce, Feasibility Study of an Accelerator-booster,
      Fast, Pulsed Research Reactor-neutron Yield Studies.
      GA-8087, Gulf General Atomic (1967).

4.16 Anders Brahme, private communication and N.N. Kausal et al.,
      Phys. Rev. 175, 330 (1968) and J. Nucl. Energy 25,91 (1971).

4.17 W.A. Lochstet and W.E. Stephens, Phys. Rev. 141 1002 (1966).

4.18 B. Arad and G. Ben-David, Rev. Mod. Phys. 45 230 (1973).

5 Interaction of Radiation with Matter

When a beam of electrons, mesons, nucleons, α-particles, γ-rays, etc., passes through matter, the intensity of the beam is attenuated, since the particles are scattered out of the beam, stopped in matter or absorbed by it. The kind of interaction depends on the nature and energy of particles. Some particles interact very strongly with matter and thus have a very short range while others interact only weakly and thus have long ranges. It is important to know in detail how the interaction occurs, since many methods of detecting particles are based on these interactions. It is also necessary to know how particles and radiation are absorbed in matter so as to be able to construct shields and reduce their interaction.

It is suitable to discuss separately the interaction of heavy charged particles, light charged particles, neutrons and γ-rays, since their stopping powers are quite different.

5.1 Heavy charged particles ($m >> m_e$)

When a charged particle passes through matter, it interacts with atomic electrons and this is the main process by which it loses energy. The interaction is a normal Coulomb interaction between charged particles and the energy transferred from the incident particle to the atoms and molecules along its path ionizes and excites them. The incident particle can also interact with the atomic nucleus losing a large amount of energy and being strongly deflected. The probability of a nuclear interaction, however, is very small and can normally be neglected in discussing the slowing down of particles.

Since the incident charged particle is much heavier than the electrons, the slowing down path will be reasonably straight.

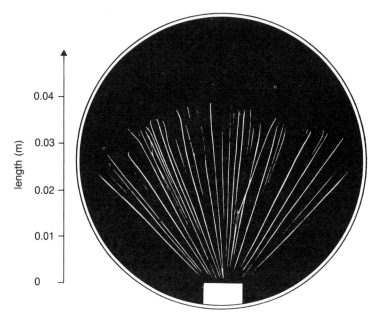

Fig. 5-1   A sketch of some cloud chamber tracks of α-particles from ^{210}Po.

The energy loss is small in any one interaction, but the number of interactions is large which gives rise to short particle ranges, e.g., the range of a 5 MeV α-particle in air is only about $3.5 \cdot 10^{-2}$m (fig. 5-1).

5.1.1 STOPPING POWER

From collision theory, it is possible to predict the specific energy loss or stopping power of a charged particle passing through matter. This is an important quantity in construc-
ting nuclear detectors and radiation shields:

$$- \frac{dT}{dx} = \frac{4\pi \, e^4 \, NZ \, z^2}{(4\pi\varepsilon_o)^2 m_e v^2} \, \ln \frac{2m_e v^2}{I} \tag{5.1}$$

Here, $-dT$ is the kinetic energy lost in a distance dx; N is
the number of atoms per unit volume in the stopping substance;
and Ze is the charge of the atom; m_e is the electron mass; ze
and v are, respectively, the charge and velocity of the particle and I is the mean excitation potential of the atoms of the

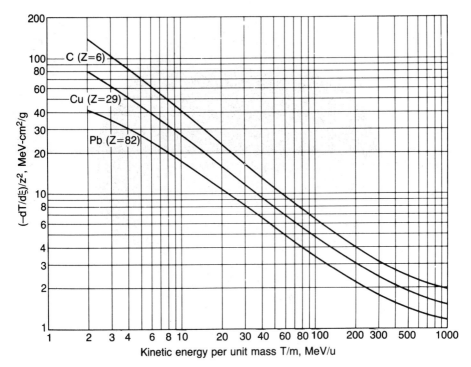

Fig. 5-2   Stopping power for protons in carbon, copper, and
             lead. The scales are given in such units that the
             graph can also be used for particles other than
             protons, provided charge exchange between the par-
             ticles and absorber can be neglected (see text).
             Particle charge = ze, particle mass = m, mass unit
             u.

stopping substance. Eq. 5.1 is an approximation which suffices
for estimations.

Since the density of atoms $N = N_A \rho A^{-1}$ it is convenient to
measure the thickness of a given absorber in kgm^{-2} or gcm^{-2}
rather than in m or cm. The relationship between mass per
unit area and thickness x is of course, $\xi = \rho x$, where ρ is
the density of the material. From eq. 5.1 it is seen that the
stopping power varies inversly with the kinetic energy T for
moderate and rather low velocities. This is also shown in fig.
5-2. For higher velocities, the stopping power passes through
a minimum and then begins to increase slowly.

It can be seen from eq. 5.1 that particles with different masses but with the same velocity have the same stopping powers in a medium, if their charges are equal. If the charges are unequal, the ratio of the stopping powers corresponds to the square of the ratio of the charges. Thus in fig. 5-2 the graphs can also be used for particles other than protons (see sec. 18.2.1).

We will also define the stopping cross section per atom ε, a quantity which has already been used in sec. 3.9.

$$\varepsilon = - \frac{1}{N} \frac{dT}{dx} \tag{5.2}$$

With the numerical values for protons, we obtain

$$\varepsilon_p = - \frac{1}{N} \frac{dT_p}{dx} = \frac{0.24}{T_p} \; Z \left[\ln \frac{T_p}{Z} + 5.24 \right] \cdot 10^{-15} \; eV \; cm^2 /atom$$

$$\tag{5.3}$$

if T_p is given in MeV.

5.1.2 RANGE

The theoretical range of a heavy charged particle is equal to its path length in the medium because scattering is negligible (fig. 5-1). The range may formally be obtained by integration of the stopping power equation:

$$R = \int_0^{T_0} \frac{dT}{dT/dx} \simeq \int_0^{T_0} T dT \simeq T_0^2 \tag{5.4}$$

In practice this integration cannot be carried out over the entire range of the particle, because of corrections necessary to perform for the explicit formula of dT/dx at low energies.

Range-energy curves are therefore constructed semi-empirically by combining experimental data at low energies with integration of the stopping power equation for moderate velocities (see fig. 18-1).

Fig. 5-3 shows the stopping power and range curves in silicon and germanium, two very important detector materials, for

different particles over a wide range of energies. The range
curves are approximately linear when plotted on a logaritmic
scale and have similar slopes. Therefore, we can write a semiempirical formula for the range

$$R = aT^b + c \qquad\qquad (5.5)$$

where $\underline{a}$ depends on the particle type and $\underline{b}$ is approximately the
same for all particles, having a value of about 1.7. The discre-
pancy from the exponential factor in eq. 5.4 comes from the
energy dependence of the logarithmic factor in eq. 5.1.

5.1.3 RANGE AND ENERGY STRAGGLING

When a charged particle is stopped in an absorber it makes
thousand to million collisions resulting in small discrete energy transfers to electrons before coming to rest. A beam of particles of initially uniform velocity, after passage through a
certain thickness of matter, will show a distribution of velocities around a mean value owing to the statistical nature of
the energy loss. The ranges of the particles will therefore
be distributed around the mean range value and since the number of collisions is large, and since also they are independent
processes, the range straggling distribution may be a normal
(or gaussian) (sec. 8.3.3). Analogously energy straggling
refers to the variations in energy noted in a beam of initially
monoenergetic particles often passing through some fixed amount
of absorber.

The energy straggling is estimated quantitatively in ref.
1.2, where it is shown that the standard deviation (eq. 8.8) σ_T
of the mean residual energy after passing an absorber with
thickness Δx is given by

$$\sigma_T^2 = \frac{4\pi e^4 N Z z^2}{(4\pi\varepsilon_o)^2} \Delta x \qquad (J^2) \qquad\qquad (5.6)$$

The standard deviation σ_R of the range distribution of heavy
particles of energy T slowed down in an absorber is given by

$$\sigma_R^2 = \int_o^R \sigma_T^2 \left(\frac{dT}{dx}\right)^{-2} dx = \int_o^T \sigma_T^2 \left(\frac{dT}{dx}\right)^{-3} dT \qquad\qquad (5.7)$$

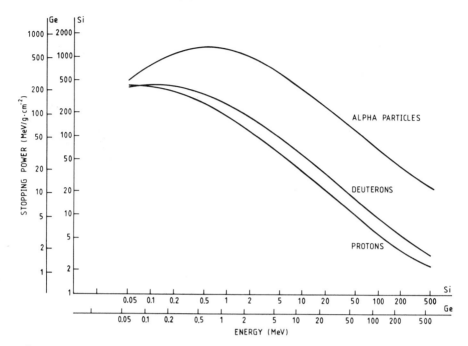

Fig. 5-3a.   Proton, deuteron and α-particle stopping powers.
            Ortec 1976, Instruments for research and industry,
            Catalog 1004, p. 16,17.

By substitution from eqs. 5.1 and 5.6 and integration, neglecting the energy variation of the logarithmic term, we find approximately

$$\frac{\sigma_R^2}{R^2} = \frac{2m_e}{M} \; \frac{1}{\ln(2m_e v^2/I)} \tag{5.8}$$

For the α-particles of polonium (T = 5.3 MeV, R = 3.84 cm of
air) we find σ_R/R = 0.9 per cent. For protons of the same initial velocity as these α-particles, eq. 5.8 shows that the straggling σ_R/R is just double that for the α-particles. This is
because the α-particle dT/dx is roughly four times greater than
for protons so that the straggling effect is relatively smaller.

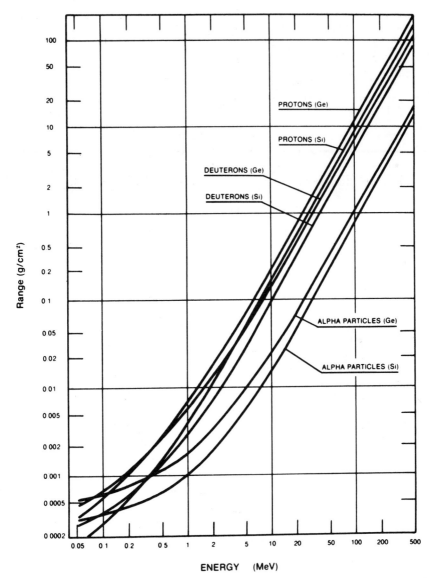

Fig. 5-3b. Proton, deuteron and α-particle ranges in Si and Ge.

5.2 Electrons

When electrons pass through matter, two processes contribute to stop them. The first is collisions with other electrons.
This process dominates for slow electrons and is the same inter-

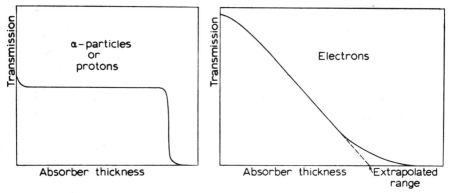

Fig. 5-4   Ranges of heavy particles and electrons.

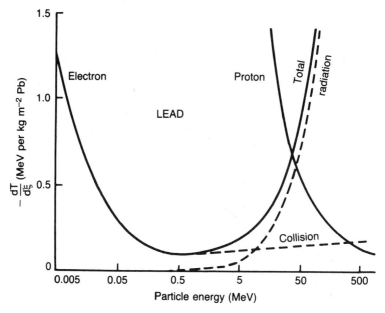

Fig. 5-5   Energy loss of electrons and protons in lead.

action as causes the stopping of heavy charged particles. Since,
however the incident electron and the electron in the stopping
material have the same mass, the incident electron is easily
scattered. Hence, the path length in the stopping material can
be considerably longer than its distance in a straight line
or its range in the material and thus a well-defined electron
range does not exist (fig. 5-4).

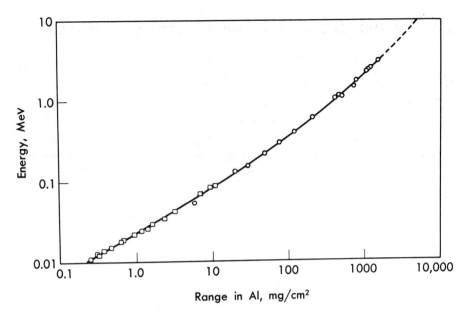

Fig. 5-6   Range of electrons in aluminium as a function of
            their energy.
            L.E. Glendenin, Nucleonics $\underline{2}$, 12 (1948).

The other process is the emission of electromagnetic radiation, which always occurs when charged particles are retarded
(bremsstrahlung) or accelerated. This radiation effect dominates
for fast electrons, but can be neglected for not too fast mesons
and protons since the effect is inversely proportional to the
square of the mass of the incident particle (fig. 5-5).

The ratio of the energy losses due to the two different
processes is for fast electrons given by

$$\frac{(dT/d\xi)_{rad}}{(dT/d\xi)_{col}} \simeq \frac{TZ}{800} \quad \text{(T in MeV)} \tag{5.9}$$

The extrapolated range of slow electrons times the density of the absorber is approximately independent of the material
of the absorber. The range-energy relationship in aluminium is
given in fig. 5-6. Between about 0.6 and 12 MeV, the extrapolated range is well represented by the linear relation (see sec.
18.2.2):

$$R(\text{g cm}^{-2}) = 0.526 \, T_e \, (\text{MeV}) - 0.094 \tag{5.10}$$

5.3 Passage of charged particles through crystals

In the discussion of the slowing-down of charged particles in matter given earlier, we have assumed that the stopping medium was isotropic. However, if a charged particle moves through a single crystal, anomalous penetration effects appear in certain directions. The phenomena is called channelling and is caused by the fact that, in a crystal , directions and planes exist for which the electron density may be particularly low.

The occurrence of channelling in a simple two-dimensional case is illustrated in fig. 5-7. When a charged particle is incident along a crystal axis, it can penetrate more deeply than is usually the case and, in addition, its passage through the crystal is determined by Coulomb interactions which tend to keep it in the low electron-density region. Obviously, such effect will be enhanced for low-energy heavy ions.

Channelling has been a useful tool in solid state physics. There will, for example, be a marked reduction of channelling in a crystal which has suffered radiation damage or has some other type of lattice dislocation. Displaced atoms will scatter particles penetrating channels which are normally open.

An effect closely related to channelling is blocking of the emission of charged particles in certain directions from lattice sites. Because of the strong Coulomb interaction, particles emitted from nuclei placed in the lattice or from nuclei which have been knocked out of the lattice will not be able to enter crystal symmetry directions, thus giving marked dips in the yields for these directions.

Blocking has been used to study extremely short nuclear lifetimes ($<10^{-16}$ s). If an excited nucleus is produced at a lattice site and a charged particle is emitted, this will not penetrate along the symmetry directions if the nuclear lifetime is short ($< 10^{-17}$ s). If, on the other hand, the lifetime is somewhat longer, the recoil velocity of the nucleus due to the production reaction may remove the nucleus from the lattice point before the particle is emitted and the particle can escape from the crystal without blocking. Hence a study of this phenomena can give information about the very short times involved in nuclear reactions.

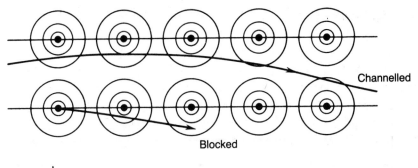

Channelled

Blocked

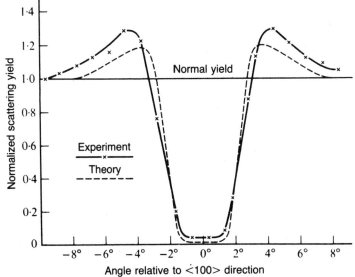

Fig. 5-7   Channelling and blocking: (a) Trajectories of ions
            between rows of atoms. A particle incident nearly
            parallell to the row is channelled; a particle
            emitted from a lattice site is scattered away from
            the channelling direction. The circles represent
            the atomic electron density. (b) Blocking dip in
            $135°$. Rutherford scattering of 480 keV protons
            incident, with an angular spread of $0.1°$ along the
            $< 100 >$ direction of a tungsten crystal.
            Adopted from J.V. Andersen, E. Uggerhöj, Can. Jour.
            Phys. 46, 517 (1968).
            W.E. Burcham, Elements of nuclear physics (Longman
            1979) p.94.

5.4 Neutrons

The main difference between neutrons and protons is that neutrons are uncharged. Even at low energies, a neutron can easily reach the edge of an atomic nucleus without being deflected by electromagnetic forces. This implies that even a slow neutron can be scattered elastically by the nuclear potential or that it reacts with the nucleus and is captured or reemitted.

Normally neutrons are produced as fast neutrons, $T_n > 0.1$ MeV, but in both physical and technical applications it is often important to thermalize them. Slowing down of the neutrons is made by elastic and inelastic scattering against nuclei.

Assuming that only elastic collisions occur (even if inelastic scattering may be of importance for neutron energies above 0.1 MeV), it is possible to calculate the number of collisions needed to thermalize a fast neutron. If the neutron has a kinetic energy T_k before collision and scatters elastically against a nucleus with mass A which is at rest, the neutron after collision has energy T (see fig. 3-3c) where

$$\frac{T}{T_k} = \frac{1}{2} (1+\alpha) + (1-\alpha) \cos \theta \tag{5.11}$$

with

$$\alpha = (\frac{A-1}{A+1})^2 = 1 - \frac{4}{A} + \frac{8}{A^2} - \frac{12}{A^3} - - \tag{5.12}$$

$$\simeq 1 - \frac{4}{A} \quad \text{when A is large}$$

and θ = the scattering angle in the centre-of-mass system. As can be seen from eq. 5.11, the largest energy loss is obtained when the scattering angle θ is π giving

$$(\frac{T}{T_k})_{min} = \alpha \tag{5.13}$$

The probability $P(T,T_k)$ dT for a neutron with initial kinetic energy T_k to have an energy in the interval (T, T-dT) after one collision is

$$P(T,T_k) = \frac{1}{(1-\alpha)T_k} \tag{5.14}$$

174

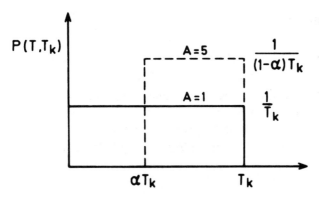

Fig. 5-8   Energy distribution of neutrons after one collision
with initial energy T_k.

It is also of interest to calculate the average energy
loss of a neutron after one collision. In practice the average logarithmic decrease in energy per collision ξ is used,
since this is independent of the neutron energy

$$\xi = (\ln T_k - \ln T)_{av} = (\ln \frac{T_k}{T})_{av} = 1 + \frac{\alpha}{1-\alpha} \ln\alpha \qquad (5.15)$$

This expression may be approximated for values A > 10 by

$$\xi \approx \frac{2}{A+2/3} \quad (\xi \approx 1.0 \text{ for hydrogen}) \qquad (5.16)$$

We see from eq. 5.14 and fig. 5-8 that there is equal probability for the neutron to reach any energy between the minimum and maximum energy. If the neutron scatters against a
proton, the minimum energy is zero. In this case, the recoil
protons have the same energy distribution as the scattered
neutrons.

The average number of collisions $\bar{c}_k$ needed to thermalize
a fast 2 MeV neutron is now

$$\bar{c}_k = \frac{\ln(\frac{T_k}{T_{th}})}{\xi} \qquad (5.17)$$

and the value of $\bar{c}_k$ for 2 MeV neutrons is given in table 5-1
for some moderators. In order for a material to act as a good

175

Table 5-1

Properties of moderators.

Source neutron T_k = 2 MeV, thermal neutrons T_{th} = 0.025 eV (v_0 = 2200 ms^{-1}).

D. Jakeman, Physics of Nuclear Reactors (The English University Press 1966) p.346

Material	H$_2$O	D$_2$O	Be	BeO	Graphite
Density g/cm^3	1.00	1.10	1.85	3.09	1.6
Atoms or molecules per unit volume (x10^{24}/cm^3)	0.0335	0.0331	0.1236	0.0728	0.0803
Scattering cross section per atom or molecule σ_s barns	103*	13.6*	7.0	6.8	4.8
Absorption cross section σ_{a0} barn	0.66	∿0.001†	0.01	0.01	0.004
Scattering mean free path λ_s cm	0.290*	2.23*	1.16	2.00	2.60
Approximate number of collisions to thermalize neutron	19	32	86	102	114
$(\cos \psi)_{av}$	0.324	0.116	0.074	0.061	0.056
ξ	0.948	0.570	0.209	0.173	0.158
Slowing down power $\xi\Sigma_s$ cm^{-1}	1.53	0.17	0.18	0.086	0.064
Moderating ratio $\xi\Sigma_s/\Sigma_a$	72	∿12.000†	159	119	170
Diffusion coefficient D cm	0.14	0.82	0.49	0.57	0.87
Diffusion length L_{th} cm	2.7	∿110†	21	28.6	53
Scattering length L_s cm	5.2	∿10.5	8.9	9.6	17.7
Slowing down time µs	6	53	50	102	140
Mean neutron lifetime $1/\Sigma_0 v_0$ µs	205	∿10^5	3460	7600	13.000

* At 0.025 ev.   † Depends on purity of D$_2$O

moderator in slowing down neutrons from fission energies to
thermal energies, it would have a high value of the average
log decrement. This alone is not sufficient; there must also
be a high probability that the material will scatter the neutrons.
   Hydrogen would appear to be the best material to use as a
moderator and it is indeed often used in the form of ordinary
ligth water, i.e. H_2O. Deuterium oxide, or water made with heavy hydrogen, should also be a good moderator. Since the atomic
mass of the deuterium nucleus is twice that of the hydrogen
nucleus, one would not expect it to be quite as effective in
slowing down neutrons. This is described in table 5-1.
   The slowing down power is the product of the average log
decrement and the macroscopic cross section for scattering

$$SDP = \xi \, \frac{\rho N_A}{A} \, \sigma_s = \xi \, \Sigma_s \qquad (5.18)$$

   The purpose of a moderator, however, is to slow down the
neutrons without absorbing them. For this reason the slowing-
down power alone is not a sufficient criterion. A better criterion for a moderator is the moderating ratio, which takes
into account the absorption of the neutrons while slowing down.
This is given by

$$MR = \frac{\xi \Sigma_s}{\Sigma_a} = \frac{\xi \sigma_s}{\sigma_a} \qquad (5.19)$$

Values of SDP and MR are given in table 5-1.
   The number of collisions required to slow down a neutron
is only a part of the description of the slowing down process.
The distance that the neutrons travel before becoming thermal
is also of importance. The paths are randomly distributed, and
there is no unique range; therefore the average (straight line)
distance travelled by a neutron in slowing down from the origin
is used and is expressed as L_s = the mean scattering length -
a function of λ_s, the scattering mean free path (eq. 3.21).
   The total path length should be $\bar{c}_k \lambda_s$ but the average
crow-flight distance L_s can be shown to be

$$L_s^2 = \frac{(r_s^2)_{av}}{6} = \int_{T_{th}}^{T_k} \frac{\lambda_{tr} \lambda_s}{3\xi} \, \frac{dT}{T} \simeq \frac{\lambda_{tr} \lambda_s}{3} \, \bar{c}_k \qquad (5.20)$$

where λ_{tr} is the mean free path for transport. This accounts for a weak preferential forward scattering of the neutrons in the laboratory system, since the scattering is isotropic in the centre of mass system. λ_{tr} is closely related to λ_s

$$\lambda_{tr} = \frac{\lambda_s}{1-(\cos\psi)_{av}} = \frac{\lambda_s}{1 - \frac{2}{3A}} \tag{5.21}$$

where ψ is the scattering angle in the laboratory system and $(\cos \psi)_{av}$ the average value of $\cos \psi$.

The mean square distance between the neutron source and the point where the neutron becomes thermal is denoted by $(r_s^2)_{av}$ and is called the second spatial moment of the slowing down density.

Analogously with the concept of mean scattering length L_s we define a mean diffusion length for thermal neutrons which is the mean distance travelled by a thermal neutron from formation to capture. For a point source, the diffusion length L_{th} is defined as

$$L_{th}^2 = \frac{(r_{th}^2)_{av}}{6} = \frac{\lambda_{tr}\lambda_a}{3} \tag{5.22}$$

where $\lambda_a = \Sigma_a^{-1}$ = free mean path of absorption and r_{th}^2 = second spatial moment of the thermal neutron flux.

We also define the migration length M, which is a measure of the distance travelled by a neutron from its source to its point of capture.

$$M = (L_s^2 + L_{th}^2)^{\frac{1}{2}} \tag{5.23}$$

The L_s and L_{th} values for some moderators are given in table 5-1.

5.5 Gamma radiation

If a collimated beam of electromagnetic radiation passes through matter, it is attenuated exponentially. This is because radiation in its interaction with matter must be considered as a beam of photons. All the dominating interaction processes remove individual photons from the beam in proportion to the total number of photons in the beam either by scattering them out of the beam or by absorbing them

$$dn = - n \ N\sigma \ dx \qquad\qquad (5.24)$$

or

$$n = n_o e^{-N\sigma x}$$

where n_o is the number of incident photons,
$\quad$ n is the number of photons at depth x,
$\quad$ N is the density of atoms and
$\quad$ σ is the interaction cross section.

$\quad$ Analogously with the macroscopic cross section defined in eq. 3.20, we introduce a linear attenuation coefficient μ_ℓ for γ-attenuation

$$\mu_\ell = N\sigma = \frac{\rho N_A}{A} \ \sigma \qquad m^{-1} \qquad\qquad (5.25)$$

giving

$$n = n_0 e^{-\mu_\ell \ x} \qquad\qquad (5.26)$$

Thus, it is seen that the attenuation of monoenergetic photons is exponential and that no definite maximum range exists (see sec. 18.2.3).

$\quad$ The γ-ray attenuation coefficients are more conveniently expressed as mass attenuation coefficients μ_m (table 5-2).

$$\mu_m = \frac{\mu_\ell}{\rho} = \frac{N_A}{A} \ \sigma \qquad m^2 \ kg^{-1} \qquad\qquad (5.27)$$

5.5.1 DIFFERENT KINDS OF INTERACTION

$\quad$ The dominating interaction processes between photons and matter are the following:

a) photoelectric effect in which the photon is completely absorbed by a bound electron,

b) Compton scattering in which the photon is scattered by a

Table 5-2
Some mass attenuation coefficients.

E (MeV)	Water 10^{-2} m^2/kg	Aluminium 10^{-2} m^2/kg	Iron 10^{-2} m^2/kg	Lead 10^{-2} m^2/kg
0.1	1.71	1.69	3.70	54.60
0.2	1.37	1.22	1.46	9.42
0.3	1.19	1.04	1.10	3.78
0.5	0.97	0.84	0.84	1.52
1.0	0.71	0.61	0.60	0.70
2.0	0.49	0.43	0.42	0.46
3.0	0.40	0.35	0.36	0.41
5.0	0.30	0.28	0.31	0.43

free electron thus losing energy which is transfered to the electron and

c) pair production in which a photon having an energy above 1.022 MeV is absorbed and an electron-positron pair created.

The photon can also be scattered by a bound electron without losing energy. This process is called Rayleigh scattering and may be of importance at low photon energies. The scattering is, however, strongly peaked in the forward direction which implies that it contributes little to the attenuation of a γ-beam and will thus not be considered in the following discussion.

Nuclear inelastic scattering, nuclear photoabsorption and photomeson production are other interaction processes which are of considerable interest, but they do not normally contribute to any significant degree to the values of attenuation coefficients observed experimentally.

If we assign cross sections σ_{pe}, σ_c and σ_{pp} to the three interactions mentioned above, the linear attenuation coefficient μ_ℓ may be written

$$\mu_\ell = N(\sigma_{pe} + \sigma_c + \sigma_{pp}) \qquad (5.28)$$

where N is the density of atoms.

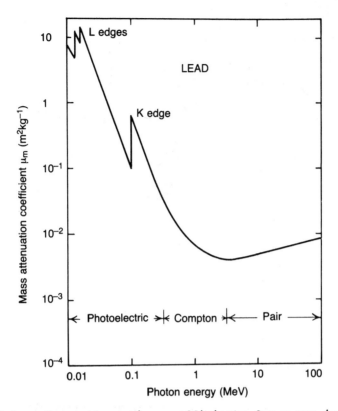

Fig. 5-9   Mass attenuation coefficients for γ-ray interaction
in lead.
R. Evans, The Atomic Nucleus· (McGraw-Hill, 1955)
p. 716.

The cross section σ_c is the atomic cross section for
Compton scattering, which is the atomic number Z times the
scattering cross section per electron, there being Z bound
electrons per atom.

Fig. 5-9 shows the variation of the mass attenuation
coefficient with energy for lead and fig. 5-10 the relative
importance of the three main processes as a function of energy
and the atomic number of absorber.

5.5.2   PHOTOELECTRIC EFFECT

An incident photon can be absorbed by a bound electron
which is then ejected. Momentum is conserved by the presence

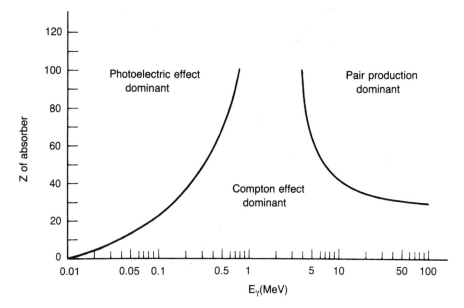

Fig. 5-10   Relative importance of the three major types of
             γ-ray interaction.
             R. Evans, The Atomic Nucleus
             (McGraw-Hill, 1955) p.716.

of the residual ion. On the other hand, a free electron cannot
absorb a photon because it is impossible to conserve both
energy and momentum simultaneously in the process. Therefore
it seems reasonable that, the more tightly-bound the electrons
are in the atom, the larger the probability of photoelectric
absorption. K-shell electrons are thus preferentially ejected
from the atom if sufficient energy is available.

The kinetic energy of the ejected photoelectron is

$$T_e = E_\gamma - E_k \tag{5.29}$$

where E_γ is the photon energy and E_k the ionization energy of
the K-electrons.

The electron hole created in the K-shell is filled by another electron probably belonging to the L-shell and at the
same time a K X-ray characteristic of the absorber is emitted.
Alternatively to X-ray emission, the atomic excitation energy
may be concentrated on another of the bound electrons which

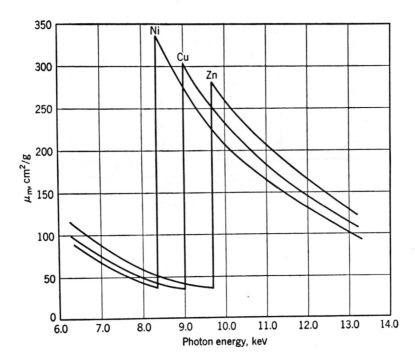

Fig. 5-11   Experimental mass attenuation coefficients for
             Ni, Cu and Zn, showing the K-absorption edges
             D. Halliday, Introductory Nuclear Physics (Wiley
             & Sons 2nd ed. 1962)p.169.

will be emitted (Auger effect). This process can be compared
with the internal conversion process in nuclear decay (sec.2-4).

The photoelectric effect predominates at low photon energies up to about 0.5 MeV. The cross section increases strongly
with the atomic number of the absorber and decreases strongly
with photon energy. It can roughly be written as

$$\sigma_{pe} \sim Z^5 E_\gamma^{-3.5} \tag{5.30}$$

The variation of σ_{pe} with energy shows abrupt increases
at the ionization potentials of successive atomic shells. The
phenomena of absorption edges can be used in analyzing unknown radiation (fig. 5-11). By measuring the absorption
coefficient in foils of two adjacent (or near adjacent) ele-

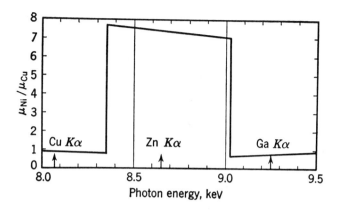

Fig. 5-12   Ratio of the mass attenuation coefficients in nickel
to that in copper for photons of various energies.
The arrows show the quantum energies of the characteristic K-radiations as marked. D. Halliday,
Introductory Nuclear Physics (Wiley & Sons, 2nd
ed. 1962)p. 169.

ments, it is possible to decide if the γ-energy falls within
or outside a particular narrow interval (fig. 5-12).

5.5.3 COMPTON SCATTERING

In the photoelectric absorption region, the whole atom
takes part in the interaction since momentum is transferred
to both the ejected K-electron and to the residual ion.

At higher γ-energies (i.e., smaller γ-wavelengths) there
is an increasing tendency for the γ-rays to interact with
individual electrons without the residual ion being involved.
Such an interaction can be regarded as the collision of a photon
with a free electron.

At the γ-energies under consideration, all the electrons
bound in the atom may be regarded as free. This interaction
process is called Compton scattering. It is the most important
process of γ-interaction for energies between about 0.5 MeV and

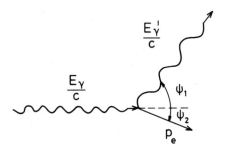

Fig. 5-13   Momentum diagram for Compton scattering.

10 MeV (fig. 5-10). During the collision, part of the γ-energy
is transferred to the electron, while the γ-ray which is now
less energetic is deflected through an angle ψ, with respect
to the incident direction (fig. 5-13).

We use the conservation laws for linear momentum and energy (relativistic case).

$$\vec{p}_\gamma = \vec{p}_\gamma' + \vec{p}_e \qquad\qquad (5.31a)$$

$$E_\gamma + m_e c^2 = E_\gamma' + E_e \qquad\qquad (5.31b)$$

$$E^2 = p^2 c^2 + m^2 c^4 \qquad\qquad (5.31c)$$

Compare with eqs. 3.1, 3.2 and 3.3 (nonrelativistic case, elastic collision).

We rearrange 5.31a and take the scalar product

$$p_e^2 = p_\gamma^2 + p_\gamma'^2 - 2 p_\gamma p_\gamma' \ \cos \psi_1, \qquad\qquad (5.32a)$$

and convert to energy

$$E_e^2 - m_e^2 c^4 = E_\gamma^2 + E_\gamma'^2 - 2 E_\gamma E_\gamma' \ \cos \psi_1 \qquad\qquad (5.32b)$$

using

$$p_\gamma = E_\gamma / c \quad \text{(the photon has zero rest mass)} \qquad\qquad (5.32c)$$

Eliminating E_e using eq. 5.31b, we obtain

$$(E_\gamma - E_\gamma') m_e c^2 = E_\gamma E_\gamma' (1 - \cos \psi_1) \qquad\qquad (5.32d)$$

185

Table 5-3

Compton effect for $\psi_1 = \pi/2$.

E MeV	λ 10^{-12}m	λ' 10^{-12}m	E_γ' MeV	$(E_\gamma - E_\gamma')/E_\gamma$ %
0.01	124	124	0.010	1
0.10	12.4	14.8	0.0837	16
1.00	1.24	3.67	0.337	66
10.00	0.124	2.55	0.486	95

or

$$E_\gamma' = \frac{E_\gamma}{1 + \dfrac{E_\gamma(1-\cos\psi_1)}{m_e c^2}} \qquad (5.33)$$

As expected, eq. 5.33 reduces to $E_\gamma' = E_\gamma$ for $\psi_1 = 0$ corresponding to no interaction. For $E_\gamma \gg m_e c^2$ and $\psi_1 > 0$ we have

$$E_\gamma = m_e c^2/(1-\cos \psi_1) \qquad (5.34)$$

This shows that photons scattered at $\psi = \pi/2$ approach the limit of $E_\gamma' = 0.511$ MeV (table 5-3).

The incident photon loses most energy when it is scattered backwards ($\psi_1 = \pi$). In this case, as the initial energy increases, the photon energy tends to $m_e c^2/2$ after the collision. The probability for such scattering is, however, very small (fig. 5-14). If $E_\gamma \ll m_e c^2$, the nonrelativistic case, eq. 5.33 reduces to $E_\gamma' \simeq E_\gamma$.

Since there are Z electrons bound to an atom with atomic number Z, the atomic Compton scattering cross section is proportional to this number:

$$\sigma_c = Z\sigma_c^e \qquad (5.35)$$

where σ_c^e is the Compton scattering cross section for an individual electron. The form of the differential cross section for different incident energies is shown in fig. 5-14.

186

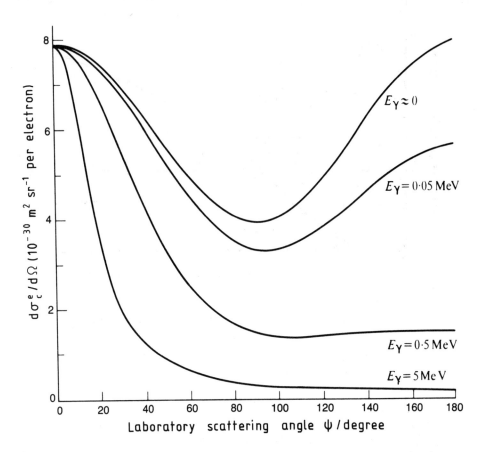

Fig. 5-14. Differential cross section for Compton scattering,
giving number of photons scattered per unit solid
angle at angle ψ, for energies 0-5 MeV.
W.E. Burcham, Elements of nuclear physics.
(Longman 1979) p.86.

5.5.4 PAIR PRODUCTION

     A γ-ray having an energy larger than $2m_e c^2$ = 1.022 MeV
can interact with the Coulomb field of an absorber nucleus producing an electron-positron pair. A positron is a particle with
exactly the same properties as an electron except that it is
positively charged. We have already discussed nuclear β^+ -decay
(sec. 2.3.2) in which a positron and a neutrino are emitted.
The existence of the positron is an example of the fundamental principle that to every elementary particle there exists an

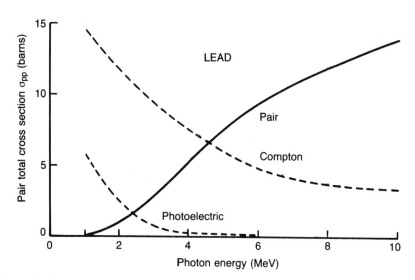

Fig. 5-15 Variation of the pair cross section σ_{pp} with energy for lead. The Compton and photoelectric cross sections are indicated for comparison
W.E. Burcham, Nuclear Physics an introduction
(Longman 2nd ed. 1973)p.136.

antiparticle. The electron-positron pair is such a particle-anti-particle couple. The antiparticles are often symbolized by a bar over the relevant particle symbol ($\bar{\nu}$, $e^+ = \bar{e}$, $\bar{n}$, $\bar{p}$ etc.).

Antiparticles are as stable as the corresponding particles. The positron is thus a stable particle. If, however, a particle and an antiparticle approach each other, they interact either by scattering or annihilation. In the latter case, energy corresponding to the total mass and kinetic energy of the pair is released. A positron and an electron at rest thus release $2m_e c^2 = 1.022$ MeV, this energy appearing as two oppositely-directed quanta of energy $m_e c^2 = 0.511$ MeV, the so-called "annihilation radiation". If a positron is annihilated by collision with a moving electron, the angle between the two annihilation quanta becomes less than $\pi/2$ and measurement of these angles is an important means of learning about the distribution of electron momenta in matter.

The energy dependence of the cross section for pair production in lead is shown in fig. 5-15. The cross section in-

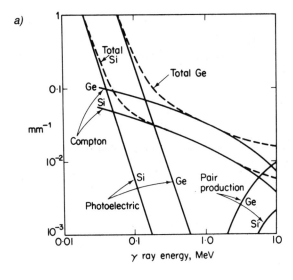

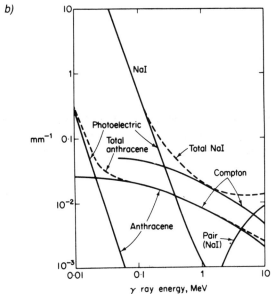

Fig. 5-16   Partial and total γ-ray linear attenuation coefficients for (a) Si and Ge and (b) NaI and anthracene.
(a) E. Storm, E. Gilbert and H. Israel, Report
LA 2237 U.S. Atomic Energy Comission (1958).
(b) J.H. Neiler and P.R. Bell, Alpha-Beta and Gamma
ray Spectroscopy, ed. K. Siegbahn, (North-Holland
1965)p.287.

creases at first with energy, the pair production taking place
at larger and larger distances from the nucleus, but a limit
is set to the increase by the screening of the nuclear charge
by the atomic electrons. In the absence of screening, the pair
cross section is proportional to Z^2

$$\sigma_{pp} \sim Z^2 \qquad\qquad (5.36)$$

Pair production is therefore the predominant effect at
high energies and for heavy nuclei, being much larger than
Compton scattering and the photoelectric effect.

In fig. 5-16 are given the partial and total linear attenuation coefficients for γ-rays in some detection materials. From
the figure, it can be seen how the various attenuation processes
compete in different energy regions.

5.6 References

5.6.1 RANGE-ENERGY AND STOPPING POWER TABLES

5.1   J.H. Hubbel, Photon Cross Sections, Attenuation Coeffi-
       cients and Energy Absorption Coefficients from 10 keV to
       100 GeV, NSRDS-NBS 29 U.S. Government Printing Office
       Washington D.C. (1969).

5.2   J.H. Hubbel, Photon Mass Attenuation and Mass Energy-Ab-
       sorption Coefficients for H, C, N, O, Ar and Seven Mixtu-
       res from 0.1 keV to 20 MeV, Radiation Research 70, 58
       (1977).

5.3   L.C. Northcliffe and R.F. Schilling, Range and Stopping-
       Power Tables for Heavy Ions ($1 \leq Z \leq 103$). Nuclear Data Tab-
       les 7A, 233 (1970).

5.4   L. Pages et al., Energy Loss, Range, and Bremsstrahlung
       Yield for 10-keV to 100 MeV Electrons in Various Elements
       and Chemical Compounds, Atomic Data 4,1 (1972).

5.5   E. Storm and H.I. Israel, Photon Cross Sections from 1 KeV
       to 100 MeV for Elements Z=1 to Z=100, Nucl. Data Tables
       A7, 565 (1970).

5.6   J.F. Ziegler (ed.), The Stopping and Ranges of Ions in
      Matter, Pergamon Press (1977).
      Vol. 1.   The theory of stopping, straggling, and
                range of ions in matter.
      Vol. 2.   Bibliography and index of experimental
                range and stopping power data.
      Vol. 3.   Hydrogen: stopping powers & ranges in all
                elements.
      Vol. 4.   Helium: stopping powers & ranges in all
                elemental matter.

Part B.
Nuclear detectors

6 Principles of Radiation Detectors

6.1 Gas detectors

6.1.1 IONIZATION CHAMBER

The ionization chamber is a gasfilled chamber in which deposited radiation causes ionization. In the chamber there are two electrodes over which a voltage is applied to create an electrostatic field across the chamber (fig. 6-1). The electrons released and the positive ions produced are accelerated toward their respective electrode and the resulting current can be measured.

This current exhibits saturation above a certain field strength E at which all of the primary ions are collected before recombination (fig. 6-1). The saturation current is therefore proportional to the total energy deposited. The current is of the order of 10^{-14} - 10^{-16} A and can be detected using a high impedance output circuit (10^{10}-10^{13} Ω). Different gases are used depending on the application. For example, high pressure argon is used for γ-rays, hydrogen for fast neutrons (recoiling protons are detected) or BF_3 for slow neutron (the $^{10}B(n,\alpha)^{7}Li$ reaction).

An ionization chamber can also be used to detect single events. Now the condition exists that the collecting time of the electron-ion pair produced should be as short as possible. This avoids the overlapping of successive signals, which results in incorrect measurement of pulse height. A short signal also improves the properties of the detector in coincidence experiments. The electron drift time in a normal chamber is of the order of 10^{-6} s and the corresponding drift time for the ions produced is about 1000 times larger. In fig.6-2a an ionization chamber is shown schematically. The potential

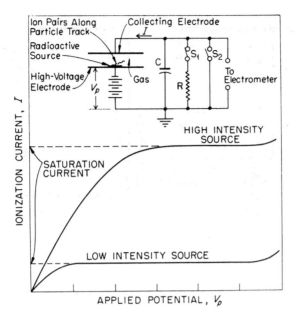

Fig. 6-1   Illustration of ionization chamber operation.
Typical current-voltage curves are shown for different source intensities. The insert shows how a
parallel-plate chamber is arranged for current measurements with the switch S_1 closed and S_2 opened.
For very small current ($<10^{-12}$A) measurements the
switch S_1 is opened and S_2 first closed and then opened.
The rate of charge collected on the capacitor is
measured by the electrometer[6.1].

of the electron collector (anode plate) is shown in fig. 6-2b
for a large value of anode resistance.

Initially, a sharp rise in potential is obtained when
the fast moving electrons are collected. The magnitude of this
initial rise is however dependent on where in the chamber
the ionization event occured. The positive ions move much
more slowly and until they have all reached the cathode, the
potential continues to increase giving an undesirable potential tail. To avoid these problems, a gridded ion chamber
is often used. In this device the collector plate is shield-
ed from the effects of the positive ions and only the fast
electron pulse is obtained. The chamber gas must be chosen

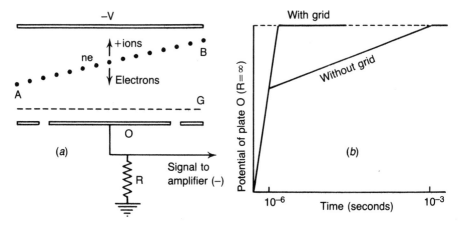

Fig. 6-2   The ionization chamber. The collector potential for
            ordinary and gridded chamber shown as a function of
            time, neglecting collector capacity due to the grid.
            A, B is a ionized track, G shows schematically the
            grid[(6.2)].

so that negative ions are not formed. This is achieved by
adding methane or carbondioxid to the filling gas which can
be argon. Counting ionization chambers have been profitably
used to study such strongly ionizing particles as α-particles
and fissions fragments, but they are not used for electron
counting, the primary ionization being too low.

6.1.2 PROPORTIONAL COUNTER

    If the voltage applied on an ionization chamber is
increased beyond the saturation interval, the primary ions
begin to produce secondary ions by collisions with the gas,
and consequently the primary ionization is multiplied by a
factor, depending on the geometry of the apparatus and on
the applied voltage (fig. 6-3). The device is now functioning
as a proportional chamber. In argon gas, the critical voltage
gradient needed to enter the proportional region is about
10^6 Vm^{-1}.

    To achieve such a large field gradient, it is usual to
use cylindrical geometry so that a large field gradient is

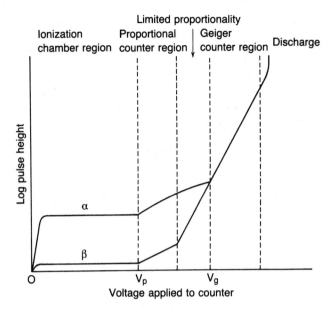

Ionization
chamber region

Limited proportionality

Proportional
counter region

Geiger
counter region

Discharge

Log pulse height

α

β

O V_p V_g

Voltage applied to counter

Fig. 6-3   Pulse site obtained from a counting chamber as a
            function of voltage applied[6.2]. Curves are drawn
            for a heavy initial ionization, e.g. from an α-partic-
            le and for a light initial ionization, e.g. from a
            fast electron. For voltages less than V_p only the
            primary ion pairs are collected; for $V > V_p$ ioniza-
            tion by collision takes place and the counter ceases
            to be a conventional ionization chamber. Very rough-
            ly, the electric field in argon at atmospheric pres-
            sure corresponding to V_p is 10^6 V m^{-1}.

set up near the axial wire electrode. In such a geometry,
the field is

$$E = \frac{V}{r \cdot \ln(b/a)} \qquad (6.1)$$

where V is the voltage applied, a is the wire radius and
b the inner radius of the cathode cylinder. For a counter
with a = 0.01 cm, b = 1 cm and V = 1000 volts, the field required for ionization by collision in argon at atmospheric
pressure is reached at a radius r = 0.02 cm. Thus, in this
small cylindrical region, we get multiplication or an ava-

197

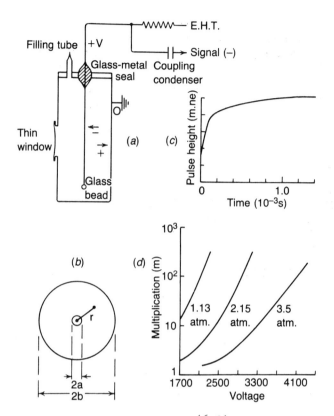

Fig. 6-4   The proportional counter[(6.2)].   (a) Construction.
The glass bead prevents sparking from points at the
end of the wire. (b) Cross section, cylindrical geometry. (c) Pulse shape. (d) Dependence of multiplication on voltage for a cylindrical counter filled
to the pressures indicated.

lanche of secondary ionization. Multiplication factors of
10-1000 are common (fig. 6-4).

The shape of the pulse from a proportional counter is
independent of the position of the ionization track in the
chamber because the main multiplication always occurs in the
high field region. The pulse rises to half its final height
in 2.5 µs but normally there is a time-lag of about 1 µs while
the initial electrons drift towards the wire.

Proportional counters are particularly useful for

lightly-ionizing particles such as electrons, mesons and fast protons whose small specific ionizations make the use of an ion chamber difficult. In many applications, however, such as in the study of charged particle spectra, as transmission detectors in particle telescope arrangements and for low-energy γ-ray and X-ray detection, the proportional counter and the ionization chamber have been superseeded by the use of semiconductor counters. Ionization chambers and proportional counters still find widespread applications as neutron detectors and also in many special arrangements as position-sensitive detectors and large-area counters.

6.1.3 GEIGER-MÜLLER COUNTER

If the voltage applied to a proportional counter is increased, the proportionality feature gradually disappears and the size of the pulse obtained, which can be several volts in amplitude, is independent of the primary ionization. A counter operated in this manner is called a Geiger-Müller counter.

The construction of a typical GM counter may be almost identical with that of a proportional counter (fig. 6-5). An argon pressure of 100 mm Hg (13.33 kPa) and an alcohol pressure of 10 mm Hg (1.33 kPa) are common. If an ion pair is produced in a GM counter, the electron moves towards the central wire just as in proportional counter action, and the positive ion of argon moves more slowly to the cathode. The avalanche produced by the electron in the high field near the wire is much more violent than in a proportional counter and secondary ultraviolet photons are produced in sufficient numbers to convey the discharge down the whole length of the wire of the counter.

When the avalanche reaches the end of the wire, a charge is collected, its size being independent of the initial ionization. Meanwhile, the sheath of positive ions moves out from the wire toward the wall and screens the wire from the cathode. The field near the wire drops below the value necessary for ionization by collision so that the discharge ceases. The alcohol gas neutralizes the argon ions by donating electrons to them and when, in this way, the

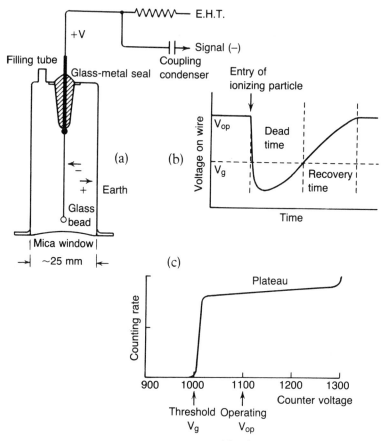

Fig. 6-5   The Geiger-Müller counter[(6.2)]. (a) Construction.
(b) Voltage on wire immediately following entry of
an ionizing particle. The dead time and recovery
time are each approximately 100-200 μs. (c) Counting rate of GM tube exposed to constant source,
as a function of applied voltage.

alcohol ions produced reach the cathode, the energy available
dissociates them rather than creates further free electrons.
The alcohol gas thus acts as a quenching gas.

  The advantages with the GM-counter are its very simple
operation, its high sensitivity and the large output pulse
signal. However, its intrinsic limitations, such as no energy
sensitivity and a long recovery time, (around 250 μs compared
with about 0.5 μs for a proportional counter) frequently

outweight its advantages. Nevertheless, GM-counters continue
to provide adequate measurements for many radioactive requirements, especially when chemical separations are part of the
procedure.

6.2 Scintillation detectors

6.2.1 NaI(Tl) AND OTHER SCINTILLATORS

The fluorescence produced in substances which are sensitive to α-particles (ZnS) or β-particles (Ba Pt(CN)$_6$)
has been utilized for the detection and counting of such
particles since the beginning of the present century. The
visual observation of scintillations with a microscope is,
however, inprecise and in the long run extremely tedious.
A modern scintillation counter, however, operates on the
same principles, except that the human eye is replaced by an
light-sensitive electronic device called a photomultiplier.
New scintillation materials have also been developed.

The most important scintillators used for particle and
photon detection are listed in table 6-1. The particular properties of these phosphors which may be emphasized are
according to Burcham[6.2] as follows:

Sodium iodide NaI(Tl): This is the most versatile of
all the phosphors and is of outstanding importance for the
study of γ-radiation, but it has the disadvantage that it is
hygroscopic and must be sealed in aluminium can (fig.6-6)
with reflecting or diffusing walls. The efficiency for γ-ray
detection is many times greater than that of a GM-counter
because of the effective thickness of the convertor. The
efficiency is also much larger than that of a germanium
counter of the same volume because of the difference in atomic
number, but the resolution is much poorer.

Zinc sulphide: This is an excellent phosphor for
particles of short range but cannot be used in thick layers
since it rapidly becomes opaque to its own radiation.

Anthracene and stilbene: Organic phosphors have a faster
decay time than inorganic phosphors but have poorer efficien-

Table 6-1
Characteristics of some scintillators .

Scintillator	Density (g/cm^3)	Photon efficiency(%)	Decay time (μs)
NaI(Tl)	3.67	20	0.25
CsI(Tl)	4.51	-	1.1
ZnS(Ag)	4.1	20	10
Anthracene	1.25	10	0.03
Stilbene	1.16	6	0.008
Terphenyl in toluene (liquid)	0.86	3.5	0.002
Terphenyl in polystyrene (plastic)		3.9	0.004
Xenon gas		10	0.01-0.1

cies, especially for heavy particles. Since they contain only light elements, they are useful for counting particles in the presence of γ-radiation. For fast counting, they have been largely superseded by

Plastic and liquid scintillators: These are readily obtainable in very large volumes and can easily be adapted to many different geometrical arrangements. In this case energy of the excitation is transferred from a solvent to a solute, which then re-emits radiation in a wavelength range for which the solvent is transparent. Plastic scintillators form the detectors in the multi-element counter used in high energy physics to define the paths of particles or to trigger the operation of a visual detector. Because of their high hydrogen content and fast response, organic scintillators are much used for the detection of neutrons by observation of the recoil protons from (np) collisions.

Gases: Xenon is particularly useful when heavy charged particles are to be counted in the presence of γ-radiation. The main emission of light takes place in the ultraviolet and a wavelength shifter (often a film of grease) is necessary to convert the energy to a wavelength suitable for detection by a standard photomultiplier.

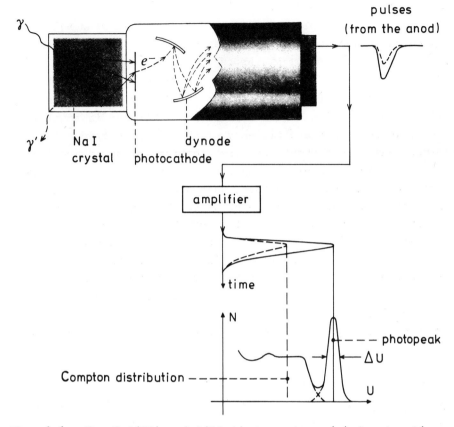

Fig. 6-6   The NaI(Tl) scintillation counter. (a) Construction,
(b) Pulse shape.

6.2.2   SCINTILLATION COUNTING SYSTEMS

A scintillation counting system consists of a scintillator, a photomultiplier, a power supply and an amplifier-
analyser-scaler system (fig. 6-6).

When an ionizing radiation passes through the scintillator,
it produces photons. The mechanism involved in the production
of light is complicated and not well understood. Also, the
mechanism is different for different types of scintillators. The
number of photons produced n_p is proportional to the energy absorbed in the scintillator. The scintillator is covered with
a reflector except on the side connected to the photomultiplier.

The photomultiplier has a photoelectric film (usually coated on to the photomultiplier tube) as its first element. When light falls on it, electrons are released. The number of electrons produced is proportional to the number of photons falling on the photocathode (which is equal to $n_p \epsilon_c$, where ϵ_c is the collection efficiency of the system and is equal to $n_p \epsilon_c \epsilon$, where ϵ is the efficiency of the photoelectric material. In most cases ϵ is about 10%. The electrons produced are focused on to the first dynode. The dynodes are coated with a material like cesium antimonide so that, when high energy electrons hit them, secondary electrons are produced and hence electron multiplication occurs. The photomultipliers have ten or more dynodes. When an electrical potential is applied between any two dynodes, the secondary electrons produced in preceding dynodes gain energy before they strike the next dynode and produce more secondary electrons. Thus multiplication occurs in each dynode and the final number of electrons N_e collected by the last electrode, called the collector, is given by

$$N_e = n_p \epsilon_c \epsilon m_1 m_2 \ldots \qquad (6.2)$$

where m_1, m_2, ... are the multiplication factors for successive dynodes. The factors m_1, m_2, ... are dependent on the potential between the dynodes and are usually independent of the number of incident electrons on any dynode. From eq. 6.2, one can see that N_e is proportional to n_p and, consequently is proportional to the energy which the incident radiation can supply to the scintillant. The electrons finally pass through a resistance, producing a voltage drop across it. The voltage drop is of short duration and is proportional to N_e. This electrical pulse can be amplified and analyzed (for more details see sec. 7.2.4).

Scintillation counters can be compared with proportional counters. Both give an output pulse proportional to the energy of the radiation. The multiplication in gas counters is due to the multiplication of ions, while in scintillation counters it is due to the production of secondary electrons at the dynodes. The duration of the pulse is shorter for scintillation counters than for proportional counters.

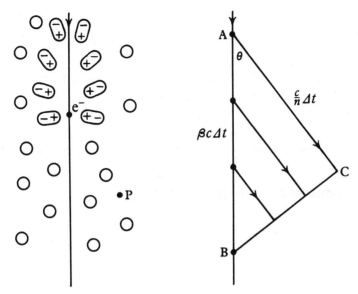

Fig. 6-7   Cerenkov radiation. (a) Polarization of atoms of a
transpararent medium by passage of a charged particle. (b) Formation of coherent wave-front.

6.3 Cerenkov counters

When a charged particle passes through a transparent
material with a velocity v greater than the velocity of
light in that material, c/n, where n is the index of refrac-
tion, a cone of electromagnetic radiation is emitted in the
material and the particle loses energy as a result of this
emission. The emission of this radiation, which is called
Cerenkov radiation, is a property of the gross structure of
the absorber material. The incoming charged particle induces
a polarisation of the medium along its path (fig. 6-7). The
time variation of this polarisation can lead to radiation.
If the charge moves slowly, the phase relations will be random
from different points along the path and no coherent wave
front will be formed. However, if the particle is moving faster
than the velocity of light in the medium, v > c/n, then a
coherent wave front can be formed and radiation emitted
(fig. 6-7). The radiation is mainly in the blue end of the

visible spectrum and in the ultraviolet. The applications of
Cerenkov detectors are mainly concerned with particle velocity
or particle mass selection in experiments involving the scattering of light relativistic particles such as pions or kaons
but they are also used in experiments where electrons must be
identified or separated out.

6.4 Semiconductor detectors

6.4.1 SEMICONDUCTOR PROPERTIES

Analogously, semiconductor radiation detectors are ionization chambers in which the usual gases have been replaced by
semiconducting solids.

A number of advantages then immediately follows. The
density of the solid dielectric is very much higher than that
of a gas, thus improving the stopping power. For a given
amount of energy lost by the ionizing particle, 5 to 10 times
as many free charge-carriers are released, which greatly
improves the energy resolution of the counter. Since both the
electrons and holes have large mobilities and the collection
distances are short, it is possible to achieve rapid collection
times (as low as a few nanoseconds).

In a semiconductor, the atoms are arranged close together
in a crystal lattice. The energy levels of the electrons are
then split up according the Pauli principle, giving rise to
energy bands as is schematically shown in fig. 6-8.

At absolute zero, the electrons in insulators and semiconductors fill up completely one or more of the lowest energy
levels, the highest filled being called the valence band.
This band is separated from the next higher one, the conduction band, by a gap of forbidden energies E_g, so that conduction cannot occur. At any other higher temperature, there
will be some thermal excitation of electrons from the valence
band into the conduction band, leaving empty places or "holes",
carrying a positive charge. With electrons falling back to
the valence band, a dynamic equilibrium is reached, which is

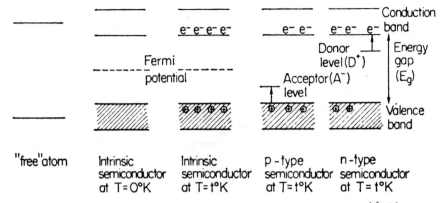

Fig. 6-8   Electron energy levels in a semiconductor[6.3].

a function of temperature. The occupation probability of a
level at energy E is given by the Fermi function:

$$f(E) = \frac{1}{1 + \exp\{(E-\zeta)/k\theta\}} \qquad (6.3)$$

where k is Boltzmann's constant, θ the absolute temperature
and ζ the Fermi potential, i.e. the potential at which the
occupation probability of an allowed state is 0.5. In an
intrinsic or pure semiconductor, the Fermi level must be in
the centre of the forbidden energy gap.

If we multiply the f(E)-function with a g(E)-function
which gives the density of electron energy states in the
region between E and E+dE in the conduction band and integrate
over all energies above E_g, we get the number of electrons
per unit volume in the conduction band

$$n_i = \int_{E_g}^{\infty} g(E)\ f(E)\ dE = A\ \theta^{3/2}\ \exp(-E_g/2k\theta) \qquad (6.4)$$

where A is a constant for a given material.

At room temperature ($\theta = 300°K$) the electron density
estimated using the above equation for Si ($E_g = 1.08$ eV,
$A = 2.8 \cdot 10^{16}$) is $\sim 10^{10}$ electrons/cm^3 and for Ge ($E_g = 0.67$ eV,
$A = 9.7 \cdot 10^{15}$) is $\sim 2 \cdot 10^{13}$ electrons/cm^3.

The electron in the conduction band is free to move and
if an electric field is applied, it moves toward the positive
electrode. The average speed, called the drift velocity, with

207

which the electron moves under the influence of an electric field is a function of the strength of the field ε, and the relationship between the two quantities is given by

$$v_e = \varepsilon \mu_e \qquad\qquad (6.5)$$

where μ_e is called the mobility of the electron. There is a similar equation for the holes. Mobility for electrons and holes is a function of the temperature and the properties of the material. The silicon and germanium detector properties are given in table 6-2. When radiation energy is absorbed in the crystal electron-hole pairs are created and the collection of these charge-carriers gives rise to an output signal, proportional to the amount of energy absorbed. To obtain the equivalent of one electron charge in the output pulse, 40 eV are needed in gas detection, 300 eV in NaI(Tl) crystals and up to 5 keV in organic scintillation detectors. In semiconductor detectors however, only a few electron volts are needed (3.6 eV in Si and 2.9 eV in Ge) for impact ionization. The fact that only a few electron volts are needed in order to obtain the equivalent of one electron charge implies that the statistical distribution of the transformation of radiation energy into electrical energy considerably improves, which accounts for the excellent resolution of the semiconductor detectors.

To obtain a satisfying detector the free charge density n_i has to be reduced. This is evident as the statistical variation in the number n_i (see eq. 8.20) is $\sqrt{n_i}$ giving for pure Si 10^5 electrons and for Ge $4.5 \cdot 10^6$ electrons. The variation thus for Si is about the same as the number of electrons produced by a 1-MeV radiation as the average energy to create an electron-hole pair in silicon is 3.6 eV. (Observe the distinction between average creation energy and the gap energy E_g required to take an electron from the top of the valence band to the bottom of the conduction band). Therefore it is difficult to distinguish pulses produced by radiations from those due to variation in electron density. We see that a small energy gap is required for better energy resolution, whereas a large energy gap is required to reduce the steady state current and its fluctuations.

Table 6-2
Properties of silicon and germanium.

	Silicon	Germanium
Atomic number	14	32
Atomic weight	28.09	72.60
Density (300 K) g cm^{-3}	2.33	5.33
Energy gap (300 K); eV	1.12	0.67
Energy gap (0 K); eV	1.17	0.75
Intrinsic carrier density (300 K); cm^{-3}	$1.5 \cdot 10^{10}$	$2.4 \cdot 10^{13}$
Instrinsic resistivity (300 K); Ωcm	$2.3 \cdot 10^{5}$	47
Electron mobility (300 K) cm^{2}/V-s	1350	3900
Hole mobility (300 K) cm^{2}/V-s	480	1900
Electron mobility (77 K) cm^{2}/V-s	$2.1 \cdot 10^{4}$	$3.6 \cdot 10^{4}$
Hole mobility (77 K) cm^{2}/V-s	$1.1 \cdot 10^{4}$	$4.2 \cdot 10^{4}$
Energy per hole-electron pair (300 K); eV	3.62	
Energy per hole-electron pair (77 K); eV	3.76	2.96

One method to reduce the charge density is to reduce the temperature of the material, which alone is not always enough because of the presence of impurities in the crystal.

Other methods also exist but before we discuss them we have to somewhat discuss the effect of impurities in semiconductor materials. The impurities can be of the donor or the acceptor type, resulting in intermediate levels in the forbidden energy region, as can be seen in fig. 6-8. A donor impurity has the tendency to give electrons to the conduction band, whereas an acceptor takes electrons out of the valence band, resulting in hole creation. A donor or n-type will thus have an excess of electrons, whereas an acceptor or p-type will show an excess of positive holes. Donor and acceptor atoms together form with other impurities and lattice imperfections "trapping-centres" where charge-carriers can be trapped for times longer than the dielectric relaxation time or can give rise to recombination with carriers of the opposite sign. Trapping and recombination both affect the proportionality between absorbed energy and output pulse amplitude.

So far, only homogeneous or bulk detectors have been considered, and we have seen that the major difficulty in obtaining such counters is to obtain material with sufficiently low carrier density. Therefore, n or p type semiconductor material can be compensated with acceptor or donor atoms in order to simulate an intrinsic material. This, of course, introduces trapping centres into the material, resulting in a loss of proportionality.

A number of different solutions have been developed to overcome these difficulties, e.g., diffused junction, surface barrier and p.i.n. detectors, as shown in fig. 6-9.

6.4.2 DIFFUSED JUNCTION DETECTORS AND BARRIER LAYER DETECTORS

The diffused junction detector consists of p and n type material. This type of detector is usually produced by diffusing a high concentration of donor impurities into a p type material. Silicon is usually used as the base material. Single crystals of high resistance are sliced into small, 1-mm pieces. Phosphorus is diffused into one surface of these slices. After diffusion proper electrical connections are made to the p and n sides.

In the junction between the p and n material there will be a resultant electric field which will clear all carriers from an intermediate layer at the interface. This is called the depletion layer. The depth of this layer can be further increased by an applied external field. Thus a region of high quality intrinsic semiconductor free of carriers is created which can then act as a bulk conduction chamber. The disadvan-

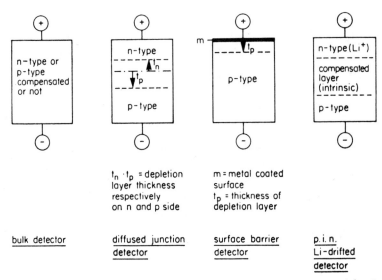

| bulk detector | diffused junction detector | surface barrier detector | p.i.n. Li-drifted detector |

$t_n \cdot t_p$ = depletion layer thickness respectively on n and p side

m = metal coated surface t_p = thickness of depletion layer

Fig. 6-9   Different types of semiconductor detectors[(6.3)].

tage of this type of counter is that the particle to be counted must penetrate the rather thick window of n-type material to reach the depletion layer.

In the surface barrier counter, one starts with n-type silicon generally of rather low conductivity. The oxide layer which forms on the surface of a fresh crystal exposed to air is p-type. Thus, between the oxide layer and the rest of the crystal a depletion layer is formed. The layer of oxide is then a very thin window. Electrical connection to both types of junction counter is made by evaporating a thin layer of gold on to the surface and making connections to it and to a similar layer of aluminium on the near surface.

Germanium can be used instead of silicon, but since it has more thermally-excited carriers than silicon, it must be used at liquid nitrogen temperature.

6.4.3   LITHIUM-DRIFTED DETECTORS

Lithium is an electron donating atom. It does not go into substitutional sites like other donor atoms, such as phosphorus. Instead, it enters interstitial sites. The diffusion coefficient is about 10^7 higher than that for phosphorus and

therefore deep diffused junctions can be prepared. Diffusion
is usually achieved in two steps. First lithium is coated onto
single crystals into which it is diffused by heating. In the
second step the sample is heated and a strong reverse bias
is applied. This helps the lithium atoms to diffuse deeper
into the crystal. Once the drifting is completed, the resulting detector has the simplified configuration shown in fig.
6-9.

The most direct way to fabricate a detector is to drift
lithium into a germanium wafer from one surface, but the size
of such detectors becomes limited to a thickness of 15-20 mm.
In order to create detectors of larger active volume, one
often drift lithium in from the outer surface of a cylindric-
ally shaped crystal, to form a coaxial detector.

Lithium-drifted germanium detectors |Ge(Li)-detectors|
must be continuously stored and operated at low temperature for reducing noise and keeping lithium-stability. All Ge(Li)
detectors are therefore enclosed in a vacuum cryostat which
provides thermal contact between the germanium crystal and a
reservoir (or dewar container) of about 20 l liquid nitrogen
at a temperature of $-196^{\circ}C$ (77 K). The dominant characteristic
of Ge(Li) detectors is their excellent energy resolution when
applied to γ-ray spectroscopy. Therefore most γ-ray spectroscopic systems that involves complex energy spectra are at
present carried out with germanium detectors (Ge(Li) or high
purity Ge).

6.4.4 HIGH-PURITY GERMANIUM DETECTORS

Ge(Li) detectors have a major disadvantage in that they
must be continuously stored at low temperature which impose a
significant cost burden of filling liquid nitrogen. A single
failure to keep the detector cold usually means that the
device must be redrifted which is an expensive time consuming
process.

Therefore in the recent development of detectors germanium
in an unusually high state of purity is used. With an impurity
concentration in germanium of as low as about 10^{10} atoms/cm^3
a depletion depth of about 10 mm can be reached using a reverse
bias of 1000 V. These large-volume germanium diode detectors

are usually called "intrinsic germanium" or "high purity germanium" (abbreviated HPGe) detectors, and have recently come into common use as γ-ray spectrometers. It is, however, still necessary to cool the detector in a liquid nitrogen dewar container due to excessive leakage current which prevents the use of the detector at room temperature. But interruption of the cooling is no longer disastrous and such detector is successfully operating again after being stored at room temperature. The γ-ray detection characteristics for a HPGe detector should be the same to those observed in a Ge(Li) detector of same size and shape.

6.4.5   SEMICONDUCTOR COUNTER CHARACTERISTICS

The advantages of semiconductor counters are, as already has been mentioned, their excellent resolution, linearity and small size and consequent fast response time permitting high counting rates. However, there are several unique problems associated with semiconductor detectors which require special consideration: (1) the signal is typically quite small so that rather sophisticated electronics are required to obtain ultimate performance; (2) noise generated in the detector itself, as a result of leakage current and random fluctuations in charge carrier density, is frequently a significant factor in the design and application selection of such devices.

Semiconductor counters have, in many applications, superseded other types of signal detectors and are now generally used for detections of charged particles, e.g., in telescope arrangements for identifying masses and charges (sec. 9.5.2), but also in the detection of neutrons, γ-rays and X-rays. Semiconductor detectors are also now widely used in detector arrangements such as are employed in time-of-flight measurements and in γ-pair spectrometry.

6.5 Track detectors

6.5.1   CLOUD CHAMBER AND BUBBLE CHAMBER

Both the cloud chamber and the bubble chamber have very limited applications in applied nuclear physics. The cloud

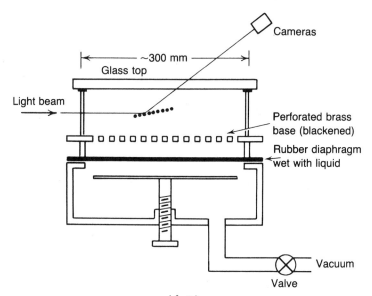

Fig. 6-10 The cloud chamber[(6.2)].

chamber works according to the following principle (see fig. 6-10). A gas mixed with a saturated vapour of, for instance, alcohol or ether is contained in a shallow cylinder. The gas can be expanded rapidly by a fast and short motion of a piston. Because of the sharp drop in temperature during the adiabatic expansion following this process, the vapour becomes supersaturated and will condense in droplets on dust particles in the gas or on other impurities acting as nuclei of condensation. If a charged particle moves through the gas immediately before or during the period of supersaturation, the ions left along the trail of the particle will act as condensation nuclei for the vapour. With proper illumination, this vapour trail can be seen and photographed. A great disadvantage of the cloud chamber is that it has a long recovery time after each expansion. This disadvantage is overcome in the diffusion cloud chamber in which a constant temperature gradient is set up giving a continuously sensitive chamber.

The bubble chamber (fig. 6-11) can be described as the inverse of a cloud chamber. A superheated liquid, normally hydrogen, is the medium through which the particles move. The trail of ions left in the path of a charged particle pro-

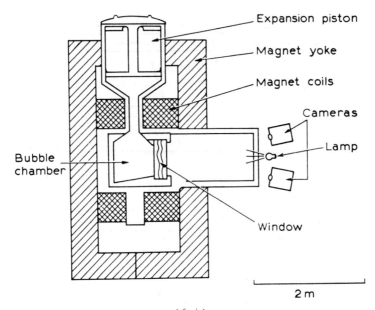

Fig. 6-11 The bubble chamber$^{(6.4)}$.

vide evaporation nuclei for the liquid hydrogen and this
trail can be seen and photographed as a chain of bubbles.

Both the cloud chamber and the bubble chamber can be
immersed in magnetic fields so that the charge and momentum
of various particles can be measured.

The bubble chamber has its greatest ·range of application
in elementary particle research.

6.5.2 SPARK CHAMBER

A spark chamber consists of two metallic plates insulated
from each other and with a noble gas in the gap. If a charged
particle passes through the chamber, a high voltage is applied
to the plates either during the passage of the particle or
immediately after and a spark appears along the trail of ions
left by the penetrating particle.

This is utilized in the spark chamber detector, which is
characterized by a high spatial resolution. The information
that is used to obtain this resolution may be derived from
the sonic wave produced by the spark or by the magnetic field
set up by the spark current. The main use of spark chambers

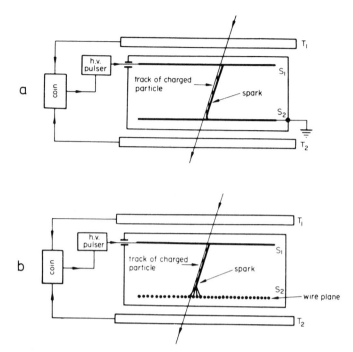

Fig. 6-12 (a) Schematic representation of a section through
a spark chamber with plane electrodes; (b) Schematic
representation of a section through a spark chamber
with a high-resolution wire plane.

has been in high-energy particle physics. In recent years,
however, their good spatial resolution has made them increas-
ingly attractive in a number of different low-energy applications.

They are used at the focal planes of magnetic spectrometers, in studying rare events over large areas, or when
measurements must be made over large solid angles, such as
e.g. in studying the low yields of fission fragments. Often
the spark chamber can be made insensitive to background
radiation such as α-particle fluxes in fission studies or γ-
radiation fluxes in neutron studies. In the latter cases, for
instance, the spark chamber must be boron-loaded (see fig. 6-12).

In applications, spark chambers are, however, not normally
preferred over other instruments because of their low count-
rate capability. Also, as a position determining detector it has
partly been superseded by the multi-wire proportional detector.

Table 6-3

Characteristics of nuclear emulsions.

Ilford Emulsion Type	D1	E1(K1)	C2(K2)	B2	G5
Mean grain diameter (μm)	0.12	0.14	0.10	0.21	0.18
Highest velocity detectable (v/c)		0.2	0.31	0.46	all
Highest detectable energy of protons (MeV)		20	50	120	all

6.5.3 NUCLEAR EMULSION

A nuclear emulsion is a modified photographic emulsion with greater thickness than usual and very fine grains. An ionizing particle passing through or stopping in the emulsion will reduce some of the AgBr molecules in the same way as photons do in the normal photographic process. After development, the nuclear emulsion can be studied under a microscope and the path of the ionizing particle can be directly examined. Different types of emulsion exist with different sensitivity. Some are sensitive only to very highly-ionizing particles, as, for instance, fission products and low-energy α-particles while others are sensitive also to minimum ionizing particles (fast electrons or other singly ionized particles with $v \simeq c$), table 6-3.

Nuclear emulsions have been extensively used as detectors of charged particles in magnetic spectrographs. The track length in the emulsion is then used to discriminate between different types of particles.

Nuclear emulsions can also be used for neutron spectroscopy, because the gelatine of the emulsion contains a considerable amount of hydrogen, so that a neutron impinging on a proton produces a proton track. If the neutron direction is known and the proton's direction and track length are measured under the microscope, the neutron energy can be calculated.

The advantage of nuclear emulsions is partly that they record continuously and store the recorded information but also that they have energy and velocity discrimination properties and high stopping power. High spatial resolution,

217

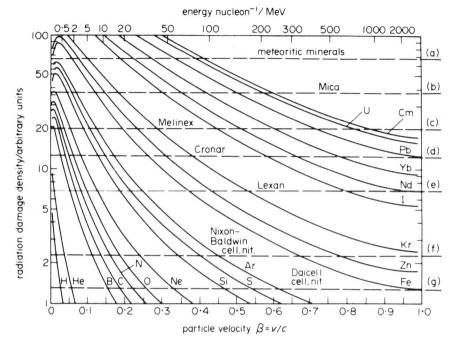

energy nucleon^{-1}/ MeV

Fig. 6-13 Variation of radiation damage density versus ion
velocity for various ions. The dashed lines indicate the approximate thresholds for track recording.
K is here taken as 16. (a) meteoritic minerals,
(b) mica, (c) Melinex, (d) Cronar, (e) Lexan, (f)
cellulose nitrate, (g) cellulose[6.5].

small size and no auxiliary apparatus are other properties
which make the detector interesting. Tedious scanning makes
it hard, however, to get good counting statistics and thus
signal detectors have in most cases superseded nuclear
emulsions.

6.5.4 SOLID STATE TRACK DETECTOR

If the electric conductivity and temperature of a solid
are sufficiently low, permanent linear tracks of radiation-
damaged material are formed by nuclear particles whose ionization rates exceed a threshold value characteristic of the
solid. These damage trails may be rendered visible under the

microscope by suitable chemical etching. This ability to
record nuclear tracks exists in all insulating solids such as
plastics, glasses and inorganic crystals as well as in natural
insulating minerals (fig. 6-13).

Glass, mica and plastic track detectors, which have quite
different threshold values, are extensively used in studying
heavy nuclear reaction products, when there may be a background
of ionizing radiation and are, for instance, well adapted for
fission product studies.

The primary advantages with solid state track detectors
are their good charge resolution for heavy nuclei and the
high event-background resolution even when there is an intense
background of less highly ionizing radiation. Other advantages
are the compactness and the easy-handling of the detectors.

Of special interest is the capability of solid state
track detectors to measure the concentration and spatial distribution of certain elements. In principle, any nuclide is
capable of being studied if it emits heavy nuclear particles
either directly because of its natural radioactivity as measurements of the radon content in air or, as a result of specific nuclear reactions when bombarded in an accelerator or
nuclear reactor. Nuclides with large cross sections for specific reactions are, of course, more suitable than others.
Here will be mentioned specifically boron mapping from the
reaction $^{10}B(n,\alpha)^{7}Li$ and also mapping from the biological
element reactions $^{17}O(n,\alpha)^{14}C$, $^{18}O(p,\alpha)^{15}N$ and $^{15}N(p,\alpha)^{12}C$.

6.6 References

6.1   N.F. Johnson, E. Eichler, and G.D. O'Kelley, Nuclear
      Chemistry in Technique of Inorganic Chemistry Vol. II,
      (eds. H.B. Jonassen and A. Weissberger), Wiley-Inter-
      science (1963).

6.2   W.E. Burcham, Nuclear Physics-An Introduction, Longmans
      2nd ed. (1973).

6.3   D. DeSoete, G. Gijbel, and J. Hoste, Neutron Activation
      Analysis, Wiley-Interscience 190 (1972).

6.4   E.B. Paul, Nuclear and Particle Physics, North-Holland
      137 (1969).

6.5   R.L. Fleischer, B.P. Price, and R.M. Walker, Nuclear
      Tracks in Solids, University of California Press (1975).

6.6   P.J. Ouseph, Introduction to Nuclear Radiation Detectors,
      Plenum Press (1974).

6.7   G.F. Knoll, Radiation Detection and Measurement, Wiley
      & Sons (1979).

7 Nuclear Electronics

7.1 Introduction

When an ionizing particle or photon hits a detector, a current or voltage pulse is produced. This takes place either directly by ionization of the active medium of the detector (e.g. proportional counter or semiconductor detector) or via complex intermediary processes, such as in the scintillation counter.

The size and time of this pulse carry the information in a continuous stochastic manner. This means that its amplitude can, in principle, have any value, and can appear at any time. The representation of data by continuously variable stochastic parameters is known as the analogue representation.

The processing of analogue signals is very difficult, since all small disturbances affect the value adversely. The analogue signals are therefore converted to digital signals. This involves the problem of registering events whose parameters are within a given range, and a "yes-no" pulse is generated. This type of data representation is also known as digital representation. The general electronic array of nuclear detectors involves the following parts:
- Detector.
- Analogue part (preamplifier, linear amplifier).
- Analogue to Digital Converter, abbreviated ADC.
- Digital part (scaler, timer, etc.).

7.2 Analogue electronic circuits

7.2.1 PREAMPLIFIERS FOR IONIZATION CHAMBERS

In an ionization chamber an ionizing particle produces
about 30 000 ion pairs or about $5 \cdot 10^{-5}$ As per MeV energy loss
This corresponds to a pulse height in the order of 0,25 mV,
with an integrating capacity, C, of 20 pF. The resulting pulse
must thus be amplified in a low-noise amplifier. The basic
preamplifier circuits are the voltage sensitive preamplifier
(fig. 7-1) and the charge-sensitive preamplifier (fig. 7-2).
(The amplifier, A, has an inverting input(-) and a non-inverting
input(+).)

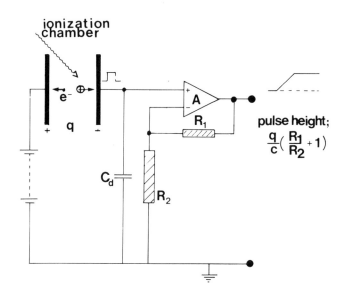

Fig. 7-1   Preamplifier circuit using a non-inverting ampli-
            fier. The closed loop gain of the circuit is $(1+R_1/R_2)$.
            An interaction process in the ionization chamber which
            releases 'q' charges causes a voltage pulse of size
            q/C. The output signal from the amplifier has thus
            the amplitude $\frac{q}{C}(1+\frac{R_1}{R_2})$

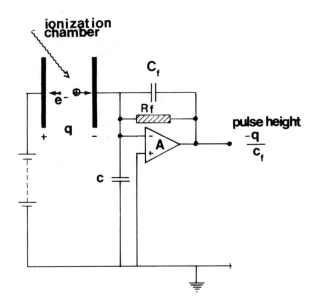

Fig. 7-2   Charge sensitive preamplifier. By means of the feed-
back capacitor C_f the preamplifier becomes charge
sensitive with gain $1/C_f$ independent of the detector and cable capacitance C_d at high loop gain 'A'.
Thus the output pulse height is q/C_f if $AC_f >> C_d$ The
feedback resistance R_f establish discharge of the
integrating capacitor C_f.

        Pulses from electron or γ-ray interactions typically corre-
spond to fewer original ion-pairs, and the number of created ion-
pairs can be as much as a factor of 100 smaller than in the
case of a charged heavy particle as mentioned above. Therefore
it becomes very difficult, if not impossible to amplify ion-
chamber pulses of electrons and photons successfully without
severel deterioration due to various sources of noise in the signal chain. In gas filled pulse detectors for electrons and photons one has to take advantage of internal gas multiplication
of the charge which occurs in proportional and Geiger Müller
counters. One of the most important applications of ion chambers
for photons is in the measurement of exposition. A determination of the ionization charge in an airfilled ionization chamber can give an accurate measurement of the exposition, and
a measurement of the ionization current will indicate exposition rate.

7.2.2 PREAMPLIFIERS FOR PROPORTIONAL COUNTERS

The maximum pulse height from a proportional counter with a capacity in the order of 10 pF becomes 10-100 mV. This pulse height is well above the noise level and no special precautions for low-noise operation normally need to be taken. The preamplifier in proportional counter assemblies therefore often consists of a simple emitter-follower, which functions as an impedance converter matching the high-resistance counter output to the low resistance (50-100 Ω) of the connection cable to the main linear amplifier. To obtain a low integrated capacity, C, the preamplifier is generally mounted directly on the electrode of the proportional counter tube.

Fig. 7-3 shows a circuit diagram of a simple emitter follower for proportional counters. The silicon diodes, D_1 and D_2, limit the pulse height and prevent the amplifier from being destroyed during the switching-on of the high-voltage supply,

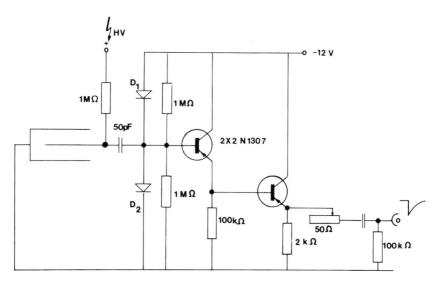

Fig. 7-3   Circuit diagram of a simple emitter follower. The signal is first differentiated with a time constant of 50 µs (1MΩ, 50 pF). The diodes D_1 and D_2 limit the pulse height and prevent the preamplifier from being destroyed during switching on the high voltage HV. The 50 ohm potentiometer matches the output impedance to the cable.

V_{HT}. The amplifier is built of two p.n.p transistors, which gives advantages for negative pulses. In order to match the output impedance to the cable, a potentiometer (~ 50 Ω) is located at the out-put.

7.2.3 PREAMPLIFIERS FOR GM-TUBES

Typical circuit for a GM-tube is shown in fig. 7-4a. The cathode is connected to a high potential (HV) through the resistors R_1 and R_2. When no discharge is present, no current passes through the resistors. The anode is on the same potential as the high voltage and the capacitance, C, is charged. When the GM-tube is discharged, a current flows through the resistors, R_1 and R_2, and the capacitance is discharged. After the discharge, the capacitor is recharged and a pulse with the shape shown in fig. 7-4b is obtained.

The integrating time constant, RC, can be chosen smaller than the ion-collection time as the discharge break-down occurs automatically. The maximum count rate is limited only by the intrinsic dead time of the counter, which is described in fig. 7-4b. During the spread of the ions towards the cathode, the field is disturbed and the counter remains entirely insensitive for a certain time (dead time) and recovers slowly with growing pulse height.

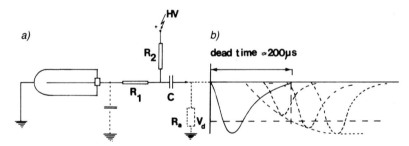

Fig. 7-4   a) GM-countercircuit. The signal is taken from the anode by means of a high value resistor R_1(2-20 Mohm) in order to minimize the influence from the cable and other parasitic capacities C_p.
b) GM-counter deadtime. The deadtime of the GM-counter in the example is 200 μs which depends on the discriminator level V_d.

7.2.4 PM-TUBES AND PREAMPLIFIERS

The photomultiplier is a light-sensitive device which is optically connected to a scintillation crystal or vial filled with a liquid scintillator. On the inside, the photomultiplier window is covered with a layer of alkali antimonide which emits electrons when it is hit by light quanta. This layer is often given a negative potential relative to the anode and is called photocathode. It is important that the wave length distribution of the light from the scintillator matches the spectral sensitivity of the photocathode. In fig. 7-5, the luminescence light distribution from a NaI(Tl) crystal is shown together with the response curve of a photocathode with S-11 response.

The PM-tube also consists of a series of ten to fifteen dynodes which are covered with a Sb-Cs and Mg-Ag alloys in order to emit secondary electrons when hit by accelerated electrons. The electrons emitted from the photocathode are focused with a focusing shield on the first dynode. A potential gradient of about 100 V is achieved between each dynode

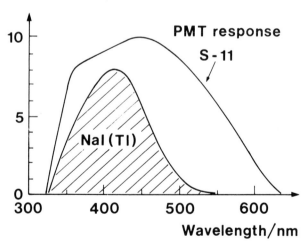

Fig. 7-5   The luminescence light distribution from a NaI(Tl) crystal. The sensitivity characteristics (S-11) of the cathod of the photomultiplier (PMT) is also given in the figure.

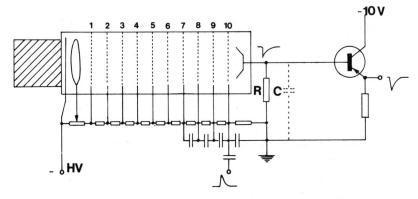

Fig. 7-6   Photomultiplier circuit with ten dynods connected
via a chain of resistors and with decoupling ca-
pacitors at the last four dynods. The parasitic
capacities are denoted with C. A single emitter
follower is connected to the output of the PM-tube.
Note that no high voltage decoupling from the anode
is needed because the anode is held at ground potential.

through a resistor chain, as shown in fig. 7-6. The shape of
the dynodes is such that electrons emitted from one dynode
are focused on the next one. The electrons from the last dynode are collected as a negative current pulse. An RC-circuit
is used as shown in fig. 7-6 to convert the current pulse to
a voltage pulse, which then will be transferred to the preamplifier.

Negative voltage pulses with rise-times of about 0.25 μs
are obtained from NaI-crystal detectors, depending on the de-
cay-time of the scintillation light. A pulse reaches its maximum value $\hat{u} = Q/C$, where Q is the charge collected by the anode
(about 10^{-4}-10^{-7} As). The decay of the pulse is exponential,
with a time constant, RC, $u=\hat{u}\ e^{-t/RC}$. In order to obtain a high
value of $\underline{u}$, the capacitance, C, in fig. 7-6 must be as low as
possible. This capacitance depends on various capacitances in
the circuits, and the connection cable to the preamplifier
usually contributes most. The preamplifier circuit, is therefore
often built directly at the PM-tube connection.

The preamplifier usually is an emitter follower which

serves as an impedance transformer between the PM-tube and
the main amplifier.

7.2.5 PREAMPLIFIERS FOR SEMICONDUCTOR DETECTORS

The semiconductor detectors are equivalent to diodes or-
iented as shown in fig. 7-7a. In this manner only leachage
currents flow through the detector.

Fig. 7-7b shows a circuit with reversed polarity on the
high voltage. The field effect transistor (FET) collects the
free charge carriers on its gate electrode.

For lownoise input stages a FET-transistor is the device which is currently almost universally used. The principle of an n-channel junction, FET, is shown in fig. 7-8.
The gate is made of p-type material and the pn-junction formed is biased so that part of the depletion layer extends
into the n-channel region. The resistance of the channel and,
hence, the current flowing between source and drain are thus
dependent on the extent of the depletion region in the channel. Since the gate-channel junction is reverse biased, the
current through the gate is very low. As shown in the I_d/U_{ds}-
diagram, the channel current, I_d, is effectively controlled
by the gate voltage (U_{gs}).

The most usual type of preamplifiers is the so-called
charge sensitive amplifier, which has an FET-transistor as
the first active element in order to get a high input impedance.

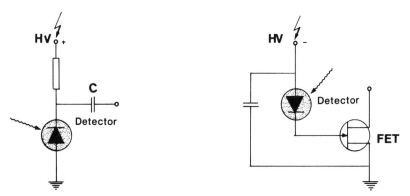

Fig. 7-7   a) Semiconductor detector.
            b) Semiconductor detector with FET-transistor.

228

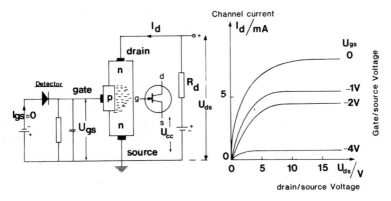

Fig. 7-8   Principles of the FET-transistor.
            a) The 'gate' is made of p-type material and the
            pn-junction thus formed is reverse biased so part
            of depletion layer extends into the channel region
            (shadowed area).
            b) The typical characteristic for a FET-transistor
            showing the drain or channel current I_d as a func-
            tion of the gate/source-voltage U_{gs} and the drain/
            source-voltage U_{ds}.

We will, however, only look at the amplifier as a black
box with a voltage amplification, $A = U_{out}/U_{in}$, as shown in
fig. 7-9. The input and output terminals are connected with
a feedback circuit consisting of a resistance, R_f, and capacitance, C_f, in parallel. The total capacitance can be given
as: $C_{tot} = C_d + (1 + A) \cdot C_f$, where C_d is the capacitance of
the detector including cable and input connection capacitances.

If a charge amount of Q released from an absorbed γ-quanta
appears on the input, the voltage, U_{in}, can be written as:

$$U_{in} = \frac{Q}{C_d + C_f(1+A)} \approx \frac{Q}{A \cdot C_f}, \qquad (7.1)$$

if A >> 1 and $C_d \lesssim C_f$

and the output voltage as:

$$U_{out} = U_{in} \cdot A \approx \frac{Q}{A \cdot C_f} \cdot A = \frac{Q}{C_f} \qquad (7.2)$$

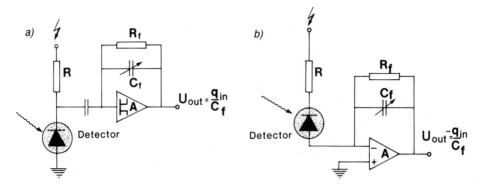

Fig. 7-9   Preamplifier circuits with FET-transistor as charge
sensitive preamplifier input via a feed-back resistor R_f.
a) AC-coupling with high-voltage across the coupling capacitor which can contribute to spurious noise.
b) DC-coupling with detector bias applied through
the feed-back resistor R_f.

The output signal is thus only determined by the passive
element, C_f, which normally is in the order of 1-2 pF. Small
variations in the amplification, A, do not influence the output signal nor variations in the capacitance of the detector
or the cable connections. In spectroscopy with semiconductor
detectors, this is very important because the applied high voltage strongly influences the capacitance of the detector. In
spite of this, smaller fluctuations in the applied high voltage do not influence the outsignal. Two different possibilities to connect a semiconductor detector to a charge sensitive
amplifier are shown in fig. 7-9. The circuit (b) in the figure
is usually to be preferred because no capacitor is needed in
the connection.

7.2.6 MAIN LINEAR PULSE-AMPLIFIERS

The primary function of the linear pulse-amplifier is to
increase the amplitude of the detector signal, but also to
introduce as little noise and nonlinearity as possible.

A detector-preamplifier-amplifier circuit is shown in
fig. 7-10. The triangular symbols represent an active ampli-

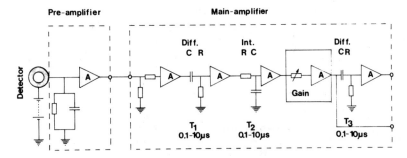

Fig. 7-10   Detector-preamplifier-amplifier circuit. The first
            differentiation circuit shortens the duration of
            "tail"-pulse from the detector and preamplifier.
            The resulting pulseform of the sequence CR-RC-CR is
            shown in Fig. 7-10d.

fying and isolating element, for example, a transistor assumed to give a frequency-independent gain, A (or -A, the minus
sign indicating inversion of polarity between input and output). The amplifier actually has a certain band-width, which
in a typical nuclear pulse amplifier is between 1 kHz-3.5 MHz,
or higher. This frequency dependence may be incorporated as
part of the first differentiator T_1 and the integrator T_2,
shown in fig. 7-10.

A differentiation is always required in order to shorten
the duration of the "tail"-pulse from the detector and preamplifier. This reduces pile-up and, thus, the possibility of
overload in the later stages, and improves the resolution of
the individual pulses for subsequent analysis. In fig. 7-11,
four different pulse-forming circuits are shown. These combination are present in the amplifier in fig. 7-10.

7.3 Analogue to digital converters

As was mentioned in the introduction of this chapter,
the registration of pulses corresponding to selected events
is performed in digital devices. It is therefore necessary
to digitize the analogue information carried by the pulse,
i.e., to decide whether or not, as well as where, a given

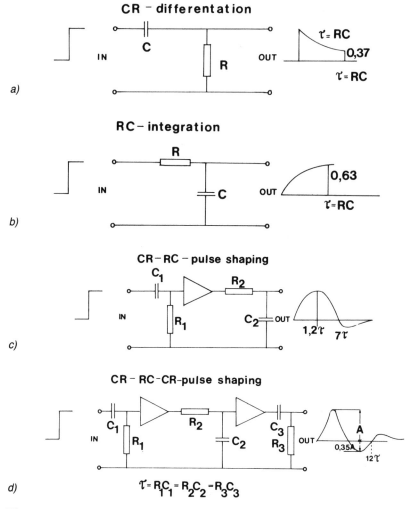

Fig. 7-11   Four different pulse forming circuits.

a) Ordinary CR-differentiation circuit where the
output is exponentially declining with a time constant RC.

b) Ordinary RC-integration circuit where the output is increasing exponentially with a time constant
RC.

c) CR-RC-pulse shaping with one differentiator C_1R_1
and one intergrator R_2C_2.

d) CR-RC-CR-pulse shaping with a chain of differentia-
ting, integrating circuits.

pulse should be registered. A circuit performing the described operation is denoted as an analogue-to-digital converter, abbreviated ADC.

7.3.1 DISCRIMINATORS

The discriminator generally has one upper and one lower level. The area between these two levels is called energy, pulse height, or amplitude window. Pulses whose pulse heights coincide with this window are labeled by a logical outsignal from the discriminator. The principle of the differential and integral discriminator is shown in fig. 7-12.

The electronic circuit for an integral discriminator is a "Schmitt" trigger. Fig. 7-13 shows the diagram for a simple integral discriminator circuit. The standing voltage, V_o, can be adjusted by means of the potentiometer, and all positive pulses with amplitudes higher than $V_1 - V_0$ trigger the circuit and an output pulse is obtained. The capacitor, C_a, compensates for the parasitic capacity, C_b, of the voltage divider, R_a, R_b, only. Except for the difference in base-emitter voltages, V_{BE} of Q_1 and Q_2, respectively, the critical voltage value, V_1, is equal to V_2.

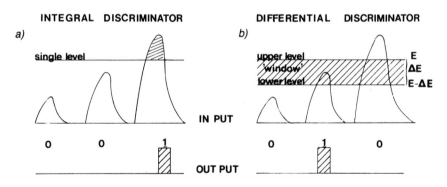

Fig. 7-12   Principle of differential and integral discrimators.
a) An integral discriminator has a single level and responds with standard output pulses for input pulses above this level.
b) A differential discriminator has two levels and responds with standard output pulses which max value lies in the window between the two levels.

233

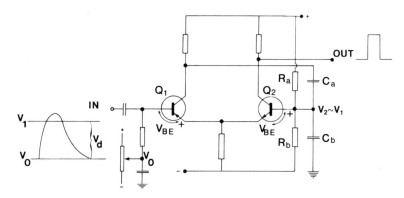

Fig. 7-13   Circuit diagram of a simple RC-coupled integral
            discriminator circuit. The standing voltage V_1 can
            be adjusted by means of the potentiometer at V_0
            which results in an adjustable discriminator threshold
            V_c. All positive pulses with amplitudes above
            $V_d = V_1 - V_0$ will trigger the cirquit.

Differential discriminators, often called single-channel
analysers, are circuits producing a normalized digital output
pulse for every input pulse, the amplitude of which satisfies
the condition $V_c < V_{in} < V_c + \Delta V_c$
    A single-channel analyser essentially consists of two
integral discriminators fixing the lower and upper thresholds,
respectively, and of a simple digital logic circuit selecting
events which trigger the lower Schmitt trigger, ST_1, but not
the upper one, ST_2, as shown in fig. 7-14. The gate system
denoted as "anticoincidence" should pass through the output
pulse of the lower Schmitt trigger, ST_1, unless it is blocked
by the output pulse of the upper one, ST_2.

7.3.2 DIGITAL CODING OF THE PULSE HEIGHT

    The high energy resolution of modern detector systems
necessitates a correspondingly high number of analyser channels. For instance, with semiconductor detectors with a relative FWHM (full width half maximum) of 0.1%, we need channel
numbers exceeding 1000 in order to display a spectrum peak in
more than one channel. Despite the difficulties connected
with the necessary stabilization of the particular channel

234

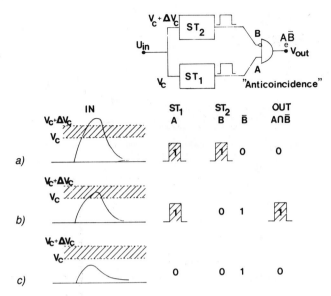

Fig. 7-14 Block diagram of a differential discriminator. The
anticoincidence circuit should pass through pulses
of the lower "Schmitt-trigger" ST_1 unless it is
blocked by the inverted output pulse of the upper
one ST_2.
a) The input pulse height is above ST_2 and output
pulses comes from both A and B. The anticoincidence
circuit is thus blocked and no pulse will pass.
b) The input pulse is in the window $\Delta(V_c)$ and no
pulse from ST_2 will block the anticoincidence circuit. Thus the pulse will pass through to the output.
c) The input pulse is below both ST_1 and ST_2 and
no pulses will come out from the "Schmitt-trigger".

widths and positions, an analogue-to-digital converter according to the multidiscriminator principle would be very complicated and expensive.

In nuclear pulse techniques, the so-called Wilkinson method
is used instead. This method is based on conversion of the pulse amplitude into a pulse length, which can then be measured
digitally by means of a frequency standard.

The operational principle of a Wilkinson type converter

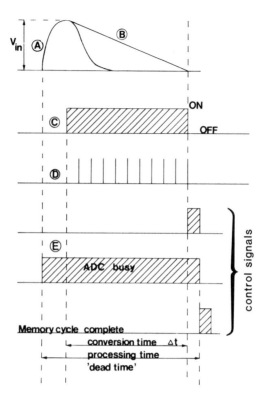

Fig. 7-15 The operational principle of a Wilkinson type of
ADC-converter.

A. The input pulse
B. A capacitor is charged to the peak voltage of
the input pulse.
C. A gate signal start discharging the capacitor
with constant current.
D. An oscillator produce pulses of a constant repetitive rate during C. The number of pulses produced, proportional to the input amplitude V_{in}
is the digital equivalent to the pulse height.
E. The busy signal goes 'on' when a input pulse reaches the ADC-converter and goes 'off' when the
ADC-convertion is finished. This signal stops
input pulses to reach the ADC when it is in 'on'
state.

is shown in fig. 7-15. A capacitor is charged by the input pulse (A) until it reaches its maximum charge corresponding to maximum value of the input pulse. The capacitor is then discharged with a constant current thus producing a voltage decrease of high linearity (B). At the same time, a gate signal (C) is produced. A comparator circuit compares (B) with zero and cuts off the gate (C) in the instant when (B=C). Hence, the length of the gate signal (C) which is a measure of the time duration Δt of B, is proportional to the input amplitude, V_{in}. In an electronic scaler, pulses (D) of constant repetitive rate (e.g. from a quartz-stabilized oscillator) are counted during Δt. The number of pulses, n, are proportional to Δt and, thus, also to V_{in}. The number $\underline{n}$ thus represents the desired digital equivalent of the pulse height and can be used directly as the address to the memory cell in which the event is to be stored.

At the input of an analogue-to-digital converter, a linear gate is always used, which is cut off during the conversion and thus prevents this process from being disturbed by following input pulses. The gate-control signal is longer than Δt by the time interval which is required for the processing of the digital output signal in the following digital equipment.

7.3.3 MULTICHANNEL ANALYSER

A pulse height multichannel analyser uses an ADC, equipped with an address-scaler which generates an address corresponding to the pulse height and transfers it to the memory unit. A schematic view of a multichannel analyzer with ADC, memory unit and display unit is shown in fig. 7-16.

The input pulse from the amplifier passes through a gate, which is open if the ADC is ready to receive a pulse. If the pulse fulfils the upper and lower discriminator settings of the linear gate, it loads the capacitor, C. When the maximum value is reached, the "linear gate" is blocked and the "address scaler" is opened. A constant current generator starts discharging the capacitor lineary and a clock oscillator loads an address-scaler until C has reached zero.

The time for discharge, C, is proportional to the pulse

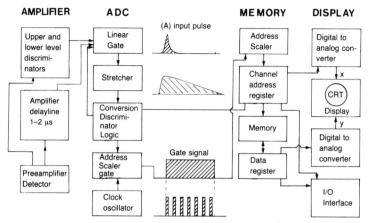

AMPLIFIER   ADC   (A) input pulse   MEMORY   DISPLAY

Fig. 7-16   Block diagram of a detector connected to multi-
             channel analyser with a Wilkinson type of ADC, memo-
             ry unit and display unit. The output pulse from
             preamplifier at the detector is feed into the main
             amplifier and discriminator. In the ADC a storage
             capacitor is charged up to a voltage equal to the
             peak height of the input-pulse. The capacitor is
             then discharged producing a ramp waveform and
             during this period a clock oscillator is switched
             on. The number of clock pulses counted in the address
             scaler until the ramp is zero gives the address and
             channel number for the pulse location in the memory.
             The display system gives possibilities to follow
             the build up of the pulse height distribution by
             fetching the information from the memory and dis-
             play the pulse-height distribution on a cathod-
             ray tube.

height. When C has reached zero, the address-scaler gate is
closed, and the address-scaler has registered a number of
pulses which corresponds to the channel number in the memory
where the information will be stored. The ADC now sends a
"ready" pulse to the memory, which answers with an "ADC transfer". The conversion cycle is finished by a "memory cycle
complete" pulse from the memory which sets the "adress-scaler"
at zero. The ADC is now ready to receive a new pulse.

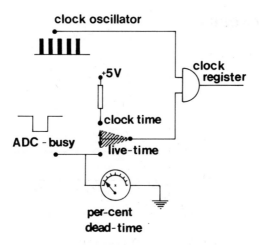

Fig. 7-17   Principle of "live-time" and "clock-time" operation.
             In "live-time" position the ADC-busy signals are
             blocking the gate and pulses from the time-base
             clock-oscillator cannot pass through. Thus the
             clock register is updated only when the ABC is free.

When the ADC is busy, it sends out an "ADC busy" signal,
which is used to determine the dead time and to compensate for
this measuring time. The dead time is determined by the construction of the ADC according to the expression:

$$\tau = K + N/f \qquad\qquad (7.3)$$

where K is a fix dead time of about 6 μs, N is the channel
number and $\underline{f}$ the frequency of the clock oscillator, i.e.
100 MHz. The dead time for storing a pulse in channel 4000 is
thus $\tau = 6 + 4000/100 \cdot 10^6 = 46$ μs. The correction of the
measuring time and recording of the dead time by an instrument
are shown in fig. 7-17. In "live time" position, the "ADC-busy"
signals block the clock pulses, and the clock register only
counts the time when ADC is free.
       An important specification for an ADC is its "integral
non-linearity" and "differential non-linearity".

7.4 Digital cirquits

Digital electronics deal with bivalent signals. The two possible signal values are denoted by the symbols 0 and 1.

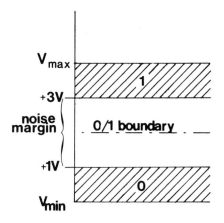

Fig. 7-18   The definition of '1' and '0' in voltage terms. The shadowed areas are forbidden zones of +/-1V. Voltage pulses above 3V correspond to '1' and below 1 V to '0'.

In any system, the special current or voltage ranges corresponding to the logical values 0 and 1 must first be defined. Fig. 7-18 shows an example of such a definition, where voltages $\geq$ 2 V correspond to "1" and voltages < 1 V to "0" When "1" corresponds to the high potential, it is called "positive logic". In order to achieve a clear distinction between "0" and "1", even in the presence of noise, spurious signals, etc., a security zone of ±1 V is left on both sides of the 0/1 boundary. Moreover, the voltage ranges corresponding to "0" and "1" are limited downwards (V_{min}) and upwards (V_{max}) by the maximum rating of the circuit components used.

A general digital circuit consists of a finite number of 0/1 inputs (m) and one output signal, (W) being a logical function of all input signals. Due to the bivalence of the input signals, there are 2^m different input-signal combina-

tions. The output from various logical functions is given in
table 7-1. These functions can be realized in electronic circuits, where symbols are also given in the table.

Table 7-1

Some basic functions of digital logic, their truth symbols
(0=false, 1=true) and below their names and symbols.

Input signals		Output signals for different logical functions W						
A	B	A∩B	A∪B	$(\overline{A \cap B})$	$(\overline{A \cup B})$	$(\overline{A}\cap B)\cup(A\cap\overline{B})$	$A\cap\overline{B}$	$\overline{A}$
0	0	0	0	1	1	0	0	1
0	1	0	1	1	0	1	0	1
1	0	0	1	1	0	1	1	0
1	1	1	1	0	0	0	0	0
Name		AND	OR	NAND	NOR	EXOR	-	NOT
Symbol								
IEC								

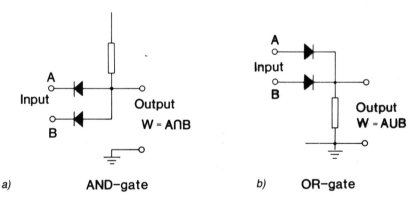

a)   AND–gate   b)   OR–gate

Fig. 7-19   a) Diagram of an 'AND'-circuit with diode logic.
When inputs A and B exhibit logical "1" a current
must flow into a load connected to the output (W=1).
b) Diagram of an 'OR'-circuit with diode logic. One
diode is sufficient to carry over a signal to the
subsequent circuit.

241

7.4.2 ELECTRONIC CIRQUITS OF DIFFERENT LOGIC

A circuit for the Logic AND and OR functions can easily be built by using diodes. In fig. 7-19a, an AND circuit is given. The potential U_W will adjust itself to the lowest of the potentials U_A and $U_B + U_{diod}$ in the AND circuit, which correspond to the logic function $A \cap B$.

In fig. 7-19b, an OR circuit is given. The potential U_W will adjust itself to the highest of the potentials U_A and $U_B - U_{diod}$, which corresponds to the logic function $A \cup B$.

The diod gate cannot be used alone because of the potential fall (0.6V for Si), which will cause the output signal to fall between the allowed levels. It is therefore necessary

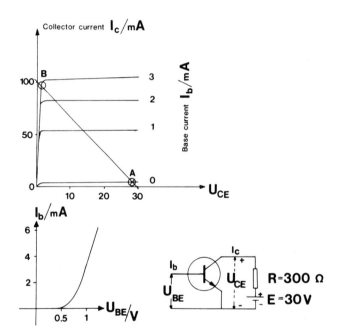

Fig. 7-20   Input current and collector current diagram of a
              transistor switch. If the base current is zero, no
              collector current can flow through the transistor,
              The whole feeding voltage is then over the collec-
              tor emitter which is cut off. Thus at point A the
              switch is cut off and at point B the switch is at
              bottom.

to use a transistor switch as output to restore the correct
potential. The function of the transistor switch can be described by its collector current diagram, shown in fig. 7-20.
If the base current is zero, no collector current can flow
through the transistor. The whole feeding voltage is then
over the collector emitter, and we say it is cut off. This
position corresponds to the point A in the collector diagram.
If the base current now starts to flow, and if it is high
enough, the transistor switches to point B. At this point,
the collector current is high and the collector emitter potential is only a few mV.

One of the most important circuits for digital electronics is the TTL-logic (transistor-transistor-logic). In this
type of circuits, the diodes have been replaced by multiemitter-
transistor. In fig. 7-21, examples of a NAND gate are shown in
a multiemitter-transistor logic performance.

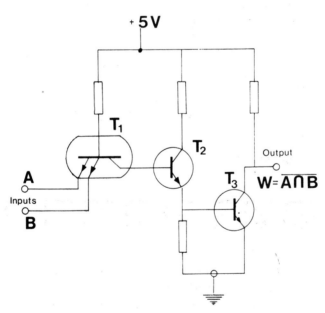

Fig. 7-21   NAND gate with transistor-transistor logic (TTL).
             The diodes in fig. 7-19b are replaced by a multi-
             emitter transistor (T1). The transistors T2 and
             T3 invert the level from T1 in order to fit the
             NAND logic.

7.4.3 SCALERS AND REGISTERS

Shift registers are in principle chains of flip-flop circuits capable of storing coded digital information, in which the information is shifted by one digit per clock pulse.

The flip-flop arises when two pulse inverters are connected in a series, as shown in fig. 7-22. The electronic circuit is a bistable multivibrator which in principle is a deamplifier with positive feedback. The bistable multivibrator has

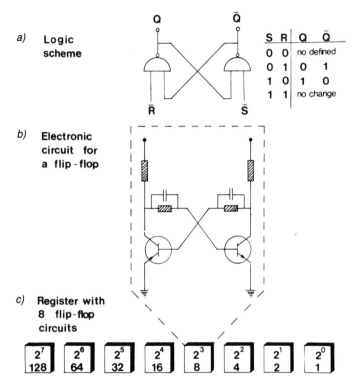

a) Logic scheme

S	R	Q	Q̄
0	0	no defined	
0	1	0	1
1	0	1	0
1	1	no change	

b) Electronic circuit for a flip-flop

c) Register with 8 flip-flop circuits

Fig. 7-22   Principle of flip-flop circuits.
a) Logic scheme of a flip-flop cirquit.
b) The electronic circuit for a flip-flop is a bistable multivibrator with two stable states either one transistor cut off and the other conducting, or th opposite.
c) A binary register is in principle a number of flip-flop circuits which can store one binary digit each.

two stable states: either one transistor cut off and the other
conducting, or the opposite. A shift takes place if it is triggered by a pulse. A flip-flop circuit is shown in fig. 7-22b.

A register is in principle a number of flip-flop circuits
which can store one binary digit each. The size of a register
is therefore given in a number of bites. In fig. 7-22c an
eight bit register is given, which in principle consists of
8 flip-flop circuits. In an eight-bit register, the lowest
position is given the weight 2^0 and the most significant bit
2^7.

The hexadecimal system uses groups of four bits for the
decimal numbers 0-15 and the letters A-F are used for the
numbers 10-15. This is simply an easier way of writing binary
numbers.

Example of Decimal:

$$212_{10} = 11010100_2 = 324_8 = D4_{16}$$
$$212_{10} = 1 \cdot 2^7 + 1 \cdot 2^6 + 0 \cdot 2^5 + 1 \cdot 2^4 + 0 \cdot 2^3 + 1 \cdot 2^2 + 0 \cdot 2^1 + 0 \cdot 2^0$$
$$212_{10} = 3 \cdot 8^2 + 2 \cdot 8^1 + 4 \cdot 8^0$$
$$212_{10} = 13 \cdot 16^1 + 4 \cdot 16^0$$

Some registers use the so-called BCD-code ('binary coded decimal'), which uses four bits for representing a decimal figure. Because four bits can be combined to 16 different figures, the BCD does not use the register efficient, but is used
because of its simple decoding.

Example:

$$212_{10} = \underset{2 \quad\;\; 1 \quad\;\; 2}{0010\ 0001\ 0010}\ \text{BCD}$$

$$4711_{10} = \underset{4 \quad\;\; 7 \quad\;\; 1 \quad\;\; 1}{0100\ 0111\ 0001\ 0001}\ \text{BCD}$$

7.4.4 MEMORIES

In principle, any element with two stable states can be
used for storing digital-coded information. The flip-flop is
thus actually a storage element, scalers are memories with
the additional capability of performing logical operation of
adding 1 and shift registers are also memories, etc. The in-

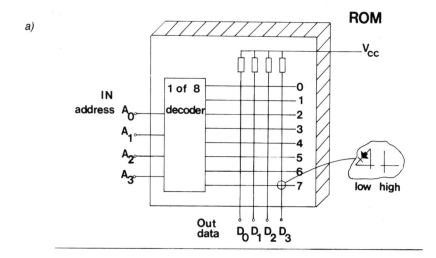

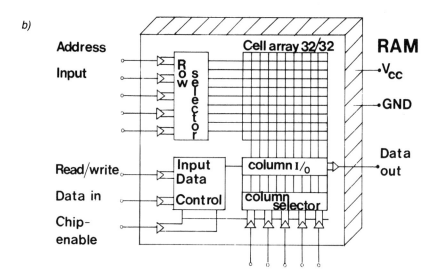

Fig. 7-23   Principle of ROM and RAM.
a) Read Only Memory can only be read by putting in
an address on A_0 - A_3.
b) Random Access Memory can be used for both data
input (write) and for reading of stored data.

formation is stored in the memory as "words". One word is
built of a number of bits, usually 8 or 16. A word with 8 bits
is called 1 byte. Each word has its given address in the memory. One can think of the memory as a shelf system and the
address is the number of the shelf where you can store what
you want.

There are at present two main types of memories available:
ROM (Read Only Memory)
RAM (Random Access Memory)
which are shown in fig. 7-23. In ROM, the information is perma-
nently stored and can only be read by putting an address on the
input. RAM has both address and data input and an input for a
control signal, which gives order for "write" or "read".

ROM is programmed by the manufacturer and cannot be altered by the user. Memories for programming by the user him-
self are called PROM (programmable ROM) or EPROM (Eraseable
programmable ROM). The latter has a quartz window, through
which one can erase the program by irradiating with intense
UV-light and use it again.

7.4.5 DATA OUTPUT

Besides the optical indication of the content of various
registers, memories or scalers, a data output device is very
often required to record the measured results in the form of
a protocol for later use or for later processing. Various elec-
tro-mechanical printers are employed for this purpose. For
intermediate data storing prior to later off-line processing
in digital computers, punched-tape recorders have proved to
be advantageous.

For monitoring digitally encoded magnitudes without high
requirements on accuracy, the digital information can be converted to analogue and displayed on a measuring instrument or
a cathode ray oscilloscope screen. In principle, all digital-
to-analogue converters are based on the addition of currents
or voltages corresponding to the weights of the particular
digital outputs. The principle is illustrated in fig. 7-24,
where digital "0" is assumed to correspond exactly to 0 V,
and the digital "1" to some well-defined voltage, V_N. The di-

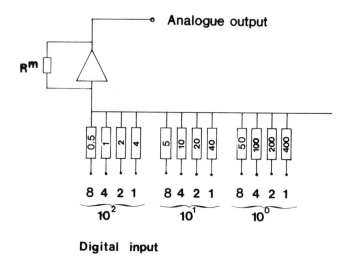

Analogue output

Digital input

Fig. 7-24 Principle of digital to analogue converter.
BCD-code of input signals is assumed.

gital input pulses are then converted to electrical currents
in proportion to their weights.

Count rates can also be recorded by means of analogue
representation where the output pulses from the amplifier are
charging a capacitor.

The operating principle of the integrating count rate
meter is shown in fig. 7-25. A standard charge, q_o, is applied to a capacitor for input pulse. Simultaneously as the particular values are added, the capacitor is exponentially discharged with a time constant, $\tau=RC$. If now $\underline{n}$ input pulses per
unit time, everyone with the charge q_o, reach the rate meter
this corresponds to a mean current nq_o. If this current is
constant during the measuring time T, the stored charge $q(T)$
over the capacitor becomes

$$q(T) = nq_o \int_o^T \exp \{ \frac{-(T-t)}{\tau} \} \, dt = nq_o \, \tau \qquad (7.4)$$

We observe that the stored charge $q(T)$ is proportional to
the count rate nq_o. The fractional standard deviation σ_q/q

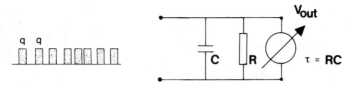

Fig. 7.25a   A capacitor C gets a standard charge q for every
              pulse caused by an event. The capacitor is dis-
              charged through the resistor R. The time constant
              τ of the system is the product between R and C.

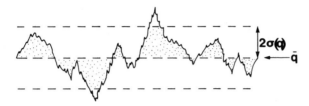

Fig. 7.25b   This figure shows the movement of the meter (V_{out}).
              For every standard charge the meter reading in-
              creases with a constant value. After every increase
              the meter is falling due to the discharge of the
              capacitor C through R. $\bar{q}$ indicates the average
              count rate.

of the stored charge for count rate fluctuations following
the Poisson distribution (sec. 8.3.2) can be shown to be

$$\sigma_q/q = (2n\tau)^{-\frac{1}{2}} \tag{7.5}$$

7.5 The microprocessor

7.5.1 STRUCTURE OF THE MICROPROCESSOR

      Electronics have undergone remarkable changes, during
the last decade, the peak of this evolution being the microprocessor. The large-scale integration (LSI) technique, that
made this technical break-through possible, can comprise se-

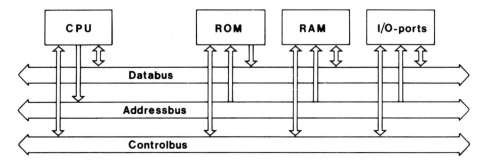

Fig. 7-26 Block-diagram of a typical microprocessor system
with central process unit CPU, memories ROM and RAM
and input/output (I/O) ports. These units communi-
cate with each other via the databus, addressbus
and controlbus.

veral tens of thousands of transistor functions in a single
chip. The function of the microprocessor is basically not
different from that of bigger computers, although several
facilities have to be omitted.

A block-diagram of a typical microprocessor system is
shown in fig. 7-26, and could be valid for most standard
microprocessors, such as the Intel 8080/8085, Zilog Z-80,
Motorola 6800, etc. The figure demonstrates the four most important blocks in a microcomputer system, central processing
unit (CPU), read-only memory (ROM), read/write-memory (RAM) and
input/output ports for communication with the outer world.

As can be seen from the figure, one CPU generates these
buses, the databus, the addressbus and the controlbus. The
databus carries information to and from the CPU, which is a
program instruction or a data-word used by the program.

The addressbus specifies where the data on the databus are
going, or where they are coming from. The controlbus, finally, directs the sequencing of the instruction being execu-
ted. To demonstrate the use of these buses, assume that the
processor needs to read the content of the memory cell, located at address 32767. The CPU then outputs the bits corresponding to the actual address on the addressbus, with a 16-bit
addressbus.

This causes one, and only one, memory cell to be selected for either read or write. The CPU will next output a read signal on the controlbus, which directs the memory to output the selected data onto the databus. In the opposite case, when data are to be stored in memory, the CPU outputs the address and the data on each bus and then sends a write signal on the controlbus to inform the memory to transfer the content of the databus into the selected memory cell.

7.5.2 FUNCTION OF THE MICROPROCESSOR

The internal architecture of a microprocessor comprises different registers, an arithmetic logic unit, a microprogram and control circuits. Fig. 7-27 shows a simplified functional block diagram of a microprocessor.

For each microprocessor type, a set of instructions are defined, i.e.

LOAD M(X), copy the content of memory cell M(X) into the
accumulator
STORE M(X), copy the content of the accumulator into the
memory cell M(X)

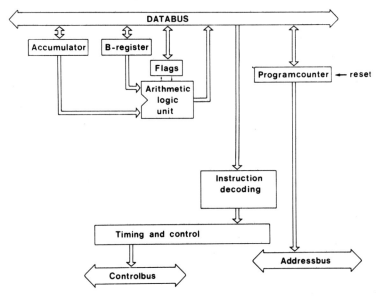

Fig. 7-27 A simplified functional block diagram of a microprocessor.

ADD B,   add the content of the B-register to the accumu-
            lator.

Assume that the programmer wants to add the content in memory address 100 to the B-register and to store the result in
memory address 101.
        The program could thus be written as follows:

LOAD 100, get data from memory address 100 into the accumu-
           lator
ADD B,   add content of B-register to the accumulator
STORE 101, store the sum in memory address 101.

        This is an example of a program written in assembly lan-
guage. Before the execution of this program, it has to be
translated into machine language, that is into pure binaries.
For an Intel 8080, the machine code corresponding to the
above program-example is:

	00111010	load
LOAD 100	01100100	lowest 16-bits representation of
		address 100
	00000000	highest 16-bits representation of
		address 100
ADD B	10000000	

	00110010	store
STORE 101	01100101	lowest 16-bits representation of
		address 101
	00000000	highest 16-bits representation of
		address 101

As is seen, the above program will occupy 7 memory locations
of 7 bytes (1 byte = 8 bits).
        Before the execution of a program can be performed, the
register denoted program-counter must be set to the start of
a program, preferably done with the rest signal which forces
the program-counter to 0. The addressbus now contains 0. The
timing and control logic then outputs a read-signal, which
effectuates that data is fetched from memory to the CPU.
Assume that the memory cell No. 0 contains the program instruc-

tion to add the content of register B to the accumulator. The
content of the databus or the ADD-instruction is next copied
on the instruction register and, finally, the content of the
program-counter is increased by one. This is called the FETCH-
cycle. The instruction register now contains the information
to ADD B to the accumulator and the program-counter and, hence,
the addressbus points to memory cell No. 1.

As shown in fig. 7-27, the instruction register feeds
data to the instruction decoder, where the ADD-construction is
encoded into a set of microoperations. First, the data in the
accumulator and the B-register are transfered to the arithme-
tic-logic-unit, ALU, which is directed to perform an add. Next,
the sum is copied, from the ALU via the databus, on the accumulator. Finally, the control logic returns for a new FETCH-
cycle, this time from memory location 1, since the program-
counter is increased by 1 for each FETCH-operation.

One important parameter of a microprocessor is its word-
length, i.e. the number of bits on the databus. Standard microprocessors like those already mentioned, 8080, Z-80 and 6800,
are 8-bits processors with a 16-bits addressbus, meaning they
can address up to 65535 memory locations. 16-bits microprocessors
are, however, also available, the marked advantage compared to
8-bits processors is a much faster execution time, and also
more detailed instructions.

7.5.3 USE OF MICROPROCESSORS

The microprocessor offers enormous advantages in complex
instruction and processing compared to conventional electronics. They can not only substantially reduce the number of
electronic circuits in an application, but also reduce the
production costs. Instrumentation equipped with microprocessors can also easily be made more favourably and extra facilities be added. As an example, the most modern multichannel analyzers often use a microprocessor for tasks like
reading the controls on the display data on the screen, etc.

Writing the program for a microprocessor application
can be performed in direct machine language, assembler or
high-level languages like FORTRAN and BASIC. The latter have

the advantages of easier programming, but suffer from producing less effective machine-code with respect to execution
time and memory need, compared to programming in assembly
language.

Numerical data-processing can be performed advantageously with a microprocessor, and several efficient operating
systems, which support FORTRAN and BASIC, have been written
for almost all 8-bits microprocessors. However, heavier number crunching is more rapidly solved on 16-bits microprocessors, not only due to their more efficient instruction re-
pertoire. Examples of well-known 16-bits microcomputers in
science are the PDP 11/03 from Digital Equipment Corporation,
and the MP 100/MP 200 micro Novas from Data General. In the
near future, 32-bits microprocessors can be expected, which
demonstrates the speed of this electronic evolution.

7.6 References

7.1 E. Kowalski, Nuclear Electronics. Springer Verlag, Berlin
     (1970).

7.2 E. Neuert, Kenphysikalische Messverfahren. (In German).
     Verlag G. Braun, Karlsruhe, W. Germany (1966).

7.3 M. Bolmsjö and O. Olsson, Elektronik-kompendium. (In Swe-
     dish). University of Lund, Sweden, Coden LUNFD6(NFRA-7003)/1-5
     (1977).

7.4 P.W. Nicholson, Nuclear Electronics. J.Wiley & Sons, (1974).

8 Statistical Methods and Spectral Analysis

8.1 Systematic and random errors

It is well known that we can never measure any physical quantity exactly, i.e., with no error. Progressively more elaborate experimental or theoretical efforts result only in a reduction of the possible error in the determination. In reporting the result of any measurement, it is therefore necessary to specify also the probability that the result is in error by some specified amount. Errors can be divided into two general classes: systematic errors and random errors.

Systematic errors may be subdivided into several classes: theoretical, instrumental and personal errors. They are all reproducible. Some examples of systematic errors in radioactivity measurements, which can occur frequently are the use of an unexact physical constant needed to evaluate the experimental data, an incorrect calibration of the efficiency of the counter as a function of the radiation energy, the neglect of the stopping power of the detector window, the use of a stop-watch, which is inclined to gain, an incorrect location of the beam target in relation to the beam in a nuclear reaction study and the tendency of an observer always to estimate a quantity to be smaller than it actually is.

Many of this kind of errors can be reduced or eliminated giving an increased accuracy of the measurements. Random errors are due to irregular causes and give fluctuations in repeated measurements under apparently fixed conditions. They can be divided in controllable and uncontrollable errors.

Controllable errors can occur if instabilites exist in the high voltage of the detector, if the room temperature fluctuates, if fluctuations exist in beam intensity of the accelerator producing the activity, if the distance between the sample and the detector varies randomly in repeated measurements, etc. These kinds of random errors can be controlled and reduced by improving the counter system and the exposure conditions.

It is possible to estimate but not to control the fluctuations in the disintegration rate of a radioactive sample. Nuclear disintegration, in common with all microscopic processes, is a random process, ultimately following a definite probability distribution (Poisson distribution). It cannot be affected by external conditions. The relative fluctuations in repeated measurements in a time interval can only be reduced by increasing the activity of the sample.

8.2 Expectation value and variance

In a mathematical sense variables which exhibit such fluctuation properties as described above are called random variables. The expected value, average value or mean value of random variable X that assumes one of a finit set of values X_1, X_2 ... X_k with corresponding probabilities P_1, P_2, P_k such that $P_1 + P_2 ... + P_k = 1$ can be defined to be

$$E(X) = X_1 P_1 + X_2 P_2 + ... X_k P_k = \sum_{i=1}^{k} X_i P_i \qquad (8.1)$$

The symbol μ or more specificly μ_X is often used for E(X), referring to the alternative terminology of mean value. The set of values P_i is called the probability function $f(X_i)$. With this notation the expectation value can be written

$$E(X) = \mu_X \qquad (8.2)$$

$$E(X) = \sum_i X_i f(X_i) \qquad (8.3)$$

In case of continuous random variables the sum becomes an integral and the probability $f(X_i)$ becomes a probability element $f(X)dX$ where $f(X)$ is the density function of X.

This leads to the following definition of expectation value in the continuous case

$$E(X) = \int_{-\infty}^{+\infty} X \cdot f(X) dX \qquad (8.4)$$

The variance is commonly employed as a measure of the dispersion of spread of the distribution and is usually denoted by σ^2. In referring to a random variable X having the given distribution f(X) the notation varX is used.

$$varX \equiv \sigma^2 \equiv E (X-\mu)^2 \qquad (8.5)$$

This expression can be expanded such as

$$E\{(X-\mu)^2\} = E(X^2)-2\mu E(X)+\mu^2 = E(X^2)-\mu^2 \qquad (8.6)$$

reminding that $E(X) = \mu$.
The following expression thus obtained is very useful for calculating the variance

$$\sigma^2 = E(X^2)-\{E(X)\}^2 = E(X^2)-\mu^2 \qquad (8.7)$$

Being an average squared deviation of X about its mean, the units of σ^2 are the square of the units of X. Thus, if X is a velocity with the units ms^{-1} the units of σ^2 are $(ms^{-1})^2$ or m^2s^{-2}. The square root of varX, however, has the same units as X and so is a bit easier to appreciate intuitively. It is called the standard deviation of X (or of its distribution):

$$\sigma \equiv (varX)^{\frac{1}{2}} = (E(X-\mu)^2)^{\frac{1}{2}} \qquad (8.8)$$

It is a kind of average deviation - what is called in physics a root-mean-square (rms) average - and can be thought of as "typical" deviation from the mean.

8.3 Probability density distributions

To understand the properties of random errors we will discuss three probability density distribution laws.

8.3.1 BINOMIAL DISTRIBUTION OR THE BERNOULLI DISTRIBUTION

The binomial distribution is a fundamental frequency distribution governing discrete random events from which other frequency distributions can be derived. If $\underline{p}$ is the probability that an event will occur and $q = 1-p$ is the probability that it will not occur, then in a random group of N independent trials the probability $W(n)$ that the event will occur $\underline{n}$ times is given by the binomial law as

$$W(n) = \frac{N!}{(N-n)!n!} \, p^n (1-p)^{N-n} \qquad (8.9)$$

The expectation value $E(n)$ of the binomial distribution is given by

$$E(n) = \sum_{n=0}^{n=N} nW(n) = pN = \mu \qquad (8.10)$$

This is equal to the true average value μ. The variance $var(n)$ is given by

$$var(n) = \sigma^2 = \sum_{n=0}^{n=N} (\mu-n)^2 W(n) = Np(1-p) = \mu(1-p), \quad (8.11)$$

giving a standard deviation of

$$\sigma = \pm(\mu(1-p))^{\frac{1}{2}} \qquad (8.12)$$

Radioactive decay is a stochastic phenomenon with a constant probability $\underline{p}$ of the nuclide to decay in a time interval dt

$$p = \lambda dt \qquad (8.13)$$

with the disintegration constant λ for the nuclide concerned. In a sample of N nuclides we can write (see eq. 2.20).:

$$p = -\frac{dN}{N} = \lambda dt \qquad (8.14)$$

(negative sign to dN is a negative number for a decaying pro-
cess). The probability $\underline{p}$ of decaying during a time period $\underline{t}$
is obtained by integrating eq. 8.14

$$-\int_{N_0}^{N_t} \frac{dN}{N} = \lambda \int_0^t dt \qquad (8.14a)$$

or

$$-(\ln N_t - \ln N_0) = \lambda t \qquad (8.14b)$$

giving

$$p = \frac{N_0 - N_t}{N_0} = 1 - e^{-\lambda t} \qquad (8.15)$$

If we have a sample containing N_0 radioactive nuclei we can
describe the probability $W(n)$ of observing $\underline{n}$ disintegrations
in the time $\underline{t}$ from the binomial distribution (eq. 8.9). The
mean value is then given by eq. 8.10

$$\mu = N_0 \cdot p = N_0(1 - e^{-\lambda t}) \qquad (8.16a)$$

and the standard deviation

$$\sigma = \{N_0 p(1-p)\}^{\frac{1}{2}} = (\mu e^{-\lambda t})^{\frac{1}{2}} \qquad (8.16b)$$

If the counting time $\underline{t}$ is short compared with the half-life
of the radioactive nucleus (eq. 2.22) as is usually the case,
then

$$e^{-\lambda t} = e^{-(t \ln 2)/t_{\frac{1}{2}}} \approx 1 - 0.693 \frac{t}{t_{\frac{1}{2}}} \approx 1 \qquad (8.17)$$

and the standard deviation is

$$\sigma = \pm \sqrt{\mu} \qquad (8.18)$$

8.3.2 POISSON DISTRIBUTION

The well-known and very important probability distribution function called the Poisson distribution can be easily derived from the binomial distribution. It is obtained when $\underline{p}$, the probability that an event will occur in a single experimental trial, is extremely small (i.e., $\lambda t \ll 1$ in our example of radioactive decay) but the number of elements involved in a single experimental trial ($N_0 \gg 1$) is so great that measurable success ($\mu = pN_0$) is assured.

Thus the Poisson distribution is deduced from the binomial distribution by assuming that N increases without bound and $\underline{p}$ decreases without bound, but in such a way that the product $\mu = pN$, the true average value, remains finite

$$W(n) = \frac{\mu^n e^{-\mu}}{n!} \qquad (8.19)$$

The Poisson distribution is characterized by a single parameter μ. The standard deviation of μ is

$$\sigma = \pm\sqrt{\mu} \qquad (8.20)$$

In all ordinary cases, radioactive decay follows the Poisson distribution law.

8.3.3 NORMAL DISTRIBUTION OR GAUSSIAN DISTRIBUTION

The normal distribution is an approximation to the binomial distribution when N is very large. It is applicable to distributions, in which the observed variable is not confined to integer values as distance, velocity, temperature, weight, but can take on any value from $-\infty$ to $+\infty$. The normal distribution law reads

$$W(x) = \frac{1}{(2\pi\sigma^2)^{\frac{1}{2}}} e^{-(x-\mu)^2/2\sigma^2} \qquad (8.21)$$

We denote the random variable $\underline{x}$ and the expectation value of the quantity, which is measured, μ. The standard derivation σ is the other parameter of the normal distribution. The distribution is shown in fig. 8-1. The amplitude of the function

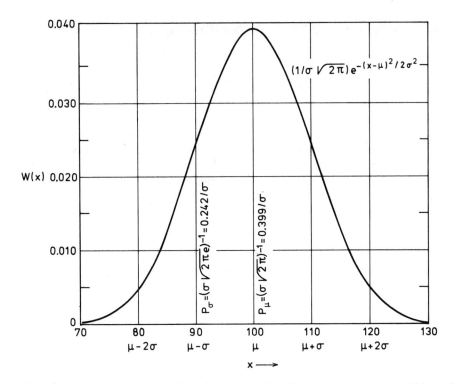

Fig. 8-1   The normal distribution. In the case shown $\mu = 100$ and
$\sigma = 10$. $P_\sigma = W(\mu \pm \sigma)$ $P_\mu = W(\mu)$.

at the values $x = \mu \pm \sigma$ is $W(\mu \pm \sigma) = e^{-0.5}\ W(\mu) = 0.607\ W(\mu)$.
The total area under the normal distribution curve is 1, and
the portion of the area between the values $\mu-\sigma$ and $\mu+\sigma$ is
0.6827.

By the substitution

$$z = \frac{x-\mu}{\sigma} \tag{8.22}$$

one obtains a variable with expectation value zero and standard
deviation 1. The standardized normal distribution function
which reads

$$W(z) = (2\pi)^{-\frac{1}{2}} e^{-\frac{1}{2}z^2} \tag{8.23}$$

is the one given in statistical tables.

8.4 Characterization of experimental data

8.4.1 ESTIMATION OF EXPECTATION VALUE

The number of counts recorded in a nuclear measurement is, of course, a statistical variable and if the experiment is repeated a slightly different number of counts is recorded each time.

In any finite series of measurements, it is impossible to find the exact value of the expectation value μ, which corresponds to the infinite population of data. Although the expectation value is constant, the individual measurements are distributed in a manner given by the particular probability density or frequency distribution of the process being studied. The best estimation of μ is simply the arithmetic average $\bar{x}$ of the N separate measurements

$$\mu \simeq \bar{x} = \frac{1}{N} \sum_{1}^{N} x_i \qquad (8.24)$$

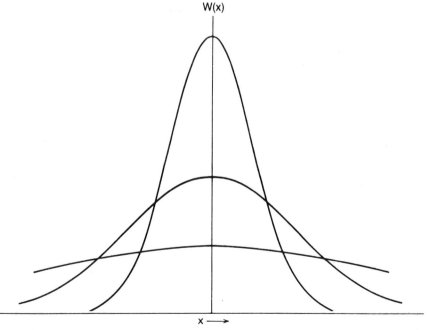

Fig. 8-2   Three normal distributions with the same mean value but with different standard deviations.

The size of the random fluctuations of the individual readings about the expectation value is expressed quantitatively by the fundamentally important parameter, the standard deviation σ. For a particular mean value μ, a small σ gives a sharply-peaked distribtuion, whereas a large σ gives a broad flattened distribution (fig. 8-2). In the normal distribution, about 32 per cent of a large series of individual observations must deviate from the mean value by more than ±σ and consequently 68 per cent of the individual observations should be within the band μ ± σ.

For any frequency distribution, the standard deviation is defined as the square root of the variance, which is the average of the squares of the individual deviations from the mean value for a large number of observations

$$\sigma^2 = \frac{1}{N} \sum_1^N (x_i - \mu)^2 \qquad (8,25)$$

In a finite serie of N observations, μ is not known exactly and hence it is impossible to determine σ exactly. The best estimate of σ in terms of the finite number N of observations can be shown to be

$$\sigma^2 \simeq \frac{1}{N-1} \sum_1^N (x_i - \bar{x})^2 \qquad (8.26)$$

There is a probability of approximately 68 per cent to find a single new observation in the interval $\bar{x} \pm \sigma$. The accuracy, with which the observed mean value $\bar{x}$ is determined after N readings, is

$$\sigma_{\bar{x}} = \frac{\sigma}{\sqrt{N}} \qquad (8.27)$$

This means that a repetition of the series of N measurements would, in general, give a different mean value, but the chance, that the new mean value would lie within $(\bar{x} \pm \sigma_{\bar{x}})$, is 68 per cent.

8.4.2   PROPAGATION OF ERRORS

Where a physical magnitude is to be obtained from the

sum or the difference of independent observations on two or
more physical quantities

$$A = a_1 \pm a_2 \pm a_3 \ldots ,$$ (8.28)

the final standard deviation σ_A of the derived magnitude is
obtained from

$$\sigma_A = (\sigma_{a_1}^2 + \sigma_{a_2}^2 + \sigma_{a_3}^2 + \ldots)^{\frac{1}{2}}$$ (8.29)

In measuring radioactivity, the counter normally has a
background which must be determined separately without sample.
Assume that the background counting rate c_b has been measured
for a time t_b giving $t_b c_b$ background counts. According to the
Poisson distribution law, the standard deviation σ_b of this
value is $(t_b c_b)^{\frac{1}{2}}$ and the background counting rate will be

$$c_b \pm (\frac{c_b}{t_b})^{\frac{1}{2}}$$ (8.30)

The total counting rate with sample and background c_{s+b}
is measured in a time t_{s+b} giving a number $c_{s+b} \cdot t_{s+b}$ of observed counts. The counting rate will then be written as

$$c_{s+b} \pm (\frac{c_{s+b}}{t_{s+b}})^{\frac{1}{2}}$$ (8.31)

and the counting rate from the sample alone is

$$c_s \pm (\frac{c_{s+b}}{t_{s+b}} + \frac{c_b}{t_b})^{\frac{1}{2}}$$ (8.32a)

where $c_s = c_{s+b} - c_b.$ (8.32b)

It will be noted that the background uncertainty enters
twice, once for the s+b measurement and once for the b measurement. By increasing the time of background measurement
t_b, it is possible to reduce the standard deviation. Often a
fixed time $T = t_{s+b} + t_b$ is available for the measurement
and has to be divided so as to make $\sigma(c_s)$ a minimum.
We then have

$$\frac{\delta\sigma(c_s)}{\delta t_b} = 0 = \frac{1}{2} \left(\frac{1}{\sigma(c_s)}\right)^{\frac{1}{2}} \left(-\frac{c_{s+b}}{(t_{s+b})^2}\frac{dt_{s+b}}{dt_b} - \frac{c_b}{t_b^2}\right) \quad (8.33a)$$

$$dt_{s+b} = -dt_b \quad \text{as} \quad t_{s+b} + t_b = T \qquad (8.33b)$$

and we get for $\sigma(c_s)_{min}$

$$\frac{t_b}{\sqrt{c_b}} = \frac{t_{s+b}}{\sqrt{c_{s+b}}} = \frac{T}{\sqrt{c_b} + \sqrt{c_{s+b}}} \qquad (8.34)$$

Thus the standard deviation with the optimum time apportioning is

$$\sigma(c_s) = \frac{\sqrt{c_b} + \sqrt{c_{s+b}}}{\sqrt{T}} \qquad (8.35)$$

If the physical quantity A is obtained by the multiplication or division of several independent observations

$$A = \frac{a \cdot b}{c} \qquad (8.36)$$

the standard deviation of A is related to the fractional errors of the observations giving

$$\frac{\sigma_A}{A} = \left[\left(\frac{\sigma_a}{a}\right)^2 + \left(\frac{\sigma_b}{b}\right)^2 + \left(\frac{\sigma_c}{c}\right)^2\right]^{\frac{1}{2}} \qquad (8.37)$$

A basic quantity in activation measurements, in which the sample has been exposed in, e.g., a reactor, is the number of counts in a specific detector channel during a measuring period, t_m.

$$K = \int_0^{t_m} c\,dt = -\int_0^{t_m} \varepsilon\,dN = \varepsilon\lambda N_e \int_0^{t_m} e^{-\lambda t}\,dt = \varepsilon N_e(1-e^{-\lambda t_m}) =$$

$$= igfbR(1 - e^{-\lambda t_e})\,e^{-\lambda t_0}\,(1 - e^{-\lambda t_m})$$

$$= igfbRST \qquad (8.38)$$

where

$c = \varepsilon\lambda N = $ counting rate (eq. 2.27a)

$\varepsilon = igfb = $ over all counting efficiency (eq. 2.27b)

$N_e = $ number of radioactive nuclei after exposure time t_e and waiting time t_0 (eq. 2.52)

265

$R = \varphi\sigma N_0 V$ = reaction rate (eq.3.26)

$S = 1 - e^{-\lambda t_e}$ = saturation factor (eq. 2.52)

$T = e^{-\lambda t_0} (1 - e^{-\lambda t_m})$ = time dependent correction factor

dependent on the waiting time t_0 between irradiation and measurement and the measuring time (eq. 2.52).

We want to know the number of stable nuclei N_0 per unit volume in the sample, of which some were transformed to be radioactive.

$$N_0 = \frac{K}{igfb\varphi\sigma VST} \qquad (8.39)$$

We assume that the saturation factor $S = 1$ and the waiting time $t_0 = 0$.

Fractional derivation gives

$$\frac{dN_0}{N_0} = \frac{dK}{K} - \frac{di}{i} - \frac{dg}{g} - \frac{df}{f} - \frac{db}{b} - \frac{d\varphi}{\varphi} - \frac{d\sigma}{\sigma} - \frac{dV}{V} -$$

$$- \frac{\lambda t_m\, e^{-\lambda t_m}}{1-e^{-\lambda t_m}} \left(\frac{d\lambda}{\lambda} + \frac{dt_m}{t_m}\right) \qquad (8.40)$$

Since all the differential quantities are independent, the fractional error in N_0 is given by

$$\frac{\sigma_{N_0}}{N_0} = \{(\frac{\sigma_k}{k})^2 + (\frac{\sigma_i}{i})^2 + (\frac{\sigma_g}{g})^2 + (\frac{\sigma_f}{f})^2 + (\frac{\sigma_b}{b})^2 + (\frac{\sigma_\varphi}{\varphi})^2 + (\frac{\sigma_\sigma}{\sigma})^2 +$$

$$+ (\frac{\sigma_v}{v})^2 + (\frac{\lambda t_m e^{-\lambda t_m}}{1-e^{-\lambda t_m}})^2 \, [(\frac{\sigma_\lambda}{\lambda})^2 + (\frac{\sigma(t_m)}{t_m})^2]\}^{\frac{1}{2}} \qquad (8.41)$$

The last term can be neglected if $\lambda t_m << 1$.

8.5 Effects of limited time resolution of counting experiment

8.5.1 COUNTING LOSSES

A particle detector and its associated electronic equipment always has a certain paralysis dead time after an event has triggered it.

For a GM counter, this dead time is of the order of 10^{-4} s, whereas fast scintillators and associated equipment may have dead times less than 10^{-8} s. When the counting rate is very high, a certain number of counts are lost because of this dead time, but a correction can be made for this loss in the following way. If the observed number of counts/ s is c' and the counter dead time is τ then the fraction of the second in which the counter is paralyzed is $c'\tau$. Of the true number c of particles which should have been recorded by the counter, only the fraction $1 - c'\tau$ is really registered. We therefore get

$$c(1 - c'\tau) = c' \qquad (8.42a)$$

which gives

$$c = \frac{c'}{1 - c'\tau} \qquad (8.42b)$$

It will be apparent that some counts are always lost due to the dead time. If the dead time is, say, 10 μs then the loss from a true counting rate of 10^4 s^{-1} is 10 per cent. This loss arises essentially from the random nature of the arrival times of the pulses.

It will be seen that the same circuit could handle a counting rate of up to 10^5 s^{-1} of regularly occurring pulses with no loss at all. Hence a useful technique in reducing counting losses is to place a de-randomizing circuit before any circuit having a large dead time. This take the form of one or more buffer stores which temporarily store the pulses until they can be processed.

8.5.2 ACCIDENTAL RATE OF COINCIDENCES

In a coincidence experiment, the aim is to establish how often two events triggering two different counters occur simultaneously. The two detector pulses are thus fed into the two inputs of a coincidence circuit having a coincidence resolving time of $\pm\tau$. We assume that the counting rate in detector 1 is c_1 c/s and in counter 2, c_2 c/s. Consider a single pulse present at input 1. If a pulse occurs at input 2 either preceding or following that at input 1 within the time interval τ, then a coincidence will be registered by the circuit. The

267

number of pulses at input 2 in any interval of duration 2τ is, on the average, $2c_2\tau$. It follows that the total accidental coincidence rate c_a is given by

$$c_a = 2c_1 c_2 \tau \qquad (8.43)$$

In an experiment designed to detect true coincidences from a single source, there will be an additional component from accidental coincidences which must be corrected for to give the true rate. The true rate c_t is obtained by first measuring the true plus accidental rate c_{t+a} and subtracting from it the accidental rate c_a which can be measured separately. If a fixed amount of experimental time T is available for performing both measurements, a situation is obtained analogous to that already discussed for the case of a source and a background. Assuming the time is apportioned in the optimum way to minimize σ_t, the fractional standard deviation in the true rate is given by

$$\frac{\sigma_t}{c_t} = \frac{\sqrt{c_a} + \sqrt{c_{t+a}}}{c_t \sqrt{T}} \qquad (8.44)$$

If the source strength is M disintegrations per second, the true coincidence rate will be equal bM where b is a constant depending on the relative intensity of coincidences in the source, the efficiencies of the detectors and the solid angles subtended by the counters at the source if the coincidences are spatially uncorrelated. The accidental rate c_a must be proportional to M^2 (eq. 8.43) and can analogously be expressed as $a M^2 \tau$, where a is another constant.

Substituting this value in eq. 8.44 we get

$$\frac{\sigma_t}{c_t} = \frac{\sqrt{aM^2\tau} + \sqrt{aM^2\tau + bM}}{bM \sqrt{T}} = \frac{1 + (1 + \frac{b}{a}\frac{1}{M\tau})^{\frac{1}{2}}}{\sqrt{b} \cdot (\frac{b}{a}\frac{T}{\tau})^{\frac{1}{2}}} \qquad (8.45)$$

As expected, the fractional error is reduced by increasing M but little is gained in using a value of $M\tau$ greater than b/a. The value $M = b/a$ corresponds to the case when the accidental rate $c_a = aM^2\tau$ is equal to the true rate $c_t = bM$ and, for practical purposes, this represents the maximum useful source strength. It is also seen from eq. 8.45 that an increase in solid angle and efficiency reduces the fractional error through the factor $\sqrt{b}$, since the ratio b/a is independent of these quantities.

8.6 Limits of qualitative and quantitative determination with application to radiochemistry

8.6.1   DETECTION AND DETERMINATION LIMITS

In the research dealing with activation analysis and other nuclear analytical methods it is necessary to determine limits of detection of radiochemical procedures, to select among alternative procedures, and to optimize given procedures with respect to certain experimental parameters. Currie[8.1] has thorougly discussed this. We follow his discussion and use the notations

blank

μ_b expected value

x_b observed value

σ_b standard deviation

gross signal

μ_{s+b} expected value

x_{s+b} observed value

σ_{s+b} standard deviation

net signal

$\mu_s = \mu_{s+b} - \mu_b$ expected value

$x_s = x_{s+b} - x_b$ value derived from an observation pair

$\sigma_s = (\sigma_{s+b}^2 + \sigma_b^2)^{\frac{1}{2}}$ standard deviation

The blank is defined as the signal resulting from a sample which is identical, in principle, to the sample of interest, except that the substance sought is absent (or small compared to σ_b). The blank thus includes the effects of interfering species.

Following an experimental observation one must decide whether or not that which was being sought was, in fact, detected. Such a binary, qualitative decision is subject to two kinds of error: deciding that the substance is present when it

is not (α, error of the first kind), and the converse, failing
to decide that it is present when it is (β, error of the second
kind) The maximum acceptable value for α, together with the
standard deviation, σ_0, of the net signal when μ_s = 0 establish
the <u>critical level L_c</u>, upon which decisions may be based.

Once L_c has been defined a <u>detection limit L_d</u> may by
established by specifying L_c, the acceptable value for β and
the standard deviation, σ_d, which characterizes the probability
distribution of the net signal when its true value μ_s is
equal to L_d.

Mathematically, the critical level is given as

$$L_c = k_\alpha \sigma_0 \qquad\qquad (8.46)$$

and the detection limit

$$L_d = L_c + k_\beta \sigma_d \qquad\qquad (8.47)$$

where k_α is related to the values of α according to

$$k_\alpha \int_{\sigma_0}^{\infty} W(s)ds = \alpha \qquad\qquad (8.48)$$

and k_β to β analogously.
The definitions and the relationships between L_c, L_d and the
probability distributions for μ_s = 0 and μ_s = L_d are de-
picted in fig. 8-3.

In a qualitative analysis neither a binary decision,
based upon L_c, nor one based on the detection limit may be
considered satisfactory. One wishes instead a result which is
satisfactorily close to the true value. Therefore for μ_s = L_q,
the <u>determination limit</u>, the standard deviation, σ_q, must be
but a small fraction of the true value. The determination limit
so defined is

$$L_q = k_q \sigma_q \qquad\qquad (8.49)$$

where L_q is the true value of the net signal μ_s, having a
standard deviation, σ_q, and $1/k_q$ is the requisite relative
standard deviation.

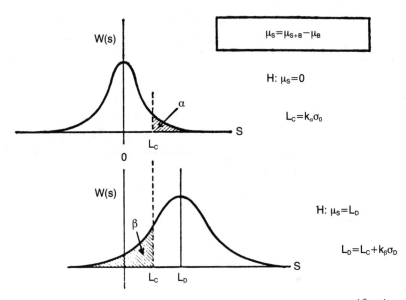

$$\mu_S = \mu_{S+B} - \mu_B$$

H: $\mu_S = 0$

$$L_C = k_\alpha \sigma_0$$

H: $\mu_S = L_D$

$$L_D = L_C + k_\beta \sigma_D$$

Fig. 8-3   Errors of the first and second kinds [8.1].

By way of summary, the levels L_c, L_d and L_q are determined entirely by the error-structure of the measurement process, the risks, α and β, and the maximum acceptable relative
standard deviation for quantitative analysis. L_c is used to
test an experimental result, whereas L_d and L_q refer to

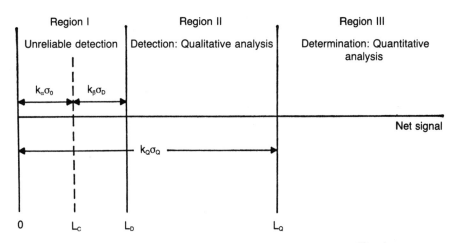

Fig. 8-4   The three principal analytical regions [8.1].

271

capabilities of measurement process itself. The relations among the three levels and their significance in physical or chemical analysis appear in fig. 8-4.

8.6.2 APPLICATION TO RADIOACTIVITY MEASUREMENTS

Application of the different kinds of limits discussed above to radioactivity involves the fact that the gross signal and blank observations are in digital form which in most cases may be assumed to be governed by the Poisson distribution. We assume that the numbers of counts are sufficiently large and thus the distributions approximately normal.

The detection limit L_d can now be written

$$L_d = L_c + k_\beta \sigma_d = L_c + k_\beta (L_d + \sigma_0^2)^{\frac{1}{2}} \tag{8.50}$$

as

$$\sigma_s = (\sigma_{s+b}^2 + \sigma_b^2)^{\frac{1}{2}} = (\mu_s + \mu_b + \sigma_b^2)^{\frac{1}{2}} = (\mu_s + \sigma_0^2)^{\frac{1}{2}} \tag{8.51}$$

if the gross signal is only measured once while there can be many observations of the blank.

A considerable simplification of eq. 8.50 takes place if we assume that the values of the two kinds of errors, α and β, are the same. This is very often an acceptable assumption. We now get $k_\alpha = k_\beta = k$ and can write:

$$L_d = L_c + k(L_d + \sigma_0^2)^{\frac{1}{2}} = L_c + (k^2 L_d + L_c^2)^{\frac{1}{2}} \tag{8.52}$$

giving

$$L_d = k^2 + 2L_c = k^2 + 2k\sigma_0 \tag{8.53}$$

The determination limit, L_q, is given by

$$L_q = k_q \sigma_q = k_q (L_q + \sigma_0^2)^{\frac{1}{2}} \tag{8.54}$$

which may be solved to yield

$$L_q = \frac{k_q^2}{2} \{1 + [1 + \frac{4\sigma_0^2}{k_q^2}]^{\frac{1}{2}} \} \tag{8.55}$$

272

Table 8-1

Working expressions for radioactivity with $\alpha=\beta=0.05$ (k=1.645) and kq = 10 (ref. 8.1).

Paired observations ($\sigma_b^2 = \mu_b$)	L_c	L_d	L_q
	$2.33\sqrt{\mu_b}$	$2.71+4.65\sqrt{\mu_b}$	$50\{1+\left[1+\dfrac{\mu_b}{12.5}\right]^{\frac{1}{2}}\}$

"Well-known" blank

$(\sigma_b^2 = 0)$	$1.64\sqrt{\mu_b}$	$2.71+3.29\sqrt{\mu_b}$	$50\{1+\left[1+\dfrac{\mu_b}{25}\right]^{\frac{1}{2}}\}$

Zero blank

$\mu_b = 0$	0	2.71	100

Asymptotic ratio*

x_s/μ_b (μ_b large)	1.64	3.29	10

* "Well-known" blank case; for paired observations, multiply by $\sqrt{2}$.

We refer to the original paper for further discussions.

Thus far, μ_b and σ_b have been used to refer to the "blank". In observations of radioactivity one frequently approaches the situation where the blank is due only to background radiation. When such is not the case, it may be desirable to decompose the blank into its separate components: background and interfering activities. Using **bg** to denote background and $\underline{i}$ to denote interference, the above quantities take the form

$$\mu_b = \mu_{bg} + \mu_i \tag{8,56a}$$

$$\sigma_b^2 = \sigma_{bg}^2 + \sigma_i^2 \tag{8.56b}$$

The detection limit L_d may be related to the minimum detectable mass m_d by means of eq.

$$L_d = q \, m_d \qquad (8.57)$$

where q represents an overall calibration factor relating the detector response to the mass present. The factor q is related to the factor K already discussed in eq. 8.38.

$$q = \frac{KN_A}{N_0 VA} \qquad (8.58)$$

where N_A = Avogadro's number A = mass number.

Currie gives in his paper a series of illustrations in order to make clear the application of the given formulaes. He also gives some working expressions for radioactivity measurements which are reproduced in table 8-1.

8.7 Regression

8.7.1 LEAST SQUARES FITTING OF A STRAIGHT LINE

Frequently in science, observations must be made, in which one quantity is unknown. For example, one often needs to determine the paramets a_0 and a_1 of a linear equation

$$y = a_0 + a_1 x \qquad (8.59)$$

with associated standard deviations $\sigma(a_0)$ and $\sigma(a_1)$. The least squares method for fitting a straight line to a series of experimental points is well-known. Suppose that n measurements have been done and that, when x has the value x_i, y is observed to have the value y_i. If the quantitiv $a_0 + a_1 x_i$ is calculated by using some pair of values for a_0 and a_1, then the difference between this and the observed value is

$$a_0 + a_1 x_i - y_i = \varepsilon_i \qquad (8.60)$$

We now add the squares of these differences ε_i for the $\underline{n}$
observations and minimize the sum by adjusting a_0 and a_1

$$\sum_i \varepsilon_i^2 = \sum_i (a_0 + a_1 x_i - y_i)^2 \tag{8.61}$$

$$\frac{\delta \sum_i \varepsilon_i^2}{\delta a_0} = 0 = 2 \sum_i (a_0 + a_1 x_i - y_i) \tag{8.62}$$

$$\frac{\delta \sum_i \varepsilon_i^2}{\delta a_1} = 0 = 2 \sum_i x_i (a_0 + a_1 x_i - y_i) \tag{8.63}$$

but since $\sum_i a_0 = n a_0$ we get from equation 8.62 and 8.63 respectively:

$$n a_0 + a_1 \sum_i x_i = \sum_i y_i \tag{8.64}$$

$$a_0 \sum_i x_i + a_1 \sum_i x_i^2 = \sum_i x_i y_i \tag{8.65}$$

These two equations have following solutions for a_0 and a_1
giving

$$a_0 = \frac{\sum y_i \sum x_i^2 - \sum x_i \sum x_i y_i}{\Delta} \tag{8.66}$$

$$a_1 = \frac{n \sum x_i y_i - \sum x_i \sum y_i}{\Delta} \tag{8.67}$$

where Δ is the equation determinant

$$\Delta = n \sum x_i^2 - (\sum x_i)^2 \tag{8.68}$$

If the observations have been made with different accuracy,
they have different weights ω_i, which are inversely proportional to the variances of the corresponding measurements,

$$\omega_i = \frac{k}{\sigma_i^2} \tag{8.69}$$

where $\underline{k}$ is a constant.
We now have to minimize the sum

$$\sum_i \omega_i \varepsilon_i^2 \tag{8.70}$$

and every term in eq. 8.61 must be multiplied with ω_i before
summation. Observe e.g. that the number $\underline{n}$ is changed to

$\sum\limits_{i} \omega_i$. The standard deviations of the two parameters can be calculated from the following formulas

$$\sigma^2(a_0) = k \; \frac{\sum \omega_i x_i^2}{\Delta} \tag{8.71}$$

$$\sigma^2(a_1) = k \; \frac{\sum \omega_i}{\Delta} \tag{8.72}$$

Fits to linear equations with more than two parameters, $y = a_0 + a_1 x_1 + a_2 x_2 \ldots a_n x_n$, can easily be developed along the same lines.

8.7.2 FITTING OF NONLINEAR EQUATIONS TO OBSERVATIONS

It frequently happens that one desires to use the method of least squares on nonlinear equations. It is simplest to treat the case of two unknowns. From that solution the extension to any number of unknowns is obvious. Assume that the unknown quantities are Z_1 and Z_2 and the observation equations are

$$f_1(Z_1, Z_2) = y_1 \quad \text{with weight} \quad \omega_1 \tag{8.73a}$$

$$f_2(Z_1, Z_2) = y_2 \quad \text{with weight} \quad \omega_2 \tag{8.73b}$$

$$f_n(Z_1, Z_2) = y_n \quad \text{with weight} \quad \omega_n \tag{8.73c}$$

where the functions f_1, f_2 ... f_n are nonlinear in Z_1 and Z_2. Solution of nonlinear equations starts with approximate values for the unknowns, which usually are obtained by graphical methods. Let us assume that the approximate values A and B have been obtained for Z_1 and Z_2 respectively. We then get

$$Z_1 = A + \zeta_1 \tag{8.74a}$$

$$Z_2 = B + \zeta_2 \tag{8.74b}$$

where ζ_1 and ζ_2 are new unknown quantities which are to be determined. Using Taylor's theorem, we can develop the function f around the point (A,B).

$$f_1(Z_1, Z_2) = f_1(A,B) + (\frac{\delta f_1}{\delta Z_1})_{A,B}\zeta_1 + (\frac{\delta f_1}{\delta Z_2})_{A,B}\zeta_2 \quad (8.75a)$$

$$f_2(Z_1, Z_2) = f_2(A,B) + (\frac{\delta f_2}{\delta Z_1})_{A,B}\zeta_1 + (\frac{\delta f_2}{\delta Z_2})_{A,B}\zeta_2 \quad (8.75b)$$

where we have neglected terms of higher order. The subscript A, B of the partial derivatives means that these partial derivatives are to be evaluated for $Z_1 = A$ and $Z_2 = B$. The quantities $f_1(A,B)$, $f_2(A,B)$, ..., are numerical values which should almost equal the values of the observations y_1, y_2, ..., if A and B have been properly chosen.

Let

$$y_1 - f_1(A,B) = \eta_1 \qquad\qquad (8.76a)$$

$$y_2 - f_2(A,B) = \eta_2 \quad etc \qquad\qquad (8.76b)$$

where the η's are small numerical quantities which now can be looked upon as new observations. We then get the observation equations

$$(\frac{\delta f_1}{\delta Z_1})_{A,B}\zeta_1 + (\frac{\delta f_1}{\delta Z_2})_{A,B}\zeta_2 = \eta_1 \quad \text{with weight} \quad \omega_1. \quad (8.77a)$$

$$(\frac{\delta f_2}{\delta Z_1})_{A,B}\zeta_1 + (\frac{\delta f_2}{\delta Z_2})_{A,B}\zeta_2 = \eta_2 \quad \text{with weight} \quad \omega_2 \quad (8.77b)$$

These equations are linear in the unknown quantities ζ_1 and ζ_2 and can be treated in the same way as in sec. 8.7.1. Often an iterative calculation must be performed where the funtion $\underline{f}$ successively is developed around the calculated (Z_1,Z_2) points.

If possible a coordinate transformation can be made which transforms the nonlinear observation equation into a linear one. This is obviously possible with the function for physical decay of radioactive nuclei (A is the activity s^{-1}).

$$A = A_0 e^{-\lambda t} \qquad\qquad (8.78a)$$

$$y = \ln A = \ln A_0 - \lambda t \qquad\qquad (8.78b)$$

In such a transformation one has to be careful to transform the weights of the observations to new weights.

Since $y = \ln A$, we get from eqs. 8.20 and 8.69

$$(dy_i)^2 = \frac{k}{\omega_i^t} = (\frac{dA_i}{A_i})^2 = \frac{k}{\omega_i A_i^2} \qquad (8.79)$$

and thus the transformed weights ω_i^t are

$$\omega_i^t = \omega_i A_i^2 \qquad (8.80)$$

As an example, let us estimate the parameters in the disintegration rate function from the following measurements on a radioactive sample, table 8-2. The time of observation was one minute in every case. Since the given rates are the number

Table 8-2

Measurements of the counting-rate on a radioactive sample.

time $t_i = x_i$	Counting rate $c_i = \omega_i^t$	$\ln c_i = y_i$
h	min^{-1}	
0	–	
0.5	9636	9.17
1.0	8310	9.03
1.5	7120	8.87
2.0	6092	8.72
3.0	4327	8.37
4.0	3355	8.12
5.0	2387	7.77
6.0	1759	7.47
7.0	1368	7.22
8.0	955	6.86
9.0	694	6.54
10.0	592	6.33
11.0	375	5.93
12.0	306	5.72

Σ 47246

of counts registered in every observation, the observation weights obtained from eqs. 8.20 and 8.69 are

$$\omega_i = \frac{1}{\sigma_i^2} = \frac{k}{c_i} \qquad (8.81)$$

and the transformed weights ω_i^t

$$\omega_i^t = \omega_i c_i^2 = kc_i \qquad (8.82)$$

The sums $\sum_i \omega_i^t$, $\sum_i \omega_i^t x_i$, $\sum_i \omega_i^t y_i$, $\sum_i \omega_i^t x_i y_i$ and $\sum_i \omega_i^t x_i^2$

have to be calculated and from formulas (8.66), (8.67), (8.71), (8.72) we get

$$\ln c_0 = 9.326 \pm 6.68 \cdot 10^{-3} \qquad (8.83a)$$

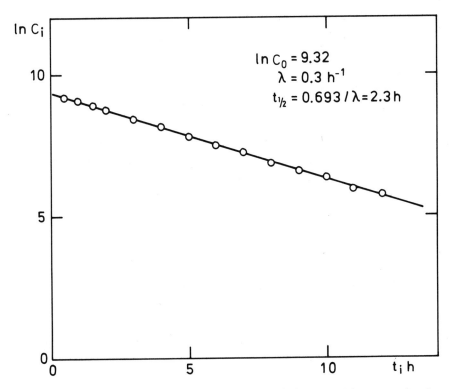

Fig. 8-5   Graphical solution of the disintegration constant.

or

$$c_0 = 11226 \pm 316 \text{ min}^{-1} \text{ as } dc_0 = c_0 d(\ln c_0) \qquad (8.83b)$$

and

$$\lambda = -0.306 \pm 0.002 \text{ h}^{-1} \qquad (8.84)$$

in close agreement to the graphical solution (fig. 8-5).

8.7.3 SMOOTHING OF DATA

One of the simplest ways to smooth fluctuating data is by
a moving average. In this procedure one takes a fixed number
of points, adds their ordinates together, and divides by the
number of points to obtain the average ordinate at the center
abscissa of the group. Next, the point at one end of the group
is dropped, the next point at the other end added, and the process is repeated.

Let us follow the description of Savitzky and Golay.[8.2]
Fig. 8-6 illustrates how the moving average might be obtained.
This description is based on the concept of a convolute and of
a convolution function. The set of numbers at the right are the
data or ordinate values, those at the left, the abscissa information. The outlined block in the center may be considered to
be a separate piece of paper on which are written a new set of
abscissa numbers, ranging from -2 through zero to +2. The ω's
at the right represent the convoluting integers. For the moving
average each ω is numerically equal to one. To perform a convolution of the ordinate numbers in the table of data with a
set of convoluting integers, ω_i, each number in the block is
multiplied by the corresponding number in the table of data,
the resulting products are added and this sum is divided by
five. The set of ones is the convoluting function, and the
number by which we divide, in this case, 5, is the normalizing
factor. To get the next point in the moving average, the center
block is slid down one line and the process repeated.

The concept of convolution can be generalized beyond the
simple moving average. In the general case the ω's represent
any set of convoluting integers corresponding to weights.
There is an associated normalizing or scaling factor, $\Sigma\omega_i$.

1800.0		705
1799.8		712
1799.6		717
1799.4		718
1799.2		721
1799.0		722
1798.8	x_o-2 ω_{-2}	725
1798.6	x_o-1 ω_{-1}	730
1798.4	x_o ω_o	735
1798.2	x_o+1 ω_1	736
1798.0	x_o+2 ω_2	741
1797.8		746
1797.6		750

Fig. 8-6   Convolution operation

Abscissa points at left, tabular data at right.
In box area the convolution integers, ω_i. Operation is the multiplication of the data points by
the corresponding ω_i summation of the resulting
products, and division by a normalizer, resulting
in a simple convolute at the point x_o. The box is
then moved down one line, and the process repeated.

The procedure is to multiply ω_{-2} times the number opposite it,
then ω_{-1} by its number, etc., sum the results, divide by the
nomalizing factor, if appropriate, and the result is the desired function evaluated at the point indicated by ω_0. For the
next point, we move the set of convoluting integers down and
repeat, etc. The mathematical description of this process is:

$$\bar{y}_j = \frac{\sum_{i=-m}^{i=m} \omega_i y_{j+i}}{N} \tag{8.85}$$

where $N = \sum_{-m}^{m} \omega_i$

The index j represents the running index of the ordinate
data in the original data table.

The moving average method to smooth fluctuating data is a

very rough method. In the general case the experimenter, if presented with a plot of the data points, would tend to draw through these points some type of curve giving as good a fit as possible.

This fitting is most commonly made with the least squares method and the curve chosen very often is a 2nd or 3rd degree polynomial. If the points are taken sufficiently close together then practically any smooth curve will look more or less like a quadratic in the vicinity of a peak, or like a cubic in the vicinity of a shoulder.

Let us discuss the fitting of a set of points to the 3rd polynomial $a_0 + a_1x + a_2x^2 + a_3x^3$. The $\underline{a}$'s are to be selected such that when each abscissa point is substituted into this equation the square of the differences between the computed numbers $\underline{y}$, and the observed numbers is a minimum for the total of the observations used in determining the coefficients.

Assuming we have 2m+1 consecutive values which are to be used in the determination of the best mean square fit through these values of our cubic polynomial. We now write (see sec.8.7.1)

$$\varepsilon^2 = \sum_{i=-m}^{i=m} (a_0 + a_1i + a_2i^2 + a_3i^3 - y_i)^2 \qquad (8.86)$$

Note that in the coordinate system being considered, the value of i ranges from -m to +m, and that i=0 at the central point of the set of 2m+1 values.

The least square criterion requires that the sum of the squares of the differences between the observed values y_i and the calculated be a minimum over the inverval being considered. We now get four equations

$$\frac{\delta\varepsilon^2}{\delta a_0} = 2\sum_{i=-m}^{i=m} (a_0 + a_1i + a_2i^2 + a_3i^3 - y_i) = 0 \qquad (8.87a)$$

$$\frac{\delta\varepsilon^2}{\delta a_1} = 2\sum_{i=-m}^{i=m} (a_0 + a_1i + a_2i^2 + a_3i^3 - y_i)i = 0 \qquad (8.87b)$$

$$\frac{\delta\varepsilon^2}{\delta a_2} = 2\sum_{i=-m}^{i=m} (a_0 + a_1i + a_2i^2 + a_3i^3 - y_i)i^2 = 0 \qquad (8.87c)$$

$$\frac{\delta\varepsilon^2}{\delta a_3} = 2\sum_{i=-m}^{i=m} (a_0 + a_1i + a_2i^2 + a_3i^3 - y_i)i^3 = 0 \qquad (8.87d)$$

These equations can be written more compactly

$$\sum_{k=0}^{k=3} a_k \sum_{i=-m}^{i=m} i^{k+r} = \sum_{i=-m}^{i=m} y_i i^r = F_r \qquad (8.88)$$

where $\underline{r}$ is the index representing the equation number of the four equations in eq. 8.87 which runs from 0 to 3,

or even more compactly

$$\sum_{k=0}^{k=3} a_k S_{k+r} = F_r \qquad (8.89)$$

where $S_{k+r} = \sum_{i=-m}^{i=m} i^{r+k}$

Note that $S_{k+r} = 0$ for odd values of $k+r$.

If 5 points (m=2) are fitted to our polynomial we get

$$S_0 = 5, \ S_2 = 10, \ S_4 = 34, \ \text{and} \ S_6 = 130$$

and our equations can be written

$$a_0 S_0 + a_2 S_2 = F_0 \qquad (8.90a)$$

$$a_1 S_2 + a_3 S_4 = F_1 \qquad (8.90b)$$

$$a_0 S_2 + a_2 S_4 = F_2 \qquad (8.90c)$$

$$a_1 S_4 + a_3 S_6 = F_3 \qquad (8.90d)$$

We want to determine a_0 which is obtained from eqs. 8.90a and 8.90c

$$a_0 = \frac{S_4 F_0 - S_2 F_2}{S_0 S_4 - S_2^2} \qquad (8.91)$$

giving

$$a_0 = \frac{-3y_{-2} + 12y_{-1} + 17y_0 + 12y_1 - 3y_2}{35} \qquad (8.92)$$

or (see eq. 8.85)

$$a_0 = \bar{y}_0 = \frac{\Sigma \omega_i y_i}{\Sigma \omega_i} \qquad (8.93)$$

283

Table 8-3

Weights in smoothing of consecutive points to quadratic and cubic polynomials.

POINTS	25	23	21	19	17	15	13	11	9	7	5
i					ω_i						
-12	-253										
-11	-138	-42									
-10	-33	-21	-171								
-09	62	-2	-76	-136							
-08	147	15	9	-51	-21						
-07	222	30	84	24	-6	-78					
-06	287	43	149	89	7	-13	-11				
-05	322	54	204	144	18	42	0	-36			
-04	387	63	249	189	27	87	9	9	-21		
-03	422	70	284	224	34	122	16	44	14	-2	
-02	447	75	309	249	39	147	21	69	39	3	-3
-01	462	78	324	264	42	162	24	84	54	6	12
00	467	79	329	269	43	167	25	89	59	7	17
01	462	78	324	264	42	162	24	84	54	6	12
02	447	75	309	249	39	147	21	69	39	3	-3
03	422	70	284	224	34	122	16	44	14	-2	
04	387	63	249	189	27	87	9	9	-21		
05	322	54	204	144	18	42	0	-36			
06	287	43	149	89	7	-13	-11				
07	222	30	84	24	-6	-78					
08	147	15	9	-51	-21						
09	62	-2	-76	-136							
10	-33	-21	-171								
11	-138	-42									
12	-253										
$\Sigma\omega_i$	5175	8059	3059	2261	323	1105	143	429	231	21	35

and we obtain the weights

$$\omega_{-2} = -3, \quad \omega_{-1} = 12, \quad \omega_0 = 17, \quad \omega_1 = 12, \quad \omega_2 = -3,$$

and $\Sigma\omega_i = N = 35.$

In a similar way it is possible to calculate the weights for smoothing more data points. In table 8-3 these weights are given for sequences up to 25 points. The same weights are valid for a quadratic polynomial which is easy to show.

The optimal value of points involved in the smoothing varies for each type of data distribution. In smoothing a pulse height distribution from a γ-detector it is a rule of thumb to choose the number of consecutive points equal to those running the full energy peak from half maximum to half maximum. In many cases one run is not enough to obtain a desired degree

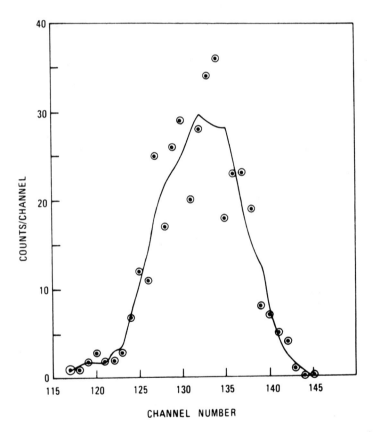

Fig. 8-7   A photopeak showing pronounced scatter. The circles are the raw data. Straight line segments connect the smoothed points. Note that the peak centroid and width are readily apparent from the smoothed data.

of smoothing. Then the smoothing process can be run several
times.

The smoothing method is exemplified in fig. 8-7, where
the raw data of a photopeak show a pronounced scatter. Here
the smoothing is made by using nine consequtive points. As
is seen the method is reduced to a running calculation and
is well adapted to small computers as it needs a relatively small
amount of progamming and a relatively little use of computer
memory or of the computer's processing capacity.

8.7.4 GOODNESS OF FIT

When a physical process is studied or when repeated measurements
are performed of a quantity, the measurement points distribute
according to some distribution law. Obviously, it is impossible
to determine exactly the true distribution since an infinite
number of measurements are then needed. Instead, one often pro-
poses some distribution that the observations appear to follow.
Now it is of interest to test from a finite number of observations if the proposed distribution may be correct, or if the
difference between the expected and the observed results is too
large to be acceptable. A χ^2-test may answer this question.
Such a test may be most simply stated as follows [8.3]:

$$\chi^2 = \sum_1^k \frac{|(\text{observed value})_i - (\text{expected value})_i|^2}{(\text{expected value})_i} \qquad (8.94)$$

where the summation is over the total number of classifications k in which the data have been grouped. The "expected
values" are computed from the frequency distribution assumed
a priori. In general, the data should be subdivided into at
least five classifications, each containing at least five
events.

The number of degrees of freedom, ν, is defined as the
number of independent classifications. Normally several constraints exist in the observations, e.g. the sum, the average
and the standard deviation of the observations may be specified. Every constraint reduces the number of degrees of freedom with one unit

If the proposed distribution is correct and with increasing number of observations the distribution of the χ^2 variable defined in eq. 8.94 approaches asymptotically a probability density function given by

$$f(u,v) = \frac{1}{2^{\frac{1}{2}v}\Gamma(\frac{1}{2}v)} \, u^{\frac{1}{2}v-1} \, e^{-\frac{1}{2}u} \qquad (8.95)$$

where v is the degrees of freedom and Γ is the gamma function for which $\Gamma(x+1) = x\Gamma(x)$ and $\Gamma(1/2) = \sqrt{\pi}$, $\Gamma(1) = 1$.

This function $f(u,v)$ is called the chi-square distribution with v degrees of freedom. It is an unsymmetric distribution from $0 \rightarrow +\infty$ which with increasing degrees of freedom slowly approaches a normal distribution.

In practice one is interested in calculating confidence intervals or testing hypothesis involving χ^2 distributed vari-

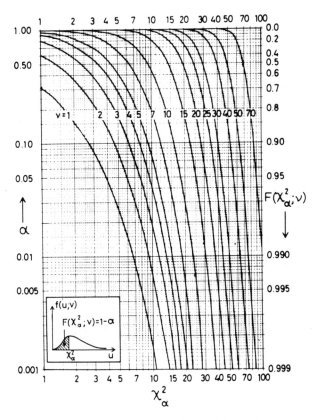

Fig. 8-8   Probability contents of the chi-square
distribution[8.3a].

ables. Fig. 8-8 gives probability contents of the χ^2 probability density function for different numbers of degree of freedom. The figure shows a double-logarithmic display of the quantities $F(\chi_{\alpha}^2, \nu)$ and α versus χ_{α}^2, as implied by the relation

$$F(\chi_{\alpha}^2, \nu) \equiv \int_0^{\chi_{\alpha}^2} f(u,\nu)du = 1 - \alpha \qquad (8.96)$$

The meaning of eq 8.96 is the following. For a given value of ν, there is a probability $F(\chi_{\alpha}^2, \nu)$ that, purely by chance, an observed value of χ^2 will be less than or equal to the value given by fig. 8-8, when the observed value is calculated from the correct distribution.

In interpreting the value of $F(\chi_{\alpha}^2, \nu)$ we may state that if F lies between 0.1 and 0.9 there is no reason to abandon the proposed distribution, while if F is less than 0.02 or more than 0.98 the proposed distribution is extremely unlikely and is to be questioned seriously.

Fig. 8-8 can also be used to calculate χ^2-values corresponding to confidence intervals. Suppose that we wish to determine from 16 classifications the value of χ^2 which has a 90% chance of being exceeded. We assume that the sum of observations is the only constraint and get 15 degrees of freedom. $F(\chi^2, 15) = 0.1$ gives $\chi^2 = 8.6$.

Suppose on the other hand that we want the value of χ^2 which has only 10% chance of being exceeded. We now get $F(22.3, 15) = 0.9$. Normally several constraints exist in the observations, e.g. the sum, the average and the standard deviation of the observations may be specified. Every constraint reduces the number of degrees of freedom with one unit.

8.8 Analysis of pulse height distributions

8.8.1 SPECTRUM STRIPPING

The statistical methods discussed above are of great use in treating experimental data. An important example in nuclear chemical analysis is to analyse a complex pulse height distribution recorded with a multichannel analyzer. The analysing technique depends of course on the kind of spectrometer used. The situation is quite different for spectra measured with a NaI(Tl) spectrometer or a Ge(Li) spectrometer. For NaI(Tl), the photopeak often overlaps other peaks or Compton edges, but in the Ge(Li) case most of the peaks are completely resolved. Independent of the detection method, there are generally two sets of numbers that are of interest, the energies and intensities of the γ-rays. The energies can be derived from the positions of the peaks and the intensities come from the peak areas. Due to the complexity of the observed peaks, either manual methods or methods involving the use of digital computers are used in the analysis.

Several such methods for spectra from NaI(Tl) detectors have been reported. In most of these, the analysis is performed with calculated or measured response functions corresponding to monoenergetic photons, individual radionuclides or other more complex components of the sample.

In practice, the peak to be determined is often superimposed on the Compton continuum of the interfering nuclide. In such cases, the Compton part under this peak has to be subtracted before the quantitative assessment of the peak can be accomplished. Such subtractions can be performed through so-called "spectrum stripping", whereby a variable amount of a pure standard spectrum from the interfering nuclide is subtracted from the complex γ-ray spectrum. Various standard spectra can be stored in the memory of the multichannel analyzer, or they can be transferred to the multichannel analyzer using tape recording.

The peak region in a complex γ-ray spectrum can further easily be integrated by means of so-called "cursors" contained in various multichannel analyzers. By adjustment of the peak region with such cursors, upper and lower discrimination levels are established and the pulse number in the peak region is revealed.

In fig. 8-9 the graphical stripping of a three-line γ-spectrum obtained with a Na(Tl) spectrometer is shown[8.4].

Similar response functions are not readily available for Ge(Li) detectors. The high resolution of the semi-conductor

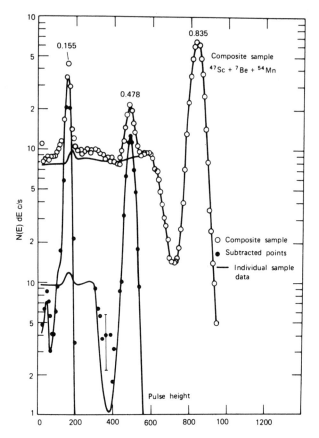

Fig. 8-9   Graphical stripping of complex γ-ray
spectrum in which the two lower energy
sources were intentionally less than 10%
of the higher energy source[8.4].

detector systems makes it possible, however, to obtain the
γ-ray energy and intensity by analysing only the data in the
vicinity of its photopeaks (full-energy as well as escape
peaks). Since no extensive response function information is
required, the method can be applied to different detector systems and experimental conditions. Also, since the errors and
uncertainties of the analysis in one part of the spectrum do
not propagate into other parts, this procedure is not subject
to the accumulation of errors which greatly reduces the precision achieved in analysing, e.g., spectra from NaI(Tl) crystals.

8.8.2 FITTING OF SHAPE FUNCTIONS (SAMPO-PROGRAM)

In most applications nowadays, Ge(Li)-detectors and HPGe-detectors supersede other γ-detectors. Various more or less complicated mathematical representations exist to describe the shape of the photopeaks and the background beneath them. Some of them use a basic gaussian function to represent the photo-peak and a simple polynomial for the background, while others use a more complex function for the peak or the background. One of the most applicable systems is that described by Routti and Prussin[8.5] and is called SAMPO.

In the SAMPO program, the central part of the peak is described by a gaussian and the tails by simple exponentials which join the gaussian so that the function and its first derivate are continuous. In this representation, the shape of a peak is defined by three parameters: the width of the gaussian and the distances from the centroid to the junction points. This gaussian-plus-exponential representation of the shape function for photopeaks has the great advantage that the defining shape parameters vary smoothly with energy and therefore the values of these parameters for any line in a spectrum may be found by interpolation between the parameters of neighbouring lines. For the purpose, intense and wellisolated lines are used as internal calibrations and their shape parameters are defined by fitting with the function described and a straightline approximation for the background continuum. This is performed by minimizing the weighted sum of the squares.

$$\sum_i \omega_i (c_i - b_i - f_i)^2 \qquad (8.97)$$

where ω_i = the weights (generally $\omega_i = \frac{1}{c_i}$ (see eq.8.81))

c_i = the counts in channel $\underline{i}$

b_i = background function and

f_i = the shape function.

The minimizing process follows the pattern sketched in sec. 8.7.

In this way, the energy-dependence of the three peak parameters are obtained and it is now possible for the program to analyze the full complex γ-spectrum. In this main fit, every peak is characterized by the three shape parameters, the peak height and the location of the peak, i.e. in all five parameters. In addition to these, the background function in this main fit is represented by a quadratic function corresponding to another three parameters. The minimization is performed in a similar way as before and the program gives the number of γ-lines, the peak heights and locations and it also determines the expected accuracy of these values from the statistics of the data and the uncertainties in the calibration procedures.

Typical data for a test spectrum are shown in fig. 8-10 where, among other things, it is seen how the program resolves a broad peak in two photopeaks. The output from the program contains in addition to tabulated numerical results, graphical information on each fit in the form of a printer plot and,

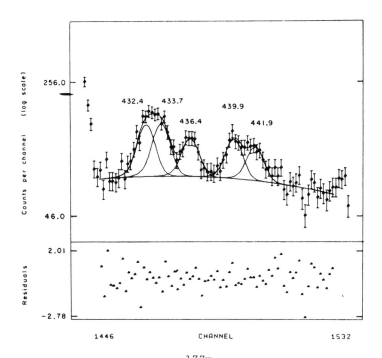

Fig. 8-10 A multiplet in the ^{177m}Lu spectrum resolved by fitting the data with the energy dependent shape functions. Residuals (lower scale) are expressed in standard deviations of the count measured$^{(8.5)}$.

optionally, pendrawn graphs as seen in the figure. More detailed information is given in the original paper. An interesting
discussion of different fitting methods for γ-data analysis is
given by Helmer et al.[8.6]

Recently Yule[8.7] has reviewed manual and computerized
data processing in activation analysis. He discusses the smoothing of data technique (sec.8.7.3) and the SAMPO-program, but
also other techniques developed from these. A comprehensive
conference "Computers in Activation Analysis and Gamma-Ray
Spectroscopy was held 1978[8.8] to which proceedings we refer
for further discussions.

8.9 References

8.1   L.A. Currie, Anal. Chem. $\underline{40}$, 586 (1968).

8.2   A. Savitzky and M.J.E. Golay, Anal. Chem. $\underline{36}$, 1627 (1964).

8.3   A.G. Frodsen, O. Skjeggestad, and H. Tofte, Probability
      and Statistics in Particle Physics, Universitetsforlaget
      Oslo (1979).
      S. Brandt, Statistical and Computational Methods in Data
      Analysis. 2 ed. Elsevier North-Holland (1976).
      E.M. Pugh and G.H. Windslow, The Analysis of Physical
      Measurements, Addison-Wesley, Reading, Massachusetts (1966).

8.4   R.L. Heath, Scintillation Spectrometry. Gamma-ray Spectrum
      Catalog. US Atomic Energy Commission AEC Report, 100-16408
      (1958).

8.5   J.T. Routti and S.G. Prussin. Nucl. Instr. Meth. $\underline{72}$, 125
      (1969).

8.6   R.G. Helmer and R.L. Heath. Nucl. Instr. Meth. $\underline{57}$, 46 (1967).

8.7   H.P. Yule, Nondestructive Activation Analysis (ed. S. Amiel),
      Studies in Analytical Chemistry 3, Elsevier Scientific
      Publishing Company (1981).

8.8   Proceedings of American Nuclear Society Topical Conference,
      in Computers in Activation Analysis and Gamma-Ray Spectro-
      scopy, (eds. B.S. Carpenter, M.D. D'Agostino and H.P. Yule)
      Mayaguez, Puerto Rico (April, 1978), CONF 780421 (1979).

9 Methods of Radiation Detection

9.1 Detection of neutrons

Neutrons, (uncharged) do not themselves produce directly ionization and their detection must therefore be accomplished by means of the ionizing products of neutron reactions. Four main types of neutron detection mechanisms are at present employed in neutron detectors:

(1) neutron-induced nuclear reactions yielding prompt charged-particle reaction products (protons, α-particles, etc) or γ-rays.

(2) induced fission in certain heavy elements.

(3) elastic scattering of the neutron from a light nucleus followed by the detection of the fast, charged, recoiling nucleus and

(4) capture of neutrons by certain stable isotopes, which are thereby converted into radionuclides emitting β- and γ-rays (activation method).

The following are brief examples of the applications of these principles in neutron detectors.

9.1.1 REACTION-PRODUCT DETECTORS

Several neutron reactions with light elements result in the formation of one or more charged particles and have large cross sections. The most common reactions are

$$^3\text{He} + n \rightarrow {}^3\text{H} + p + 0.76 \text{ MeV}$$

$$^6\text{Li} + n \rightarrow {}^3\text{H} + \alpha + 4.79 \text{ MeV}$$

$$^{10}\text{B} + n \rightarrow {}^8\text{Li} + \alpha + 2.79 \text{ MeV}$$

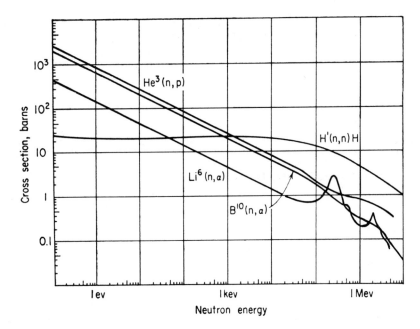

Fig. 9-1   Cross section versus neutron energy for several
common neutron-detection processes[9.12].

The cross sections for all three reactions varies for slow
neutrons as 1/v, where v is the neutron velocity (fig. 9-1).
The significance of the 1/v relationship (sec. 3.8.1) appears,
if a detector is considered, which is placed in a neutron flux

$$\varphi = \int n(v)v\,dv \tag{9.1}$$

where n(v) is the neutron velocity distribution and thus n(v)dv
is the number of neutrons per unit volume with velocity between
v and v+dv. In this discussion, the restriction is made that
there are no neutrons outside the 1/v limit for the reaction
(fig. 9-1). The reaction rate is obtained from the following
equation

$$R = \int_0^\infty n(v)v\sigma(v)NV\,dv \tag{9.2}$$

Because the product v·σ(v) is constant, R can be written

$$R = N \ V \ [v\sigma(v)] \int_0^\infty n(v) \ dv = NV[v\sigma(v)]n \qquad (9.3)$$

where N is the number of reaction nuclei per unit volume and n is the neutron density which is independent of velocity. A reaction product detector of the type discussed here thus gives, for slow neutrons, an output reading proportional to the neutron density.

The best-known of the detectors is the boron counter for slow neutrons. This is a gas counter in which either the inside wall is lined with boron or the counter is filled with BF_3 gas, preferably enriched in ^{10}B. The resulting α-particle is detected from its ionization with about 100% efficiency.

Boron- and lithium-loaded scintillators are widely used for the detection of slow neutrons. Another technique is to use the sandwich principle, in which the neutron target foil is sandwiched between a pair of low-resistivity silicon detectors.

9.1.2 FISSION DETECTORS

Fissile materials can be deposited as thin films on metallic surfaces and then placed in a gas chamber. The fissile nucleus is split by the neutron into two fission fragments which in turn cause ionization in the gas. A number of fissile materials are available, some of which have a well-defined neutron energy threshold below which the fission reaction does not take place. Suitable combinations of detectors involving these fissile materials can be used to yield a measurement of the neutron spectrum. Table 9-1 gives details of fissile materials in common use.

Fission chambers can be made very small (diameter 0.1 cm) since the range of the fission fragments is so short. On the other hand, spark chambers for neutron detection through fission are now widely used in applications in which very low levels of neutron intensity are studied over large areas.

Table 9-1
Useful data for fission detectors.

Nuclide	Fission cross section 0.025 eV neutrons (barns)	Fission cross section 3 MeV neutrons (barns)
233_U	530	1.9
235_U	582	1.3
239_{Pu}	750	2.0
	Threshold (MeV)	
232_{Th}	1.3	0.14
238_U	1.2	0.55
237_{Np}	0.4	1.5
231_{Pu}	0.6	1.1
234_U	0.4	1.5
236_U	0.8	0.85

9.1.3 NUCLEUS-RECOIL DETECTORS

By measuring the number of recoiling nuclei and particularly their energy distribution, it is possible to transform the measured distribution into the energy distribution of the impinging neutrons. This has been utilized in a series of detectors.

For direct detection of fast neutrons, hydrogen-containing scintillators or organic crystals or plastics are used. Most of them are operated with pulse-shape discrimination systems in order to increase their resolution. A counter filled with gas containing hydrogen or helium and operating in the proportionality region can also be an efficient neutron detector even at lower neutron energies. Hydrogen in nuclear emulsions or cloud chambers has also been used for detection.

9.1.4 ACTIVATION METHODS

The general principle of the activation method is that a thin foil of some suitable material is exposed to the neutron

beam for a known interval of time and it is then removed from
the beam so that measurements of its induced radioactivity can
be carried out. The main requirements for the material are,
first, that it shall have a sufficiently large capturing power
for neutrons so that it can produce an adequate number of radioactive nuclei during the exposure to the beam. Secondly, the
lifetime of the radionuclide must be sufficiently long to acco-
modate the inevitable time interval between exposure and measurement, but not so long that the rate of decay is too low to
be measured accurately. It will often be found that the capturing power of the nucleus is a sensitive function of neutron
energy and may be exceptionally large at some resonance energy.
This can be used to advantage when measurement of neutrons within some particular energy range is required, but it can otherwise be an inconvenient feature. Table 9-2 lists the properties
of a number of metals which have proved useful in the activation method of detecting neutrons.

Table 9-2

Properties of some metal foils suitable for the detection of
slow neutrons.

Element	Isotope	Abundance %	Cross section (for 0.025eV) (barns)	Half-life of product
Manganese	^{55}Mn	100	13	2.6 h
Rhodium	^{103}Rh	100	11	4.4 min
			139	42 s
Silver	^{107}Ag	51.8	45	2.3 min
	^{109}Ag	48.2	3.5	270 d
			89	24 s
Indium	^{113}In	4.3	8	50 d
			3	72 s
	^{115}In	95.7	157	54 min
			42	13 s

298

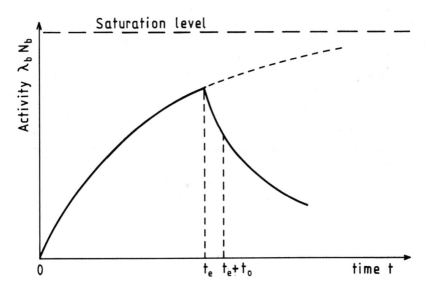

Fig. 9-2   Build up and decrease of activity during and after exposure.

Fig. 9-2 shows how the produced activity $\lambda_b N_b$ is built up during exposure and how it diminishes after stopping the accelerator. The activity after an exposure time t_e and with a delay time t_o before measurement is (eq.2.52).

$$-\frac{dN_b}{dt} = \lambda_b N_b = R(1-e^{-\lambda_b t_e})\, e^{-\lambda_b t_o} \qquad (9.4)$$

where the nuclear reaction rate $R = \varphi\sigma NV$.

By measuring $\lambda_b N_b$ with a counter, the neutron flux φ can be deduced if σ, N, V, λ_b and the efficiency of the detector are known.

9.2 Detection of photons

9.2.1 CRYSTAL- AND MAGNETIC SPECTROMETERS

Depending on the energy of the photon, different processes contribute to the absorption of photons in matter, different detection methods are thus used. For quite low energies and very concentrated and extemely intense photon sources, crystal diffraction is useful. A very good energy resolution $\frac{\delta E}{E} \simeq 10^{-4}$ is obtained with this method. Good resolution but low efficiency is also characteristic for magnetic spectrometers in which the electrons resulting from photoelectric absorption, Compton scattering and pair production are magnetically analyzed. This method is widely used and has found particular application in studying energy shifts in electron orbits due to chemical bondings. These shifts are of the order of 1-10 eV which implies that an energy resolution of about 0.1 eV is needed to study such effects carefully. Resolutions of this order have been obtained in the ESCA method[9.1] (Electron Spectroscopy for Chemical Analysis) (figs. 9-3 and 9-4).

For photon energies above $\simeq$ 3 MeV, pair spectrometers can be used to measure photon energies. In this instrument, the photon is converted into an electron-positron pair, both of

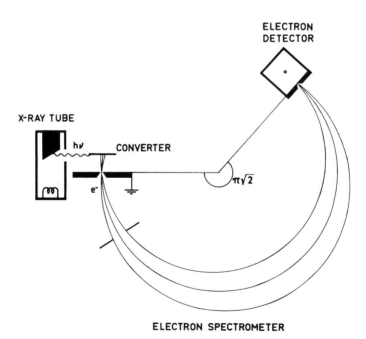

ELECTRON SPECTROMETER

Fig. 9-3   Schematic view of an ESCA arrangement for the study of electrons expelled by X-rays[9.1].

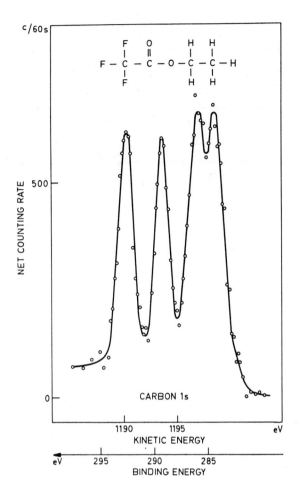

Fig. 9-4   Electron spectrum from carbon in ethyl trifluoro-
            acetate. All four carbon atoms in this molecule
            are distinguished in the spectrum. The lines
            appear in the same order from left to right as
            do the corresponding carbon atoms in the struc-
            ture that has been drawn in the figure[9.1].

which are measured in coincidence after deflection in a magnetic field (fig. 9-5).

9.2.2 SCINTILLATION SPECTROMETERS

    A very widely-used device for the detection and measure-

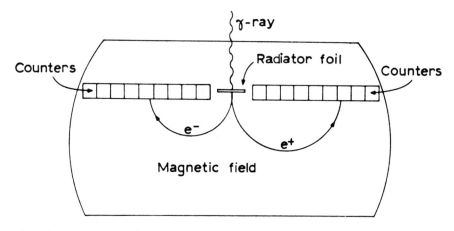

Fig. 9-5   The pair spectrometer.

ment of photons is the NaI(Tl) scintillation counter. All of
the three dominating interactions of photons with matter contribute to the pulse height spectrum. In each of the processes,
part of the primary photon energy goes into the kinetic energy
of electrons and this energy is converted to light in the scintillator, except in cases when the interaction occurs near the
surface of the crystal and the electrons escape the scintillator before being brought to rest. The secondary photons produced in the interaction may or may not release further electrons,
increasing the light production.

In the photoelectric process, all the photon energy is converted into light since the secondary is an X-ray which will be
absorbed in the scintillator. In the Compton process, a relatively large fraction of the primary energy may go into the scattered photon which often escapes the scintillator. In pair production, an energy larger than 1.02 MeV is required to produce
an electron-positron pair. Hence, the total kinetic energy of
the pair is 1.02 MeV less than the photon energy. However, when
annihilation of the positron occurs, two 0.51 MeV photons are
produced which leave in opposite directions.

The probability for the secondary photons to be absorbed
in the scintillator increases with its size. The pulse height
spectra from ^{137}Cs decay shown in fig. 9-6a illustrate this
effect.

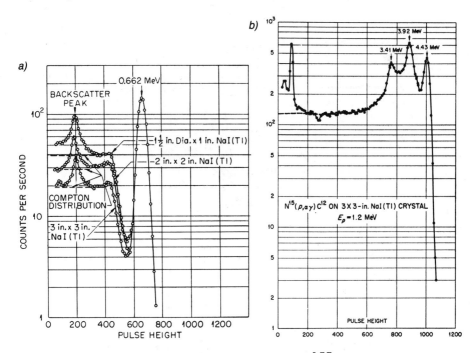

a)

b)

Fig. 9-6a   Pulse height distribution of ^{137}Cs recorder from
NaI(Tl) scintillators of different size[(9.13)].
J.H. Neiler and P.R. Bell, The Scintillation Method,
in Alpha-, Beta- and Gamma-Ray Spectroscopy (ed.
K. Siegbahn), North Holland (1965) p.259.

Fig. 9-6b   Pulse height distribution of 4.43 MeV γ-rays
from the reaction N[15] (p,αγ) [12]C[(9.13)].
J.H. Neiler and P.R. Bell Ibid. p.261.

For higher photon energies, the spectrum contains the full
energy peak when both the annihilation quanta are absorbed but
also single and double escape peaks 0.51 and 1.02 MeV below
this one. These escape peaks are super-imposed on the Compton
escape distribution (fig. 9-6b).

Different ways of improving high energy photon measurements
with scintillation counters have been tried. The three-crystal
spectrometer in which a small central crystal is placed between two large crystals is such a device (fig. 9-7). The side
crystals may be arranged to detect only 0.51 MeV radiation

303

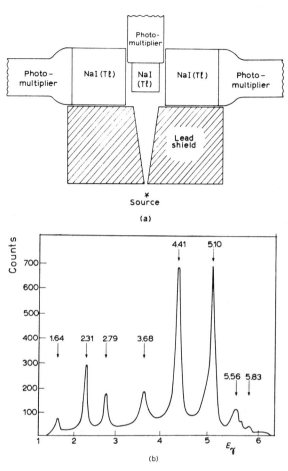

Fig. 9-7   (a) Three-crystal γ-ray spectrometer, (b) Typical
            spectrum obtained with the three-crystal spectro-
            meter[9.13].
            D. Alburger, Gamma-Radiation from Charged Partic-
            le Bombardment; Coulomb Excitation, in Alpha-, Be-
            ta- and Gamma-Ray spectroscopy (ed. K. Siegbahn),
            North Holland (1965) p.760, 761.

resulting from pair production escape. In a triple coincidence
arrangement, the spectrum from the central crystal will only
show the double escape peak.

     The amplitude of the pulse from the scintillation counter
evidently depends on the number of electrons ejected from the

photocathode of the photomultiplier tube. The resolution or
line-width of the pulse is governed by the fluctuations in this
number and, since the number is rather small, it is difficult
to get high resolution with scintillation spectrometers. Energy
resolutions of 5-8% and efficiencies of about 50% are common.
In a three-crystal spectrometer the efficiency is, of course,
reduced largely.

9.2.3 SEMICONDUCTOR SPECTROMETERS

The lithium-drifted germanium detector has completely revo-
lutionilized almost all fields for which quantitative γ-ray
data are of crucial experimental importance. It combines the
high resolving power advantage of the bent crystal spectrometer with the multichannel advantage of the NaI(Tl) scintillation spectrometer. Although its efficiency, dependent on the
energy of the photon and the size of the detector, is still
lower than that of the NaI(Tl) detector, it can be anticipated

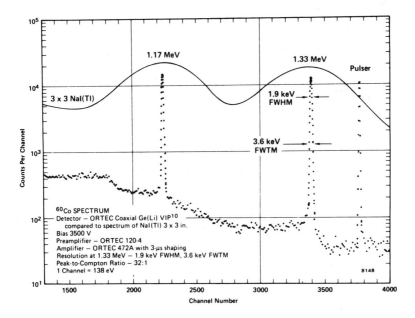

Fig. 9-8   Pulse height distribution of ^{60}Co γ-rays
from a Ge(Li) and a NaI(Tl) detector respectively[9.2].

that larger detectors will be fabricated, thus increasing the efficiency.

Normally, a photon-detector system consists of a semiconductor element, a cryostat, a liquid-nitrogen dewar and a charge-sensitive preamplifier. Depending on the photon energy region of interest, different types of crystal detectors can be chosen. In the energy range from 30 keV to over 10 MeV, a coaxial germanium detector will be useful. The efficiency of such a detector is referenced to that of a 3" x 3" NaI(Tl) detector, using a ^{60}Co source placed 25 cm from the crystal face. Typical values of efficiency for the best coaxial germanium detectors are 10-20%, with a resolution of about 2.0 keV. The spectral response of such a detector is shown in fig. 9-8.

In the photon energy region from about 3 keV to about 1 MeV, a thin-window hyper-pure germanium detector is useful.

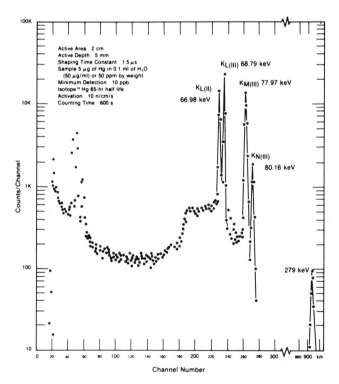

Fig. 9-9   Pulse height distribution from activated Hg (ie[197]Hg) recorded with a Ge(Li) detector[9.3].

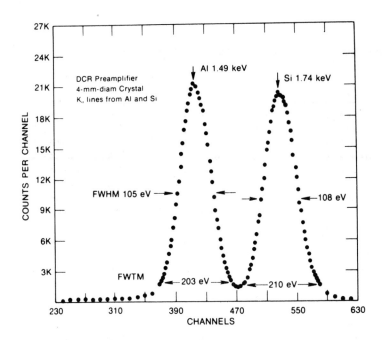

Fig. 9-10   Pulse height distribution from a Si(Li) detec-
            tor[9.3].

It is designed to take measure only of the low energy background.
The resolving power is usually superior to that of coaxial germanium detectors. A typical spectrum is shown in fig. 9-9. The
lithium-drifted silicon detector is particularly useful for
detecting photons in the low energy region of about 1 keV to
40 keV and is a highly sensitive and versatile research tool
for detecting X-rays produced in a nuclear accelerator, radioactive source or X-ray tube. A spectrum from such a detector
is shown in fig. 9-10.

A combination of NaI(Tl) detectors and a Ge detector can
be very useful in coincidence and anticoincidence measurements.
One example may be in the study of cascade γ-rays and their
correlations. Another is the method of suppressing the Compton contribution to the spectrum by placing the germanium
detector inside an annular sodium iodide detector and connec-
ting the latter in anticoincidence. A third example is the method of separating out the double escape peaks of the spectrum

Table 9-3

Recommended γ-ray calibration energies arranged by energy[9.4].

Source	E_γ (eV)	Source	E_γ (eV)	Source	E_γ (eV)
^{182}Ta	67 750.0 ± 0.2	^{108}Ag^m	614 281 ± 4	^{182}Ta	1 231 016 ± 5
^{153}Gd	69 673.4 ± 0.2	^{110}Ag^m	620 360 ± 3	^{56}Co	1 238 287 ± 6
^{170}Tm	84 255.1 ± 0.3	^{124}Sb	645 855 ± 2	^{182}Ta	1 257 418 ± 5
^{182}Ta	84 680.8 ± 0.3	^{110}Ag^m	657 762 ± 2	^{182}Ta	1 273 730 ± 5
^{153}Gd	97 431.6 ± 0.3	^{137}Cs	661 660 ± 3	^{22}Na	1 274 542 ± 7
^{182}Ta	100 106.5 ± 0.3	^{198}Au	675 887.5 ± 1.9	^{182}Ta	1 289 156 ± 5
^{153}Gd	103 180.7 ± 0.3	^{110}Ag^m	677 623 ± 2	^{59}Fe	1 291 596 ± 7
^{182}Ta	113 672.3 ± 0.4	^{110}Ag^m	687 015 ± 3	^{124}Sb	1 325 512 ± 6
^{182}Ta	116 418.6 ± 0.7	^{144}Ce	696 510 ± 3	^{60}Co	1 332 502 ± 5
^{152}Eu	121 782.4 ± 0.4	^{94}Nb	702 645 ± 6	^{56}Co	1 360 206 ± 6
^{57}Co	122 061.4 ± 0.3	^{110}Ag^m	706 682 ± 3	^{124}Sb	1 368 164 ± 7
^{192}Ir	136 343.4 ± 0.5	^{124}Sb	713 781 ± 5	^{24}Na	1 368 633 ± 6
^{57}Co	136 474.3 ± 0.5	^{124}Sb	722 786 ± 4	^{182}Ta	1 373 836 ± 5
^{182}Ta	152 430.8 ± 0.5	^{108}Ag^m	722 929 ± 4	^{110}Ag^m	1 384 300 ± 4
^{182}Ta	156 387.4 ± 0.5	^{95}Zr	724 199 ± 5	^{182}Ta	1 387 402 ± 5
^{182}Ta	179 394.8 ± 0.5	^{110}Ag^m	744 277 ± 3	^{124}Sb	1 436 563 ± 7
^{182}Ta	198 353.0 ± 0.6	^{110}Ag^m	763 944 ± 3	^{110}Ag^m	1 475 788 ± 6
^{192}Ir	205 795.5 ± 0.5	^{124}Sb	790 712 ± 7	^{144}Ce	1 489 160 ± 5
^{182}Ta	222 109.9 ± 0.6	^{110}Ag^m	818 031 ± 4	^{110}Ag^m	1 505 040 ± 5
^{182}Ta	229 322.0 ± 0.9	^{54}Mn	834 843 ± 6	^{110}Ag^m	1 562 302 ± 5
^{228}Th	238 632 ± 2	^{56}Co	846 764 ± 6	^{228}Th	1 620 735 ± 10
^{152}Eu	244 698.9 ± 1.0	^{228}Th	860 564 ± 5	^{124}Sb	1 690 980 ± 6
^{182}Ta	264 075.5 ± 0.8	^{94}Nb	871 119 ± 4	^{207}Bi	1 770 237 ± 10
^{203}Hg	279 196.7 ± 1.2	^{192}Ir	884 542 ± 2	^{56}Co	1 771 350 ± 15
^{192}Ir	295 958.2 ± 0.8	^{110}Ag^m	884 685 ± 3	^{56}Co	1 810 722 ± 17
^{192}Ir	308 456.9 ± 0.8	^{46}Sc	889 277 ± 3	^{88}Y	1 836 063 ± 13
^{192}Ir	316 508.0 ± 0.8	^{228}Th	893 408 ± 5	^{56}Co	1 963 714 ± 12
^{51}Cr	320 084.2 ± 0.9	^{88}Y	898 042 ± 4	^{56}Co	2 015 179 ± 11
^{152}Eu	344 281.1 ± 1.9	^{110}Ag^m	937 493 ± 4	^{56}Co	2 034 759 ± 11
^{198}Au	411 804.4 ± 1.1	^{124}Sb	968 201 ± 4	^{124}Sb	2 090 942 ± 8
^{192}Ir	416 471.9 ± 1.2	^{56}Co	1 037 844 ± 4	^{56}Co	2 113 107 ± 12
^{108}Ag^m	433 936 ± 4	^{124}Sb	1 045 131 ± 4	^{144}Ce	2 185 662 ± 7
^{110}Ag^m	446 811 ± 3	^{207}Bi	1 063 662 ± 4	^{56}Co	2 212 921 ± 10
^{192}Ir	468 071.5 ± 1.2	^{198}Au	1 087 691 ± 3	^{56}Co	2 598 460 ± 10
^{7}Be	477 605 ± 3	^{59}Fe	1 099 251 ± 4	^{228}Th	2 614 533 ± 13
^{192}Ir	484 577.9 ± 1.3	^{65}Zn	1 115 546 ± 4	^{24}Na	2 754 030 ± 14
^{207}Bi	569 702 ± 2	^{46}Sc	1 120 545 ± 4	^{56}Co	3 009 596 ± 17
^{228}Th	583 191 ± 2	^{182}Ta	1 121 301 ± 5	^{56}Co	3 201 954 ± 14
^{192}Ir	588 585.1 ± 1.6	^{60}Co	1 173 238 ± 4	^{56}Co	3 253 417 ± 14
^{124}Sb	602 730 ± 3	^{56}Co	1 175 099 ± 8	^{56}Co	3 272 998 ± 14
^{192}Ir	604 414.6 ± 1.6	^{182}Ta	1 189 050 ± 5	^{56}Co	3 451 154 ± 13
^{192}Ir	612 465.7 ± 1.6	^{182}Ta	1 221 408 ± 5	"^{16}O"	6 129 270 ± 50

by using a pair spectrometer (fig. 9-7) with two large sodium
iodide side crystals and a central Ge detector. The signals
from the three detectors are measured in coincidence.

9.2.4 CALIBRATION OF GAMMA-SPECTROMETERS

The γ-spectrometer must be calibrated in order to be
a useful instrument. Both an energy calibration and a photopeak efficiency calibration must be carried out. The energy
calibration is performed by means of nuclides emitting γ-
rays of known energies. Sets of calibration sources are commercially available or can be prepared by irradiation. A list
of convenient calibration sources is given in table 9-3.

The photopeak efficiency is a function of the intrinsic

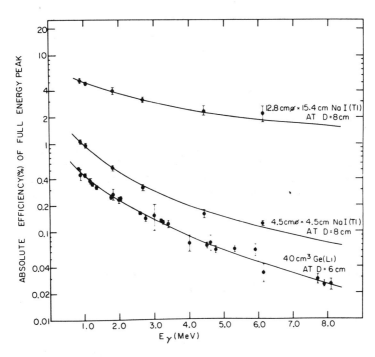

Fig. 9-11   The absolute efficiencies of the full energy
peak for NaI(Tl) and Ge(Li) spectrometers[9.5].
C. Rolfs and A.E. Litherland, Gamma Rays from
Capture Reactions, in Nuclear Spectroscopy and
Reactions (ed. J. Cerny) Academic Press (1974)
Part B.

detector efficiency, which depends on the γ-energy and the
source detector geometry, and on the peak/total ratio which is
the ratio of the area measured under the photopeak (full energy,
single escape or double escape peak) to the total area of the
spectrum. Many attempts have been made to calculate the peak
detection efficiencies of Ge(Li) detectors, but it is still
preferable to measure them experimentally. Fig. 9-11 shows typical efficiencies of full energy peaks for NaI(Tl) and Ge(Li)
spectrometers.

9.3 Detection of electrons

The highest accuracy in measuring β-spectra is obtained
with magnetic spectrometers in arrangements similar to that
shown in fig. 9-3. Since the magnetic field has to be changed
in measuring electrons with different energies, the method is
a single channel one. In measuring nuclear activities, it is
preferable to have a multichannel method in which the whole
energy distribution of the electrons is measured in one run.
In many cases, there is also a need for large detection solid
angles. The most common detectors for nuclear chemistry work
are for these reasons the flowing gas proportional counter,
the organic scintillator detector and the silicon semiconductor detector.

Some problems arise in measuring the disintegration rate
of a β-emitter. These are due to the effects of scattering
and absorption of electrons in matter combined with the distribution in energy of the β-particles. A number of different
factors influence the detection efficiency. In addition to the
geometry of detection and the intrinsic efficiency of the detector, these include: absorption in the air between source
and detector, absorption in the detection window, correction
for air scattering, for backscattering by the source support,
correction for scattering by environment, for self-absorption
and self-scattering by the finite mass of the source. Steinberg[9.14] has discussed the magnitudes and dependences of
the various factors on the experimental situation.

Many chemical experiments lead to very small amounts of radioactive samples, either because of the low yield of the reaction under study or because of the small amount of sample that is available. In such cases, methods are needed in which the background is suppressed. In figs. 9-12 and 9-13, two different low-background counters are shown. In the first case, a proportional counter is covered by an anti-coincidence guard counter and in the other a very thin GM-counter is placed in coincidence in front of a plastic scintillator.

Liquid-scintillation counting is frequently used for the direct measurement of high-energy β-emitters such as ^{32}P and ^{90}Y, but many low-energy β-particles escape detection. The amplitude of output pulses from the photomultiplier will therefore give a spectrum, which deviate from the true β-particle spectrum on the low energy side.

For radionuclides of low and intermediate β-particle energy (H^3) or pure Auger electron emitters (^{55}Fe), it is necessary to consider the distribution of pulses from the phototube arising from decay events in the scintillator in the range of 0-10 keV.

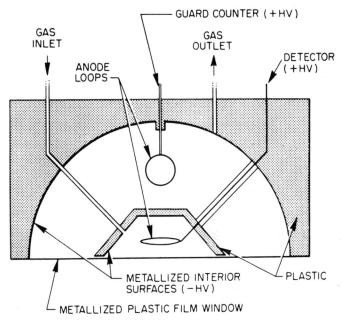

GUARD COUNTER (+HV)

GAS INLET

GAS OUTLET

DETECTOR (+HV)

ANODE LOOPS

METALLIZED INTERIOR SURFACES (-HV)

PLASTIC

METALLIZED PLASTIC FILM WINDOW

Fig. 9-12   Low-background counters$^{(9.5)}$.

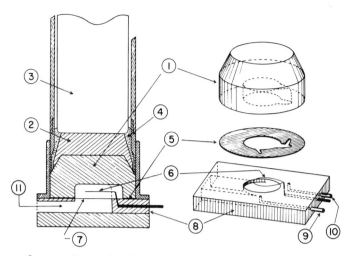

Cross-section and exploded view showing the detector assembly.

(1) Plastic scintillator;
(2) Lucite light pipe;
(3) Photo-multiplier tube (EMI 6097B);
(4) MgO powder light reflector;
(5) Vinyl packing;
(6) Anode wire of the G-M counter;

(7) Window of the G-M counter (Gold plated Mylar film of 1.3 mg/cm² thick);
(8) Lucite plate;
(9) Counting gas inlet tube;
(10) Counting gas outlet tube;
(11) Sample inlet.

Fig. 9-13   Cross section of Tanaka detector assembly[9.7].

Because there is a cut-off or threshold to the sensitivity of the system in the region of 1 to 2 keV (corresponding to the emission of only one electron from the photo cathode), it is necessary to calculate the non-detection probability in this region in order to arrive at a more accurate level of the radioactivity concentration of the β-emitter in the scintillator.

9.4 Alpha particle detection

In the past the most widely used method for activity measurement of α-emitting nuclides was that of 4π counting. The 4π counter is a defined solid-angle counter, with the highest possible geometric efficiency. In the method of 4π counting, the radioactive source, on a thin plastic film, is located at

the center of symmetry of two detectors, each subtending a
solid angle of essentially 2π steradians at the source. Provided that the effective thicknesses of the source and source mount are below the range of the α-particles in those media, good results can be obtained in the assay of α-particle
sources.

Often great difficulties are involved in producing even
distributed thin α-sources on plastic films. Therefore α-
emitting radionuclides are more often electrodeposited onto
stainless steel discs.

The measurement of α-emitting radionuclides in 2π geometry requires essentially the measurement with one half of a
4π system. As α-particles are much heavier than electrons,
they are less easily scattered from the backing, so that the
fraction of the α-particles scattered into the 2π counter is
far less than the case of β-particles.

9.4.1 SCINTILLATION COUNTING

One of the earliest methods for detecting α-particles
used a ZnS screen and a low-power microscope to view the tiny
flashes of light. This principle is still in use but discs
of polyester film coated with silver-activated ZnS and photo-
tubes with appropriate electronic circuitry are now used.

Other inorganic scintillators that are useful for the detection of α-particles are CsI(Tl) and CaF_2(Eu). CsI(Tl) is
slightly hygroscopic while CaF_2(Eu) is not hygroscopic and
can be cleaned with water. Convential liquid scintillation
counting systems are suitable for the assay of α-particle
emitting nuclides in some samples, but the background count
rates may be to high for low-level sources. The coupling of
a multichannel analyser to a liquid-scintillation counter
with a single phototube can yield α-particle spectra with an
energy resolution of several per cent.

9.4.2 ALPHA PARTICLE SPECTROMETRY

The two major types of α-particle spectrometers are the
ionization chamber and the semiconductor detector. Pulse-ionization chambers of cylindrical shape have the advantage in

low-level activity measurements of providing a large surface area for the sample. The resolution may be improved by employing a Frisch-type grid that shields the collecting electrode.

Silicon surface barrier semiconductor detectors are now probably the most widely used in α-particle spectrometry. A 1 cm^2 active area detector typically has a resolution of 15 to 20 keV FWHM for 5.5 MeV α-particles emitted from a source 1.5 detector-diameters away. Placing the source closer to the detector increases the detection efficiency at the expense of degraded resolution. Compared to pulse-ionization chambers with very large areas (15000 cm^2), larger semiconductor detectors (20 cm^2) are still very small. Therefore often more extensive sample preparation prior to counting is more necessary in semiconductor detection than in pulse-ionization detection.

9.5 Particle identification techniques

9.5.1 TIME-OF-FLIGHT TECHNIQUES

The time-of-flight technique was developed to distinguish between neutrons and the γ-rays produced in neutron interactions in matter. The most effective method of removing a background of photons in a neutron experiment is to time the arrival of the signals at the detector relative to the arrival of the burst of neutrons at the target.

If the maximum neutron energy is, say, 10 keV, then over a flight path of 10 m there will be a difference in the time of arrival of the prompt γ-ray burst and the scattered neutron burst of approximately 7 μs. Thus the neutron signal can easily be distinguished from the γ-ray signal and the neutron cross section under study be determined. For higher neutron energies, the problem of timing becomes critical. If a good time spectrum is to be obtained in, say, the 1 to 10 MeV neutron energy range, then time resolutions in the nanosecond region are required, even for long flight paths.

In modern neutron spectroscopical work, the time-of-flight technique is no longer used to separate neutrons from photons. This can more easily be done by pulse-shape discrimination of

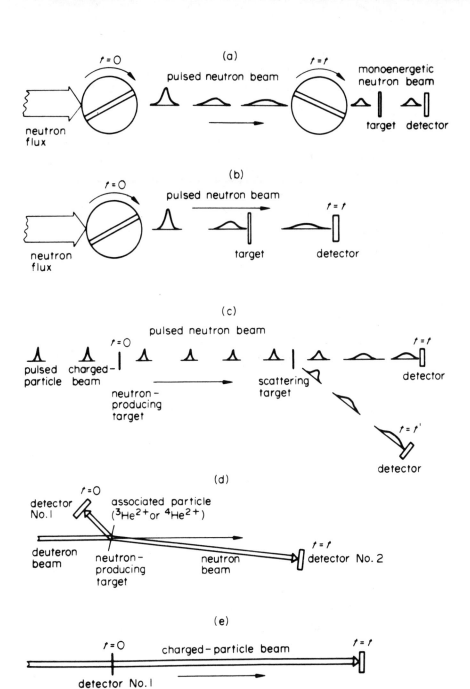

Fig. 9-14 Schematic representation of different methods
used in time-of-flight (ref. 9.9 p.421).

the outpulse from the detector. The technique, however, is common in measuring the neutron energy spectrum. Thus, identification techniques using time-of-flight are now mainly confined studying charged particles, for which the energy can be very conveniently measured by other means.

A precise definition of the start time for the flight path is a necessity for the use of the technique. Four methods for producing a well defined start pulse are generally used: these are schematically illustrated in fig. 9-14.

The twin-roter system illustrated in fig. 9-14a selects a pulse of neutrons of precisely known energy from a broad-spectrum continuous-output source such as a reactor. By varying the absolute speed and relative phase of the rotors, a wide range of monoenergetic pulsed neutron beams can be obtained. The more usual singlerotor chopper system is shown in fig. 9-14b. Here, the burst of neutrons allowed through the rotor is within a well-defined velocity range and the full time-of-flight spectrum must be established. For energetic neutrons, the system illustrated in fig. 9-14c is the most common one in operation. A pulsed charged-particle beam from an accelerator is allowed to fall on a target producing as a result either monoenergetic neutrons or a wide range of neutron energies, depending on the particular application.

If the source of neutrons is the $T(d,n)^4He$ or the $D(d,n)^3He$ reactions at low or medium energies, then the possibility exists of deriving the neutron start signal from the associated ^4He or ^3He recoiling nucleus. This associated particle technique is illustrated in fig. 9-14d. By carefully selecting the angle at which the recoil particle is detected, the angle of emission and thus to some extent the energy of the emitted neutron can be determined.

The final system, which is used exclusively with charged particles, is shown in fig. 9-14e. It is the basic system for a large number of charged-particle time-of-flight measurements, including those at high energies, even when the particle beam is pulsed. The particle passes through a very thin transmission detector in which it loses some small amount of energy, ΔE. The detector provides the start signal for the flight path. The particle then passes down the flight path to the second detector which provides the stop signal for the flight path. Particularly for light medium-energy particles, this technique works very well.

9.5.2 PARTICLE IDENTIFICATION BY TELESCOPE TECHNIQUE

Detector telescope techniques are confined to the detection and identification of charged particles. The simplest system uses two detectors which are either operated in a coincidence mode, in which only those particles which pass through the first detector and enter the second detector are of interest or are operated in anticoincidence mode when only those which stop in the first detector are of interest. When operated in these ways, the detector telescope effectively makes use of the range of a charged particle as means of identification. However, a more usual technique is to use the energy signals from the two, or three, detectors in a telescope to identify the mass and charge of the particles which traverse the first one, or two, detectors and stop in the final detector (fig. 9-15). In the most usual arrangement, the amount of energy deposited in the passing detector is denoted by ΔT and the amount of energy deposited in the stopping detector is denoted by T. The total energy is then $T_T = \Delta T + T$. From section 5.1 we know that an empirical relation exists between the range R of a particle and its energy T of the form (eq. 5.5):

$$R = aT^b + c \qquad (9.5)$$

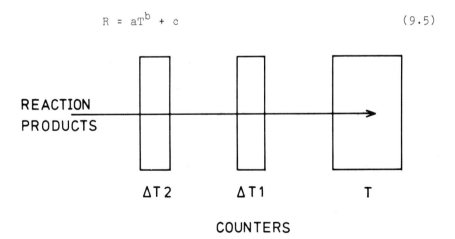

Fig. 9-15   Detector telescope.

where a, b and c are constants. It is found that over a wide energy range for light particles b is close to 1.7 and a is a function of $(MZ^2)^{-1}$. If the first, traversed detector has a thickness L, then the following relationship is true

$$L = a(T_T)^b - a(T)^b$$

or

$$(T_T)^b - (T)^b = \frac{L}{a} \simeq LMZ^2 \tag{9.6}$$

A number of methods have been used to compute this quantity, which uniquely identifies the various charged particles. Analogue pulse multiplication techniques are used to provide a signal proportional to LMZ^2 from the two detector outpulses

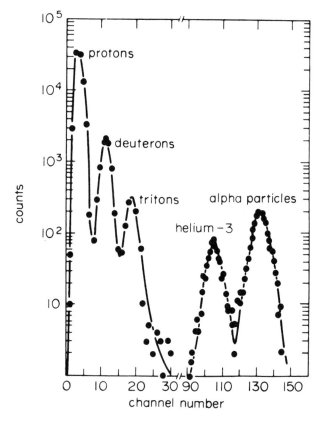

Fig. 9-16   Typical mass separation obtained with particle detector telescopes (ref.9.9 p.480).

318

by generation of the identification function eq. (9.6). In
fig. 9-16 is shown a typical mass separation obtained with
a particle detector telescope based on the identification
function discussed.

9.6 Coincidence techniques

In many nuclear counting procedures, it is necessary to
decide whether two events are time-correlated. Such information may be required for investigations of nuclear decay
schemes, where it may be necessary to know whether two radiations
are emitted at the same time. Also in many types of counting,
imposing the condition that two events must be coincident in
time will serve to discriminate effectively against noise pulses that are randomly distributed in time. Measurements of times-
of-flight or of short-lived nuclear states also need coincidence requirements. Electronic circuits which make such decisions
are called coincidence circuits, and produce an output pulse
only if inputs to the device receive a pulse simultaneously.
    In nuclear chemical analysis coincidence techniques are
very useful. As well β-γ as γ-γ cascades of the disintegration which occur within very short time intervals, are suitable for these purposes. Also both annihilation γ-rays can be
measured in coincidence. We have already seen many applications of coincidence techniques in this chapter (figs. 9-7,
-12, -13, -14, -15).

9.6.1 COINCIDENT MULTICHANNEL SPECTROMETRY SYSTEM

In the two-detector system shown in fig. 9-17 a multichannel spectrum analysis is made of only those events that are in
time coincidence with events in the alternate detector. Two
channels of detector, preamplifier and amplifier are each followed by a timing single-channel analyser, where energy criteria
may be selected by an amplitude window. Each single-channel
analyzer derives an output pulse at some point having a very
small time variation with changes in amplitude. When these two
logic signals occur within the resolving time of the coinciden-

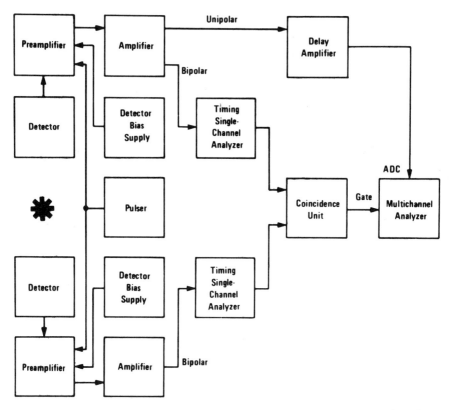

Fig. 9-17 Coincident multichannel spectrometry system[9.3].

ce unit, an output logic pulse is generated that operates the
analyser gate and permits analysis of the desired linear signal.

9.6.2 MULTICHANNEL TIME ANALYSERS

Where it is only necessary to detect coincidences between
a pair or more of detector pulses, the coincidence circuit already described will suffice. In other types of experiment however, the particles may not be emitted simultaneously and it
may be necessary to acquire information about the distribution of their relative times of detection. Examples of this
are in time-of-flight experiments or in measurements of the
lifetimes, or of radioactive nuclear states, or of unstable

particles. While information about the relative time of detection of a pair of particles may be obtained with a simple coincidence circuit by taking a delay curve, this is a time consuming procedure and has other disadvantages.

A better arrangement is often to use a multichannel time
analyser which senses the relative time of arrival of the pair
of pulses from the two detectors and accordingly directs the
output pulse into one of several time channels, each of which
has an associated scaler. Thus each time channel may be seen
as equivalent to a coincidence circuit of resolving time equal
to the width of the time channel.

The relationship of a time analyser to the simple coincidence circuit may be seen to be analogous to that of a multichannel pulse height analyser to the single channel pulse
height analyser. The time analyser can be arranged so that the
time channels cover the overall range of the time intervals
of interest and so the whole of the time interval histogram
may be obtained simultaneously by examining the contents of
the scalers. This has the addition advantage that changes in
the source intensity will affect all channels equally, the
same will be true for some sources of apparatus drift.

It is, however, important that the widths of the time
channels should be stable. In many designs all the time channels
are arranged to have the same width and to be contiguous, but
this is not always necessary or desirable. In practice the
type of experiment will dictate the range of time intervals
of interest and the number of time channels required to span
this range. For slow neutron time-of-flight measurements the
range may be of the order of 10 µs, whereas for the measurement of short-lived states the lower limit will be set by the
time jitter of the detector-time pick-off and may be of the
order of 100 ps. The corresponding channel widths for these
two extremes will therefore be of the order of 100 ns and
1 ps respectively.

Three main techniques are used in time analysis. The
first consists of feeding the pair of input pulses to several
coincidence circuits, one for each time channel, with a
successively increased relative delay prior to each circuit.
The second method consists of using one of the input pulses

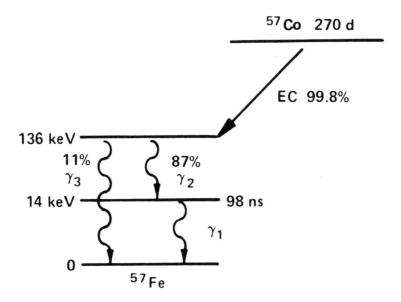

Fig. 9-18   Decay scheme of ^{57}Co.

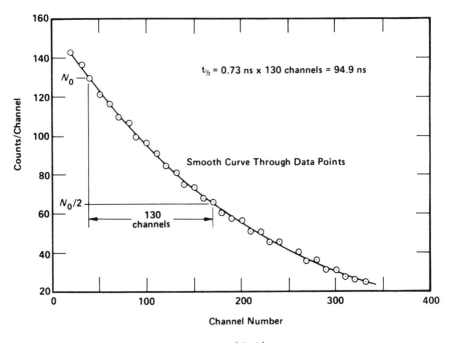

Fig. 9-19   Half-life measurement[9.3].

322

to start a high frequency clock, which is stopped by the other input pulse, and counting the number of clock cycles which occurred in the interval. The third method, which is probably the most common, is to use a time-to-amplitude converter which deliveres an output pulse of height proportional to the time interval between input pulses. The output pulse heights can then be analysed by means of a multichannel pulse height analyser and the resulting pulse height spectrum gives the time interval histogram.

As an example we will discuss a nuclear lifetime measurement on the nuclide ^{57}Co. The decay scheme for ^{57}Co is shown in fig. 9-18. The lifetime of the 14 keV state can be measured by determining the time distribution of coincidence between γ_2 and γ_1. In an experiment a 2"x2". NaI(Tl) detector was used to detect the γ_2 events and to start the time measurement in the time to amplitude converter. The γ_1 events were detected in a thin window NaI(Tl) detector which pulses were used to stop the time measurements. The output pulses were then analyzed in a multichannel analyzer. Typical results are shown in fig. 9-19.

9.7 References

9.7.1 SPECIAL

9.1 K. Siegbahn et al., ESCA Atomic, Molecular and Solid State Structure Studied by Means of Electron Spectroscopy, Almqvist & Wiksell (1967).

9.2 Ortec Application Note 34, Experiments in Nuclear Science, p.40 (1976).

9.3 Ortec Instruments for Research and Industry Catalogue, 1004, p.3, 57 (1976).

9.4 R.G. Helmer, P.H.M. van Assche, and C. van der Leun, Recommended Standards for Gamma-Ray Energy Calibration (1979), Atomic Data and Nuclear Data Tables 24 39 (1979).

9.5 C. Rolfs and A.E. Litherland, in Nuclear Spctroscopy and Reactions (J. Cerny ed.) Academic Press (1974).

9.6 N. Johnson, E. Eichler and D.G. O'Kelley Nuclear Chemistry
     p. 167 Wiley-Interscience (1963).

9.7 E. Tanaka, Nucl. Instr. Meth., $\underline{13}$, 43 (1961).

9.7.2 GENERAL

9.8 G. Dearnaley and D.C. Northrop, Semiconductor counters for
     Nuclear Radiations, J. Wiley & Sons 2 nd ed. (1966).

9.9 J.P.A. England, Techniques in Nuclear Structure Physics,
     Mc Millan (1974).

9.10 G.F. Knoll, Radiation Detection and Measurement, J. Wiley
     & Sons (1979).

9.11 P.J. Ouseph, Introduction to Nuclear Radiation Detectors,
     Plenum Press (1975).

9.12 W.J. Price, Nuclear Radiation Detection, Mac Graw-Hill
     2nd ed. (1964).

9.13 K. Siegbahn (ed.), Alpha-, Beta- and Gamma-ray Spectro-
     scopy, North Holland (1965).

9.14 E.P. Steinberg, Counting Methods for the Assay of Radio-
     active Samples in Nuclear Instruments and Their Uses
     (ed. A.H. Snell) J. Wiley & Sons (1962).

Part C.
Activation Analysis

10 Neutron Activation Analysis

10.1 Thermal neutron activation analysis

More than 2/3 of the elements in the periodic table include nuclides which possess activation and decay properties such that thermal NAA can be useful (neutron activation analysis is abbreviated NAA). A variety of matrixes can be analysed with the NAA technique, from ultra pure materials to tiny medical biopsy specimens.

10.1.1 NEUTRON PRODUCING DEVICES

In the activation processes, neutrons - and especially neutrons of thermal energies - have been more often used than other elementary particles. Two reasons for this may be given. A large number of nuclei have high cross sections for thermal neutrons and the inducement of activity by neutrons does not in general present any special difficulties.

Neutron sources with known spectral distributions and intensities are now readily available. In this respect, nuclear technology has progressed very far from the radium-beryllium sources of early days to modern reactors and other neutron-producing devices such as generators and californium - 252 isotope sources. Neutron generators and isotopic neutron sources are treated separately in sec. 4.2 .

Reactors providing neutron fluxes in the range 10^{12} - 10^{14} n cm^{-2}s^{-1} are now in operation in several countries. High flux reactors are in operation in Grenoble, France and in Oak Ridge, USA. Analysis of trace constituents in the ORNL high flux reactor in thermal fluxes of $5 \cdot 10^{14}$ n cm^{-2}s^{-1} has been reviewed by Ricci et al[10.1].

The Triga reactor, which employs uranium-zirconium-hydride fuel elements, provides in one model about 10^{13} n

$cm^{-2}s^{-1}$ at a steady state operating level. During very short
times, it is possible to obtain activations with neutron pulses with fluxes of about $2 \cdot 10^{17}$ n $cm^{-2}s^{-1}$. The pulse duration is about 14 ms.

Such high-intensity pulses have been shown to be of
value in NAA work, when the desired induced activity is very
short lived. According to studies of Yule and Guinn[10.2],
it has been shown that, with 1 000 MW pulses of the TRIGA
Mark 1 reactor, a pulse provides about a 700-fold improvement
in sensitivity for an element that forms a product with a
half-life of 0.1s, 70-fold improvement in sensitivity for an
element that forms a product with a half-life of 1s, 7-fold
if the half-life is 10s, and no improvement if the half-life
is 70s. In this connection it should also be mentioned that
short-lived nuclides could suitably be determined through
repeated activations and measurements, i.e. cyclic techniques,
described e.g. by Spyrou and Matthews[10.3].

10.1.2 IRRADIATION

The specimen to be irradiated is usually wrapped in pure
aluminium foil, or inserted in a plastic vial or a quartz
ampoule of high purity material which in turn is transferred
to a standard aluminium can (rabbit) to be introduced pneumatically in the selected reactor irradiation position.

Sample containers composed of synthetic materials such
as polyethylene are often used for short irradiation periods
or in medium high flux reactor positions. During prolonged
irradiation such materials become brittle, due to the absorbed γ-radiation and to interaction with fast neutrons; consequently aluminium containers are preferentially used in such
cases.

Let us now consider the selection of a suitable reactor
irradiation position for the specific analytical case. The
following factors have to be taken into account:
1.   The magnitude of the thermal neutron flux
2.   The contribution of fast neutrons in the irradiation
     position.

Fast neutrons can result in interferences in the analysis,

inasmuch as a given nuclide formed through (n,γ) process can also be formed through e.g. an (n,p) or an (n,α) process through fast neutron interactions.
Example:

Mn $^{55}Mn(n,\gamma)^{56}Mn$

Fe $^{56}Fe(n,p)^{56}Mn$

Thus consider the analysis of manganese in blood samples at trace level; the high amounts of iron in haemoglobin may interfere through the production of ^{56}Mn. This interference can be adjusted by making a correction for the ^{56}Mn-contribution produced in the latter reaction, with knowledge of the magnitude of the fast neutrons as well as the cross section data.

3.   The thermal neutron flux gradient in the irradiation position. Within the volume of the aluminium can (∿30mm diameter, and height ∿70mm) the thermal flux may be uniform in certain reactor positions. Usually the thermal neutron flux changes somewhat from the bottom to the top of the container in the irradiation position. The magnitude of this effect can easily be measured by means of e.g. copper foils. The thermal flux gradient in a good thermal irradiation position is illustrated in fig. 10-1.

The γ-flux in the irradiation position should further be known. Exposure to γ-radiation results in radiolytic processes in organic and aqueous materials, giving rise to destruction, evolution of gases and heating effects, which can complicate the analysis in several ways (see sec. 13.1).

10.1.3 DETECTION LIMITS

In table 10-1 detection limits are given for various elements referring to thermal neutron irradiation in a flux of 10^{13} n cm^{-2}s^{-1} for a period up to 1 hour[10.4, 10.5].

Thermal neutron flux

n·cm^{-2}·s^{-1}

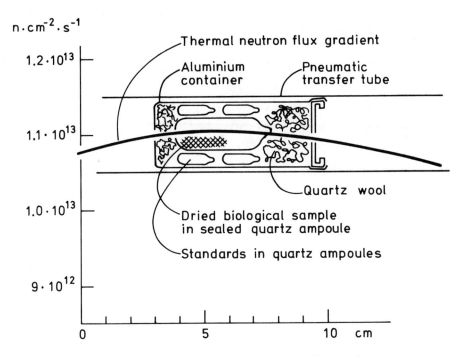

Fig. 10-1 A dried biological sample irradiated in a reactor position with low thermal flux gradient.
The sample is inserted in a sealed quartz ampoule.
The standards are irradiated in the same container,
also in quartz ampoules.

A consequence of the high sensitivity of NAA is the
possibility of accomplishing the analysis of a small sample
for elements in minor or trace quantities. Thus, in biopsy
specimens of muscle tissue of 8-30 mg wet weight, Bergström[10.6] measured Na, K, Cl, and P with an NAA technique;
this is very important in studies of mechanisms of the
disturbances in intracellular fluids and electrolytes[10.6, 10.7].

10.1.4 STANDARD COMPARATOR METHODS

The way of performing activation analysis usually com-

Table 10-1

Limits of detection for 71 elements in a thermal neutron flux of 10^{13} n cm^{-2} s^{-1}(1 h irradiation) (10.4,10.5).

µg Limit of Detection	Elements
$1-3 \cdot 10^{-6}$	Dy
$4-9 \cdot 10^{-6}$	Mn
$1-3 \cdot 10^{-5}$	Kr, Rh, In, Eu, Ho, Lu
$4-9.10^{-5}$	V, Ag, Cs, Sm, Hf, Ir, Au
$1-3 \cdot 10^{-4}$	Sc, Br, Y, Ba, W, Re, Os, U
$4-9 \cdot 10^{-4}$	Na, Al, Cu, Ga, As, Sr, Pd, I, La, Er
$1-3 \cdot 10^{-3}$	Co, Ge, Nb, Ru, Cd, Sb, Te, Xe, Nd, Yb, Pt, Hg
$4-9 \cdot 10^{-3}$	Ar, Mo, Pr, Gd
$1-3 \cdot 10^{-2}$	Mg, Cl, Ti, Zn, Se, Sn, Ce, Tm, Ta, Th
$4-9.10^{-2}$	K, Ni, Rb
$1-3 \cdot 10^{-1}$	F, Ne, Ca, Cr, Zr, Tb
$4-9 \cdot 10^{-1}$	
$1-3$	
$4-9$	
$10-30$	Si, S, Fe

prises the simultaneous irradiation of sample and a standard, the latter a known amount of the element to be assayed. The activities induced in sample and standard, denoted A_s and A_{st}, respectively, are related to weight of the sought element in the sample, W_s, and the standard, W_{st}, according to the following simple relationship:

$$\frac{A_s}{A_{st}} = \frac{W_s}{W_{st}} \qquad (10.1)$$

The activity induced through thermal neutron irradiation followed with a certain decay period has been described in eqs. 3.80 and 3.82, and is:

$$A_{th} = NV \, \varphi_0 \sigma_0 (1-e^{-\lambda t_e}) e^{-\lambda t_0}$$

NV denotes the number of nuclei irradiated.

It is assumed here that the sample and the standard are submitted to equal neutron doses and that no shielding effects occur. In multielement analysis comprising the simultaneous irradiation of standards together with the samples, the procedure may be simplified using one or a few elements as standards (comparators). The technique requires accurate knowledge of nuclear disintegration schemes of the nuclides involved, accurate data for thermal and epithermal cross sections and knowledge of the spectral neutron distribution for the subsequent calculation procedure. In practice, the comparator is a small piece of a foil or a few millimetres of a wire of a selected element or an alloy of the element. The comparator technique has been described e.g. by Girardi et al.[10.8] and by Simonits et al[10.9]. Another standard method comprises the use of an element present in the sample as the standard; this is an internal standard[10.10].

10.1.5 BIOLOGICAL AND GEOLOGICAL STANDARD REFERENCE MATERIALS

Standard reference material for application in a variety of fields, e.g. in the quality control of steel or glass and ceramic products, are generally available from institutions such as the National Bureau of Standards, Gaithersburg, US.

In the last decade special interest has been focused on biological standard reference materials to be used in trace element analysis of biological specimens for analytical control purposes or for facilitating the analysis.

Bowen has paid special attention to a kale standard, which has been distributed among laboratories around the world for intercomparison purposes. In Bowen's kale 35 elements were studied for certification purposes, i.e. the following elements: Al, As, Au, B, Ba, Br, Ca, Cd, Cl, Co, Cr, Cs, Cu, F, Fe, Hg, K, La, Mg, Mn, Mo, N, Na, P, Pb, Rb, S, Sb, Sc, Si, Sr, U, V, W and Zn[10.11]. Bowen's kale was recommended as a reference material for 12 elements[10.12].

In this context it should also be mentioned that the National Bureau of Standards has issued two biological matrix standards, Orchard leaves and Bovine liver[10.13] which are useful in the NAA of biological specimens. Reference rock samples comprising a large number of elements have been described e.g. by Nadkarni and Morrison[10.14], and by Nakahara et al[10.15].

Further, the establishment of recommended values for various trace elements in the IAEA milk powder is in progress[10.97].

10.1.6 SELF-SHIELDING DURING NEUTRON IRRADIATION

Interferences in the irradiation process can result from inhomogeneous neutron flux effects such as neutron flux gradients, neutron self-shielding by the sample matrix, or from other neutron absorbers. In aqueous solutions thermal-neutron enhancement can occur due to moderation of epithermal and fast neutrons in the water matrix.

Neutron self-shielding result in a neutron flux gradient in which the flux decreases toward the centre of the sample. This results in an activation gradient throughout the sample. Such effects can be minimized by making the samples thin or by preparing standard samples in a form comparable to the sample.

The magnitude of the self-shielding effects is estimated by means of the self-shielding factor, f, defined as follows:

$$f = \varphi_s / \varphi_\infty \tag{10.2}$$

where φ_s is the average flux inside the sample and φ_∞ is the (constant) flux in the same position if the sample were absent. The factor, f, for thermal neutron activation may be calculated for foils, spheres and wires by means of the following formulas[10,16, 10.17, 10.18]:

Foil: $f = 1 - \frac{\tau}{2} (0.923 + \ln \frac{1}{\tau})$ (10.3)

Sphere: $f = 1 - \frac{9}{8} \tau$ (10.4)

Wire: $f = 1 - \frac{4}{3} \tau$ (10.5)

where τ = shape parameter = Nσt for foils, Nσr for wires and $\frac{2}{3}$ Nσr for spheres

N = atom density (atoms cm^{-3})
σ = thermal neutron cross section (cm^2)
t = thickness (cm)
r = radius (cm)

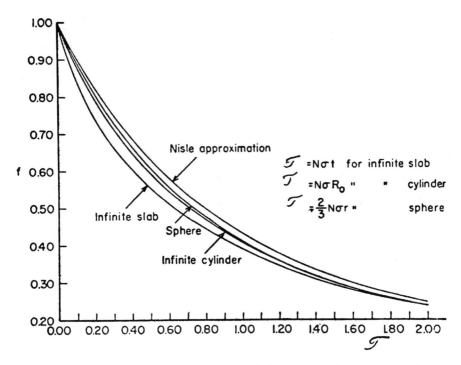

Fig. 10-2 Self-shielding factor (f) versus shape parameter
(τ).

In fig. 10-2 the self-shielding factor (f) is presented
as a function of the shape parameter. The self-shielding
effect has been treated in detail e.g. by Zweifel[10.16]
and by Reynolds and Mullins[10.17]. Further, activation
analysis of substances with large neutron capture cross
sections like boron and cadmium has been investigated by
Alimarin et al[10.19].

10.1.7 FLUX PERTURBATION DUE TO MODERATION

When irradiating large samples e.g. metals or water
various flux perturbation effects may occur and influence
the activation rate of a nuclide to be determined as was
previously pointed out. In the case of the metal pieces of
20 - 50 g used for oxygen analysis with fast neutrons, the
matrix may affect the neutrons so that they are, to some

extent, scattered out of the sample and also slowed down below threshold energy, (this is discussed in sec. 4.2.2).

The irradiation of large water samples, i.e. of volume up to 1 litre, with thermal neutrons in a reactor position which also contains a high proportion of epithermal and fast neutrons, may result in a thermal flux perturbation in the water sample, for two reasons: As with fast neutrons, thermal neutrons may be lost out of the sample through scattering processes. However, another perturbation effect leads to an increased thermal flux; in this effect epithermal and fast neutrons undergo collision processes, mainly with hydrogen and are then thermalized so that the thermal flux in the water sample is elevated.

In NAA, knowledge of the total perturbation effect is of interest. Thus, the extent of the total flux perturbation effect depends on the water volume and the neutron spectrum in the irradiation position, which can be determined experimentically. In the well-moderated Swedish reactor R1, presently closed, irradiation in a central position resulted in a thermal flux increase of about 25% for a 40 ml sample and an increase of about 40% for a 400 ml water sample[10.20].

10.2 Epithermal neutron activation analysis (ENAA)

In neutron activation analysis performed with reactor neutrons the analytical assay of a specific nuclide is not solely achieved with thermal neutrons.

Under certain circumstances, however, so called epithermal neutrons prove extremely useful for such purposes. This is the case when an element contains a nuclide which possesses high resonances in the epithermal part of the cross section curve. Two main advantages then follow from the use of epithermal neutron activation analysis:

- The activity of a desired nuclide can be optimized to that of an interfering nuclide.
- The activity level of the sample as a whole can be reduced, thus facilitating the handling procedure.

With regard to biological materials such as blood and tissue specimens[10.21] the activity of ^{24}Na normally dominates the γ-spectrum for a period of several days after the end of neutron irradiation[10.22]. The production of ^{24}Na is, however, small when epithermal rather than thermal neutrons are used. Accordingly, their use facilitates favourable analytical determination of a nuclide that possesses a high resonance cross section in such organic materials. The same consideration applies to water, especially sea-water, in which the ^{24}Na-activities dominate. The application of ENAA to rock samples has been studied extensively by Steinnes and collaborators[10.23].

10.2.1 EPITHERMAL NEUTRONS

The following is a brief summary of the general theory underlying ENAA (see also sec. 3.8.5).

The neutron flux in a reactor exhibits a broad energy distribution from about 0.001 eV to more than 10 MeV. This interval is often divided for convenience into three energy regions as follows:

The thermal region: $T_n < 0,5$ eV   (10.6)

The epithermal region: 0.5 eV $< T_n < 0.1$ MeV   (10.7)

The fast region: $T_n > 0.1$ MeV   (10.8)

Under normal circumstances, activation employs neutrons in the thermal region while accepting a certain contribution from epithermal and fast neutrons. The latter are also to some extent applied in activation processes that concern threshold reactions, although for this purpose it is more convenient to use fast neutrons produced in generators rather than in reactors. Fig. 10-3 shows the curve for a typical resonance cross section[10.24]. More particularly it represents the neutron cross section for dysprosium of normal isotopic composition as a function of energy. The individual resonance peaks are assigned to the different dysprosium isotopes. The 1/v dependence of the cross section is also evident from the figure.

Practical use is made of the resonance effect by placing

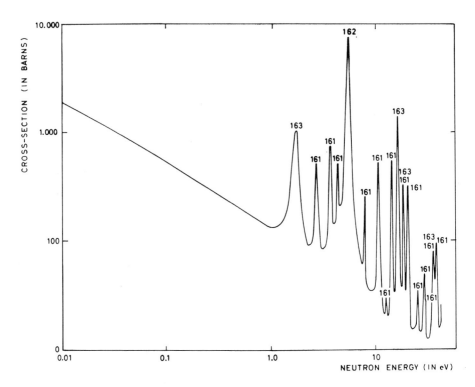

Fig. 10-3   Excitation function for (n,γ)reactions in dysprosium[10.24].

thin cadmium shields around the sample to remove thermal neutrons. It should be noted that the lower limit of the epithermal region, 0.5 eV, coincides approximately with the effective cadmium cut-off energy T_{cd} for cadmium foil thickness of about 0.5 - 1 mm[10.25]. Above 0.5 MeV, the induced activity is, in general, negligible except for threshold reactions. Also cadmium-boron filters are used in ENAA as described by Randa[10.26].

10.2.2 CALCULATION OF EPITHERMAL ACTIVITY

The activity obtained by the epithermal activation technique can be calculated from a knowledge of the various parameters relating to the epithermal part of the excitation function such as positions of resonances, their widths etc.,

followed by application of the Breit-Wigner formulae[10.27].
Borg et al.[10.28] and Racoviĉ and Prouza[10.29] have previously used this formulae to perform optimization calculations. Resonance parameters are conveniently tabulated in BNL-325[10.24] (see sec. 3.7).

Most ENAA studies are, however, simply based on the absorption of the thermal neutrons in cadmium (^{113}Cd). This nuclide is present at a level of $\sim$ 12% in natural cadmium and possesses a thermal cross section of $2 \cdot 10^4$ barns. Accordingly, the epithermal activation rate obtained by this simple approach is an integral expression of the epithermal flux and the cross section which lies above the cadmium cut-off energy of about 0.5 eV, and is described by:

$$R_{epi} = NV \int_{T_{cd}}^{0.1 \text{ MeV}} \varphi_{epi} \, \sigma(T) \, \frac{dT}{T} = NV\varphi_{epi} \, I \qquad (10.9)$$

where

NV = denotes the number of atoms in the sample
φ_{epi} = the epithermal flux
I = the resonance cross section, i.e. the so-called resonance integral, and σ_0 the thermal cross section.

The resonance integral is defined as (see eq. 3.70):

$$I = \int_{T_{cd}}^{0.1 \text{ MeV}} \sigma(T) \, \frac{dT}{T} \simeq \int_{T_{cd}}^{\infty} \sigma(T) \, \frac{dT}{T} \qquad (10.10)$$

where T_{cd} is the cadmium cut-off energy.

The resonance integral is listed in various tabulations (10.31, 10.46, 10.47) where it is given with or without the content of the 1/v cross section component which corresponds to a value of $\sim$ 0.45 σ_0. Thus in a compilation by Albinsson (10.30) the 1/v contribution has been included in the values of the resonance integral.

Consider the activity generated in a central position of a reactor that possesses thermal, epithermal and fast neutrons. The activation rate is given by the following formula (neglecting the fast contribution above 0.1 MeV) which is valid for (n,γ)-reactions (see eq. 3.79).

$$R_{tot} = NV(\varphi_o \sigma_o + \varphi_{epi} I) \qquad (10.11)$$

where σ_o represent the absorption cross section at the energy of 0.0253 eV.

10.2.3 ADVANTAGE OF ENAA

In practice the analyst is faced with the choice between applying the thermal (usually with an epithermal contribution) or the epithermal activation procedure.

If it is known which nuclide (or nuclides) interferes in the measurement of the desired activity (activities), a rough estimate of the yields can in the first instance be made by calculating the ratio: I/σ_o.

Consider the following expression:

$$\frac{(I/\sigma_o)_d}{(I/\sigma_o)_i} \gtreqless 1 \qquad (10.12)$$

where d and i designate the nuclide to be determined and the interfering nuclide, respectively. An indication is obtained of whether performance of ENAA is suitable or not. If the expression is essentially exceeding 1, let us say by a factor of 5, the ENAA technique seems favourable.

Values of I/σ_o have been tabulated by van der Linden et al[10.31].

Alternatively, the advantage to be gained from using **epithermal neutrons,** as opposed to thermal neutron activation can be assessed quantitatively by means of the cadmium ratio, D_{cd} (the cadmium ratio is in the literature often denoted R_{cd} but here D_{cd} as the letter R here denotes macroscopic rate). The cadmium ratio is given by:

$$D_{cd} = \frac{R_{tot}}{R_{epi}} \qquad (10.13)$$

For information regarding R_{tot} and R_{epi} see e.g. eqs. 3.71 and 3.79.

Thus consider a situation where a nuclide which is to be assayed is subject to interference from a second nuclide; the enhancement in activity can then be expressed simply as the ratio of their two cadmium ratios[10.32].

$$F = \frac{(D_{cd})_i}{(D_{cd})_d} \qquad (10.14)$$

The factor F may be described as the "advantage" factor. In general, this "advantage" factor can be calculated from a knowledge of such parameters as the thermal and epithermal neutron cross sections, the epithermal index, the neutron temperature, and the cadmium cut-off energy. Compilations of thermal cross section data have been given by e.g. Sher[10.33].

The cadmium ratio for gold is often known for a certain irradiation position in the reactor. It is then possible to calculate the cadmium ratio for another nuclide by means of the following equation;

$$D_{cd} = I + \frac{\sigma_0}{\sigma_0(Au)} \cdot \frac{0.45\sigma_0(Au) + I(Au)}{0.45\sigma_0 + I} \quad (D_{cd}(Au) - I)$$
$$(10.15)$$

The equation was deduced using the Westcott nomenclature[10.34]. Alternative approaches have been published by Macklin and Pomerance[10.35] and Högdahl[10.36].

10.2.4 EXAMPLES OF PRACTICAL APPLICATIONS OF ENAA

Various practical applications of ENAA are summarized in table 10-2.

10.2.5 EPITHERMAL NEUTRONS AT SELECTED ENERGIES

The performance of ENAA with a high intensity neutron beam, possessing a narrow energy range which corresponds to a major resonance in the excitation function of a specific reaction, provides the "ideal case" for analytical purposes.

A close approach to these conditions can be made by using beams transmitted through resonance window filters or resonance scatterers inserted in a reactor channel. If the filter for the neutron beam is chosen to correspond to a selected energy the γ-background can be reduced considerably.

Table 10-2
Applications of E N A A.

Nuclide measured	Interfering nuclide	Material studied	"Advantage" factor	Reference
^{56}Mn	^{38}Cl(^{24}Na)	Biological sample	6.5	10.28
^{99}Mo(^{99m}Tc)	^{59}Fe	Steel	15	10.32
^{98}Au	^{24}Na	Biological sample	6	10.29
^{80}Br	^{38}Cl(^{24}Na)	Sea water	17	10.37
^{239}U	^{56}Mn(^{24}Na)	Mineral	19	10.38
^{128}I	^{38}Cl(^{24}Na)	Pharma-ceutical	11	10.39
^{86}Rb	^{46}Sc	Mineral	24	10.23
^{124}Sb	^{134}Cs	Mineral	34	10.23
^{122}Sb	^{134}Cs	Mineral	46	10.23
^{239}U	^{233}Th,^{233}Pa	Mineral	3	10.40
^{116}In	^{56}Mn	Mineral	11.8	10.41
^{115}Cd	^{24}Na	Biological sample	77.7	10.42
^{195m}Pt	^{46}Sc,^{59}Fe,^{51}Cr	Mineral	43.5	10.26
^{182}Ta	^{46}Sc,^{59}Fe,^{51}Cr	Mineral	18.1	10.26
^{110m}Ag	^{46}Sc,^{59}Fe,^{51}Cr	Mineral	17.8	10.26

Table 10-3

Epithermal neutron intensities at selected energies[10.43, 10.44].

Energy		Intensity $(n \cdot cm^{-2} \cdot s^{-1})$	Energy Resolution	Filter, Scatterer	
19	eV	$2 \cdot 10^4$	0.6 eV	W	-Scatterer
139	eV	$3 \cdot 10^4$	6.0 eV	Co	-Scatterer
2	keV	$5 \cdot 10^6$	0.7 keV	Sc	-Filter
2.8	keV	$1 \cdot 10^4$	0.5 keV	NaF-	-Scatterer
24.8	keV	$6 \cdot 10^5$	1.8 keV	Fe	-Filter
144	keV	$1 \cdot 10^7$	30-50 keV	Si	-Filter

The technique seems therefore promising for application in prompt analysis[10.43].

A list of epithermal neutron intensities at selected energies is given in table 10-3. It should also be stressed that epithermal neutrons at selected energy intervals can be produced in accelerators such as the Van de Graaff using reactions of the type $^7Li(p,n)^7Be$ (see sec. 4.2.3).

Currents of about 20 µA can be obtained by such techniques. The neutron beams thus formed are focused in a forward direction. Below 50 keV, however, the energy resolution is poor.

Table 10-4

Neutron beams produced in the $^7Li(p,n)^7Be$ reaction.

Energy (keV)	Intensity $(ncm^{-2} s^{-1} \mu A^{-1})$	Resolution (keV)
50	$2 \cdot 10^6$	15
100	$5 \cdot 10^6$	25
200	$2 \cdot 10^7$	25

These high-energy epithermal neutrons might be applied
to in vivo analysis of elements located at various depths
in the human or animal body, after being moderated to appropriate energies by hydrogen and carbon present in tissues and
biological fluids[10.96].

10.2.6 SHIELDING EFFECTS IN ENAA

The effects of shielding and self-shielding in ENAA are
in principal similar to those that prevail in thermal neutron
activation analysis (see e.g. Zweifel[10.16]). These effects
have been given special attention by Högdahl[10.36] who also
studied the effects of closely spaced resonances in the
analysis of copper and gold in silver spheres.

A comprehensive study of the effects of overlapping
resonances upon the generation of active nuclides has further
been made by Connally et al[10.45]. The various isotopes
of cadmium (e.g. ^{106}Cd-^{114}Cd) exhibit a large number of
resonances in the epithermal neutron spectrum. Accordingly,
for a sample enclosed in cadmium these resonances will
produce strong minima in the neutron flux resulting in a
considerable reduction in the activity of a nuclide whose
resonances coincide with those of cadmium.

10.3 Fast neutron activation analysis (FNAA)

Fast neutron activation analysis has mainly been exploited in the analysis of light elements. In the following,
the analysis of oxygen and nitrogen will be given more
detailed consideration refering to industrial or agricultural
applications.

10.3.1 OXYGEN DETERMINATION

The bombardment of oxygen with fast neutrons leads to
the formation of the nuclide ^{16}N by the reaction:

$$^{16}O(n,p)^{16}N$$

^{16}N decays with a half-life of 7.3 seconds, emitting

γ-quanta with energies of 6.13 and 7.11 MeV which can be measured with a scintillator detector. Since the high energy γ-radiation is specific for a decay of ^{16}N the measurement is, in practice, performed by integrating all pulses with an energy in excess of about 4.5 MeV.

The application of this type of analysis has been greatest in samples where oxygen in trace quantities is of significance. High purity metals provide a general field of application while the speed of analysis is particularly valuable to the steel industry where rapid oxygen analysis in connection with production is required. Steel samples are generally presented for analysis in the form of cylindrical discs 10-25 mm in diameter and 5-15 mm thick. The level of detection in such samples is often quoted as being about 5 μg·g^{-1} sample.

This reaction has been successfully utilized for the determination of oxygen in metals by several investigators, e.g. Guinn[10.48], Strain[10.49], Mott and Orange[10.50], Fuji et al.[10.51] and Aude and Laverlochere[10.52]. The oxygen analysis of organic materials such as mineral oils has also been performed with this technique[10.53].

At the steel manufacturing company Cockerill à Seraing in Belgium, a Sames neutron generator (sec. 4.2.2) was installed in 1967 for the purpose of the fast assay of the oxygen content in steel. The facility has been used in routine operations and reliable results have been obtained as reported by Colette and Lacombe[10.54].

Fast neutron activation analysis is at present characterized by a high degree of automation. Samples to be irradiated are transported in pneumatic tubes from target to detector within about 1 second and the irradiation, transfer and counting sequence may be automatically controlled. In the routine analysis of oxygen in metals the application of the dual sample irradiation technique, implying the simultaneous irradiation of the standard and the sample in a rotating device in front of the target, gives a high degree of precision in the analysis. Thus at higher oxygen levels a standard deviation of less than 1% has been obtained with

this technique[(10.50)]. Generally, this figure amounts to
1-5%. At lower concentrations the precision is considerably
reduced. At an oxygen level of 10 ppm in steel samples of
about 10 g a standard deviation of 30% has been reported[(10.51)].

Fuji et al.[(10.51)] compared the results obtained in the
oxygen analysis of steel accomplished by the fast neutron
activation method and the vacuum fusion technique. The two
methods seemed to be in good agreement in the concentration
range investigated, 10-2 000 ppm oxygen. Neutron activation
analysis had in this case proved superior to the vacuum fusion
method with regard to rapidity. The reliability of the nuclear
technique for oxygen analysis has further been confirmed by
Bruck et al.[(10.55)], who compared fast neutron activation analysis and the vacuum fusion technique.

In aluminium the detection limit of oxygen is estimated
to be 10 ppm. In the aluminium analysis interfering activities
occur from fluorine through the reaction:

$$^{19}F(n,\alpha)^{16}N$$

Fluorine produces somewhat less than half the amount of
^{16}N activity through this reaction that the same amount of
oxygen produces through the corresponding (n,p) reaction[(10.56)].
The extent of this "extra" ^{16}N activity may be estimated by
determining the fluorine content of the sample by means of another reaction, e.g. by $^{19}F(n,p)^{19}O$, or by $^{19}F(n,2n)^{18}F$. The
^{18}F nuclide is identified by means of the 0.51 MeV annihilation
peak.

Fast neutron activation analysis of oxygen and silicon
has further been accomplished satisfactorily in diamond
matrices, as reported by Bibby and Sellschop[(10.57)]. For
oxygen a detection limit of about 5 µg was established while
the detection limit of silicon was estimated to be of the
order of 25 µg using the reaction $^{28}Si\ (n,p)^{28}Al$. It should
also be mentioned that FNAA technique has advantageously been
utilized in the chemical assay of oxygen in coal conversion
liquids as demonstrated by Khalil et al[(10.58)].

10.3.2 FLOW-RATE MEASUREMENTS

Using the reaction $^{16}O(n,p)^{16}N$, flow-rate measurements have been accomplished by fast neutron activation of oxygen in various solutions, as demonstrated by Peck and Pierce [10.59]. The technique can even be applied to slurries. For such analysis small sealed-tube neutron generators are used (sec. 4.2.2).

10.3.3 NITROGEN ANALYSIS OF GRAIN AND OTHER AGRICULTURAL PRODUCTS

The applicability of the fast neutron activation technique to the analysis of nitrogen in grain for purposes of protein evaluation has been industrially exploited at the Ralston Purina Co, St Louis, USA[10.60, 10.61]. This method is considered to afford an alternative to the wellestablished Kjeldahl technique; its advantages include high capacity, shorter analysis time, no grinding procedures, the lack of environmental risks due to pollution with acids or with mercury compounds used as catalysts, and the abscence of health risks due to mineral acid vapours.

The fast neutron technique has also been used for the analysis of nitrogen in eggs, hay and organic liquids such as beer. According to studies at the Ralston Purina Co, grain samples weighing between 15 and 20 g have been analyzed at a rate of 30 per hour.

Further, a large scale system for the analysis of nitrogen and various other elements in agricultural products using the fast neutron activation method has been developed in a Franco-Russian project[10.62].

Nitrogen is determined by means of the reaction:

$$^{14}N(n,2n)^{13}N$$

The radionuclide formed, ^{13}N, is a positron emitter with a 10 min half-life, which can easily be measured using scintillation spectrometry. The nitrogen content of grain samples usually lies in the range 1-3%, giving rise to distinct full energy peaks of ^{13}N in the irradiated sample.

The above reaction has been used by various investigators for the nitrogen analysis of grain[10.60-10.65] Various reactions interfere in the analysis[10.67].
Interfering reactions:

$$^{28}Si(n,p)^{28}Al \qquad (1)$$

$$^{31}P(n,2n)^{30}P \qquad (2)$$

$$^{31}P(n,\alpha)^{28}Al \qquad (3)$$

$$^{39}K(n,2n)^{38}K \qquad (4)$$

$$^{16}O(p,\alpha)^{13}N \qquad (5)$$

$$^{13}C(p,n)^{13}N \qquad (6)$$

These nuclides are either positron or γ-emitting nuclides.

The interference due to the presence of silicon and phosphorous in grain, from the nuclides ^{28}Al and ^{30}P with half-lives of 2.3 min and 2.6 min respectively, can be reduced to negligible levels by allowing the samples to decay for a period of 12 min. It is even possible to carry out the analysis omitting the initial decay period of 12 min, by making corrections for the induced short-lived activities[10.67].

The effect of potassium interference is due to ^{38}K, a positron emitter with a half-life of 7.7 min; this can be corrected for in the analysis by assuming a constant value for the potassium content (about 0.5% K). In reactions (5) and (6), ^{13}N is produced from oxygen and carbon respectively through recoil proton interactions. The recoil protons are produced through fast neutron reaction with hydrogen. However, this effect can also be corrected for by irradiating a blank sample, with the same composition as the grain sample as regards the elements carbon, oxygen and hydrogen (but with no nitrogen content). "Apparent" nitrogen contents of about 0.30% produced by reactions (5) and (6) have been measured[10.61, 10.64].

10.3.4 ANALYSIS OF NITROGEN IN EXPLOSIVES

The reaction $^{14}N(n,2n)^{13}N$ has further proved satisfactory for the determination of nitrogen in explosives. In a study of Semel et al.[10.68], high precision analysis of nitrogen in such samples was carried out using a multisample triple axis rotator where the flux gradient effects are eliminated.

10.3.5 ANALYSIS OF LUNAR SAMPLES

Fast neutron activation analysis has been used with advantage in the determination of the composition of the lunar surface. Several elements e.g. oxygen, magnesium, aluminium, silicon, titanium and iron have been analyzed with the aid of fast neutrons. In this connection the advantage of a nuclear technique for accomplishing non-destructive analysis is clearly demonstrated, since the expensive lunar samples can be studied in other ways after nuclear analysis[10.69]

10.3.6 EXPLOITING MINERAL RESOURCES

The fast neutron activation technique can further be used for prospection of mineral resources. Small, transportable neutron generators are used for bore-hole logging experiments. A review of the technique has been given by e.g. Czubeck[10.70] and Santos et al[10.71].

Using the fast neutron activation technique Gorski et al.[10.72] determined the copper content in minerals using γ-γ coincidence methods.

In various logging experiments, pulsed neutrons can be used to some extent to establish the elemental composition of the geological formation. The processes of inelastic neutron scattering may conveniently be utilized in such cases, since the γ-radiation associated with inelastic scattering can be separated from the capture radiation[10.73].

With a pulsed neutron technique the two following nuclear methods can be used for elemental characterization:
1.   Inducement of neutron pulse, then characterization of the neutrons with respect to energy and life-time.

2.   Inducement of neutron pulse, and registration of γ-
      radiation, i.e. prompt and delayed.

Using neutron life-time logging methods it is possible
to determine various parameters such as diffusion coefficients (d) and average neutron life-times (τ) in the vicinity
of the bore-hole. These parameters depend on the composition
of the rock material, so that the composition can then be
evaluated.

Further, the pulsed-neutron technique can be used in
oil exploiting technology. Thus, by carrying out measurements of life-times oil-bearing formations can be distinguished from water-bearing ones; the average neutron life-
time in oil is about 200 μs and in water 50-200 μs, depending
on the salt concentration.

10.3.7 INDUSTRIAL PROCESS-CONTROL

On-line analysis of oxygen, silicon and aluminium in
coal has been accomplished with fast neutron activation
analysis. Prompt γ-radiation and the γ-radiation of induced
radionuclides are measured for the quantitative assay of
these elements[10.74].

Silicon can also be analyzed in several matrices, e.g.
in rocks and minerals, by means of the reaction[10.75].

$$^{28}Si(n,p)^{28}Al$$

However, various interfering reactions limit the applicability of this method.

10.3.8 MULTIELEMENTAL ANALYSIS

With the aid of the neutron generator, multielemental
analysis of various minor components, i.e. F, Al, Si, P, K,
Mn, Mo and W in minerals and steel has been carried out
according to study of Gangadharan et al.[10.76] Detection
limits in the range of a few μg to some mg has been reported.
With the intense 14 MeV neutron facility at Lawrence Livermore
National Laboratory up to 20 elements of standard fly ash,

Table 10-5

Optimum detection limits for various elements using fast neutrons$^{(10.78)}$.

Limit of detection (ug)	Element	Radionuclide	Half-life	Energy of main peak (MeV) 14-MeV neutrons:10^9n cm^{-2} s^{-1},	Interference
From 1 to 10	Ag	Ag-106	24.o min.	0.51	Sb,In,Cl,Sn.
	Al	Mg-27	9.45 min.	0.84	
	Ba	Ba-137m	2.60 min.	0.66	Ag
	Br	Br-78	6.4 min.	o.51	Cu,N,K......
	Cd	Cd-111m	48.6 min.	0.25	
	Cu	Cu-62	9.80 min.	0.51	Br,Sb,N,In,F
	Ga	Ga-68	68 min.	0.51	Cd,Se,Cr....
	Hg	Hg-199m	44 min.	0.158	Cd,Cl,Sn
	Sb	Sb-120	16.4 min.	0.51	Ag,In,Mo....
	Si	Al-28	2.30 min.	1.78	P
	Sr	Sr-87m	2.80 hr.	0.39	
	Zn	Zn-63	38.3 min.	0.51	Cl,Sn,Se....
	Zr	Y-94	16.5 min.	0.92	Sr
From 10 to 100	Cl	Cl-34m	32.4 min.	0.51	Zn,Sn.......
	Cr	V-52	3.76 min.	1.44	Mn
	F	O-19	29.4 sec.	0.20	Pd,Ag
	Fe	Mn-56	2.58 hr.	0.85	Co
	Ge	Ge-75m	48 sec.	0.139	As,Pd
	K	K-38	7.75 min.	0.51	Br,Cu,N....
	Mg	Na-24	15.0 hr.	1.37-2.75	Al
	Mo	Mo-91	15.5 min.	0.51	Sb,In,Ag...
	N	N-13	10 min.	0.51	Cu,Br,K.....
	Na	Ne-23	37.6 sec.	0.44-1.65	Mg
	O	N-16	7.4 sec.	6.13	F,B
	P	Al-28	2.30 min.	1.78	Si
	Pd	Pd-109m	4.8 min.	0.188	
	Se	Se-81m	56.8 min.	0.103	
	Sn	Sn-123	40.0 min.	0.153	Hg,Cd,Cl
	Te	Te-129	74 min.	0.46	
	W	W-185m	1.7 min.	0.130	Fe,Pd
From 100 to 1000	As	Ge-75m	48 sec.	0.139	Ge,Pd
	Au	Au-197m	7.4 sec.	0.28	
	Nb	Y-92	3.60 hr.	0.21-0.94	
	Ni	Co-62	13.9 min.	1.17	Cu
	Pt	Pt-197m	80 min.	0.34	Au
	Ta	Ta-180m	8.15 hr.	0.093-0.102	
	Ti	Sc-46m	19.5 sec.	0.142	Se
From 1000 to 10000	B	Be-11	13.7 sec.	5.60-6.76	O,F
	Ca	K-44	22.0 min.	1.13	Cl
	Pb	Pb-207m	0.84 sec.	0.57	
	S	P-34	12.4 sec.	2.13	Cl,B

Irradiation and counting times=4 half-lives or maximum 20 min.

orchard leaves or bovine liver has been determined according to Williams et al[10.77].

10.3.9 DETECTION LIMITS

Lists of detection limits for various elements using the fast neutron activation technique have been compiled e.g. by Perdijon[10.78] and Cuypers and Cuypers[10.79]. Values from the first mentioned study are given in table 10-5. In this connection it should, however, be stressed that the values presented should only be used as indications in practical analysis, due to the complexity of interfering reactions.

For further details about fast neutron activation analysis reference is made to the comprehensive textbook in this field by Nargolwalla and Przybylowicz[10.80].

10.4 Instrumental neutron activation analysis

10.4.1 VARIOUS APPLICATIONS

Instrumental neutron activation analysis, abbreviated INAA, comprises the use of purely instrumental techniques in the assay of the nuclide through measurements with for example high resolution Ge(Li) detectors or the use of coincidence or anticoincidence spectrometry after appropriate activation.

The INAA technique is widely used in the assay of trace elements in air, i.e. of elements accumulated on various air filters[10.81, 10.82]. Thus, through the use of automated γ-ray spectrometric procedures, magnetic tape recording of spectral data and computerized processing of data it is possible to measure more than 30 elements in an air filter sample with a median detection limit for airborne concentrations of an element of $4 \cdot 10^{-9} \mathrm{g} \ \mathrm{m}^{-3}$ [10.81].

In this context it should, however, be mentioned that so-called energy-dispersive X-ray spectrometry has found valuable applications for multielement pollution analysis, as described by Rhodes[10.83].

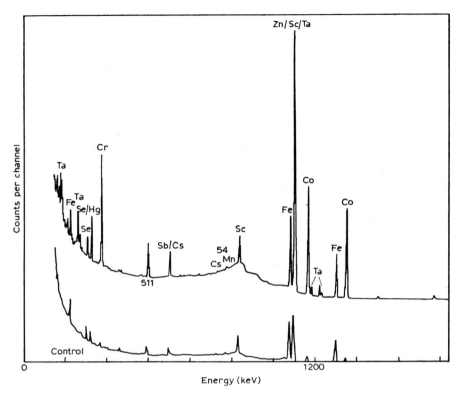

Fig. 10-4   Gamma-spectrum Ge(Li) of neutron activated lung
from metal worker compared to that of control[10.86].

Further, the INAA technique has found valuable applications in the analysis of a large number of minor or trace
elements in geological specimens[10.23] and in quartz[10.84].

Instrumental neutron activation analysis has also found
applications in the biological field. Thus, this technique
has made possible the quantitative assay of 43 elements in
kidneys of rabbits[10.85]. A typical Ge(Li) γ-ray spectrum
of lung tissue is given in fig. 10-4[10.86]. Further, 25 minor
and trace elements in human blood serum have been determined
using this technique[10.87].

The application of INAA in the control of the working
environment has been surveyed by Hewitt[10.86].

10.4.2 MEASUREMENT OF PROMPT GAMMA RADIATION

In cases where various charged particle reactions such as (p,γ), (d,p) or (d,n) are utilized, the promptly emitted reaction products such as γ-rays, protons or neutrons of defined energies can be measured and the results utilized for quantitative analysis (see chapter 11). The prompt γ-ray radiation resulting from the thermal neutron capture process may also be utilized for analytical chemical procedures.

Thus, using the neutron capture procedure, the specimen is irradiated in a neutron beam emerging from a reactor channel or from an isotope neutron source. The γ-radiation from nuclear states excited by neutron capture, i.e. the prompt γ-radiation is then analyzed while the sample is being irradiated.

In order to obtain a pure capture γ-ray spectrum the contributions from background radiation and from long lived radionuclides must be subtracted from the total spectrum. For optimization purposes a method has been developed by Isenhour and Morrison to modulate the neutron beam and gate the analyzer, permitting the rapid, accurate subtraction of interfering radiation[10.88].

Tabulations of prompt γ-rays emitted in the thermal neutron capture process are found in e.g.[10.89, 10.90]. The prompt γ-rays usually predominate in an energy range as large as 1-10 MeV. Typical half-lives of the excited states are 10^{-12} - 10^{-20} s. In experimental studies the neutron beam is often modulated with a cadmium chopper. Various filters are also used to enhance the thermal neutron flux with respect to the γ-radiation. A sum coincidence arrangement for the analysis of capture γ-rays for the purpose of eliminating interfering radiation has been developed by Lussie and Brownlee[10.91].

The thermal neutron capture method has been utilized for the measurement of nitrogen in cereals using a reactor irradiation facility, as described by Bäcklin et al.[10.92] With an isotope neutron source Tiwari et al.[10.93] examined the nitrogen content in cereal for protein evaluation in connection with plant breeding.

Table 10-6

Detection limits of various elements in neutron capture γ-ray analysis[10.95].

Element	Low-energy region		High-energy region	
	E_γ, keV	Detection limit, µg	E_γ, keV	Detection limit, µg
Boron	477.0	$0.2 \cdot 10^{-3}$	–	–
Chlorine	1163.2	$46 \cdot 10^{-3}$	6111	0.5
Scandium	227.6	$24 \cdot 10^{-3}$	8175	1.6
Titanium	1378.0	0.1	6759	1.0
Manganese	212.5	0.28	7243	2.4
Iron	353.1	0.8	7631	6.0
Cobalt	229.6	$26 \cdot 10^{-3}$	6877	1.5
Nickel	465.4	0.3	8999	2.0
Copper	278.6	0.22	7914	4.3
Rhodium	180.8	$11 \cdot 10^{-3}$	5917	7.0
Silver	198.4	$16 \cdot 10^{-3}$	–	–
Cadmium	558.6	$0.7 \cdot 10^{-3}$	5824	0.2
Indium	275.8	$5.3 \cdot 10^{-3}$	–	–
Lanthanum	217.8	0.23	5098	18
Neodynium	696.4	0.11	6502	4.8
Samarium	333.9	$0.15 \cdot 10^{-3}$	–	–
Europium	89.9	$0.2 \cdot 10^{-3}$	–	–
Gadolinium	182.1	$0.1 \cdot 10^{-3}$	6749	$80 \cdot 10^{-3}$
Dysprosium	184.2	$3.2 \cdot 10^{-3}$	5607	0.8
Holmium	136.6	$12 \cdot 10^{-3}$	–	–
Erbium	185.3	$4.5 \cdot 10^{-3}$	–	–
Thulium	148.8	$22 \cdot 10^{-3}$	–	–
Lutetium	149.6	$7.1 \cdot 10^{-3}$	–	–
Iridium	217.4	$12 \cdot 10^{-3}$	5958	5.4
Gold	215.7	$65 \cdot 10^{-3}$	6252	3.2
Mercury	371.5	$2.9 \cdot 10^{-3}$	5967	0.25

Further, the thermal neutron capture γ-ray method has
been used for the quantitative assay of the elements Cr, Mn,
Fe, Co, Ni and W in refined steel at percentage levels or
higher, as described by Zwittlinger[10.94].

According to a study by Henkelmann and Born[10.95] in
the high flux reactor in Grenoble, using a thermal neutron
beam with a flux of about $1.5 \cdot 10^{10} n \, cm^{-2} s^{-1}$ with low
γ-background, the detection limits for various elements given
in table 10-6 were obtained.

10.5 References

10.1   E. Ricci et al., J. Radioanal. Chem., 19, 141 (1974).

10.2   H.P. Yule and V.P. Guinn, Radiochemical Methods of Ana-
        lysis, 2, 111, (1965)IAEA, Vienna.

10.3   N.M. Spyrou and I.P. Matthews, J. Radioanal. Chem.,61,
        1 (1981).

10.4   H.R. Lukens, GA-5073 (1964).

10.5   V.P. Guinn, Article in Advances of Activation Analysis,
        (eds. J.M.A., Lenihan and S.J. Thomson) Academic Press,
        59 (1969).

10.6   J. Bergström, Diss., Karolinska Institutet, Stockholm
        (1962).

10.7   G.J. Batra and D.K. Bewley, Radioanal. Chem., 16, 275,
        (1973).

10.8   F. Girardi et al., Anal. Chem., 37, 1085 (1965).

10.9   A. Simonits et al., J. Radioanal. Chem., 24, 31 (1975).

10.10   G. Leliaert et al., Anal. Chem. Acta, 19, 99 (1958).

10.11   H.J.M. Bowen, J. Radioanal. Chem., 19, 215 (1974).

10.12   R.E. Wainerdi, Pure and Appl. Chem. 51, 1183 (1979).

10.13   P.D. La Fleur, J. Radioanal. Chem., 19, 227 (1974).

10.14   R.A. Nadkarni and G.H. Morrison, J. Radioanal. Chem.,43,
        347, (1978).

10.15   H. Nakahara et al., J. Radioanal. Chem., 59, 245 (1980).

10.16   P.F. Zweifel, Nucleonics, 18 (11), 174 (1960).

10.17   S.A. Reynolds and W.T. Mullins, Int. J. Appl. Rad. Iso-
         topes 14, 421 (1963).

10.18   J. Gilat and Y. Gurfinkel, IA 756, (1962).

10.19   I.P. Alimarin et al., J. Radioanal. Chem., 16, 79 (1973).

10.20   D. Brune and K. Jirlow, Radiochim. Acta, 8, 161 (1967).

10.21   C. Turkowsky et al., Radiochim. Acta, 8, 27 (1967).

10.22   D. Comar, Article in Advances in Activation Analysis
         (eds. J.M.A. Lenihan and S.J. Thomson), Academic Press
         163 (1969).

10.23   E. Steinnes, Some Neutron Activation Methods for the
         Determination of Minor and Trace Elements in Rocks. Diss.
         Oslo University (1972).

10.24   Neutron Cross-Sections. BNL-325 3d ed Vol. I (1973),
         Vol. II (1976).

10.25   K.H. Beckurts and K. Wirtz, Neutron Physics. Springer
         Verlag, Berlin, 272 (1964).

10.26   Z. Randa, Radiochem. Radioanal. Letters 24, 157 (1976).

10.27   E.B. Paul, Nuclear and Particle Physics, North Holland
         Publ. Comp., Amsterdam 207 (1969).

10.28   D.C. Borg, et al., Int. J. Appl. Radiation Isotopes II,
         10, (1961).

10.29   M. Raković and Z. Prouza, Isotopenpraxis 4, 11 (1968).

10.30   H. Albinsson, in Handbook on Nuclear Activation Cross-
         Sections, IAEA, Vienna, Tr-no 156 (1974).

10.31   R. Van der Linden et al., Nucl. Data in Science and
         Technology, Proc. of a Symp., Paris 12-16 March, 241,
         (1973).

10.32   D. Brune and K. Jirlow, Nukleonik 6, 242 (1964).

10.33   R. Sher, in Handbook on Nuclear Activation Cross-Sec-
         tions, IAEA, Vienna, Tr-no 156 (1974).

10.34   C.H. Westcott, AECL-1101 (1960).

10.35 R.L. Macklin and H.S. Pomerance, Proc. Int. Conf. Peaceful Uses of Atomic Energy, Geneva 1955, 5, UN, New York, 96 (1956).

10.36 O.T. Högdahl, Proc. Symp. Radiochemical Methods of Analysis, Salzburg 19-23 Oct 1964. IAEA, Vienna, 1, 23 (1965). (STI/PUB/88).

10.37 H.W. Nass, Int. Conf. on Modern Trends in Activation Analysis. Gaithersburg, USA, Oct. 7-11, 1968. Proc. Washington D.C, 1, 563 (1969). (NBS Spec. Publ. 312).

10.38 E. Steinnes and D. Brune, Talanta 16, 1326 (1969).

10.39 D. Brune, Anal. Chim. Acta 46, 17 (1969).

10.40 L.T. Atalla and F.W. Lima, J. Radioanal. Chem., 20, 607 (1974).

10.41 I.M. Cohen, Radiochem. Radioanal. Letters 15,379 (1973).

10.42 D. Brune and B. Bivered, Anal. Chim. Acta 85, 411 (1976).

10.43 R.M. Brugger and O.D. Simpson, Irradiation Facilities for Research Reactors. Proc. of a Symp., Teheran 6-10 Nov., 1972. IAEA, Vienna 131 (1973). (STI/PUB/316).

10.44 W. Dilg and H. Vonach, Nucl. Instr. Meth. 100,83 (1972).

10.45 T.J. Connally et al., (KFK-718; EUR-3716; EURFNR-527). (1968).

10.46 E.M. Gryntakis and J.I. Kim, J. Radioanal. Chem., 42, 181 (1978).

10.47 L. Moens, J. Radioanal. Chem., 54, 377 (1979).

10.48 V.P. Guinn, Proc., U N Intern, Conf. Peaceful Uses of Atomic Energy, Geneva 3, 1964, Geneva 15, 433 (1965).

10.49 J.E. Strain, Progr. Nucl. Energy, Ser IX, 4, Pergamon Press, New York, 137 (1965).

10.50 W.E. Mott and J.M. Orange, Proc. Conf. Modern Trends in Activation Analysis, College Station, Texas, USA, April 19-22, 115(1965).

10.51 I. Fuji et al., Anal. Chim. Acta 34, 146 (1966).

10.52 G. Aude and J. Laverlochere, 36th Congr. Intern. de Chimie
       Ind., Brussels, 10-21 Sept. 1966, (CEA-CENG-DR SAR-G/66-10/
       J1/NC).

10.53 R.A. Stallwood et al., Anal Chem., 45, 6(1963).

10.54 F. Collette and M. Lacombe, Cong. Recent Developments in
       Neutron Activation Analysis, 4-7 Aug. 1975, Churchill
       College, Cambridge.

10.55 J. Bruck et.al., IBID

10.56 K. Perry, et al., 1st International Symposium on Trace
       Characterization - Chemical and Physical (NBS), Gaithers-
       burg, Md, USA, Oct. 3-7 (1966).

10.57 D.M. Bibby and J.P.F. Sellschop, J. Radioanal. Chem., 22,
       103 (1974).

10.58 S.R. Khalil et al., J. Radioanal. Chem., 57, 195 (1980).

10.59 P.F. Peck and T.B. Pierce, 1st International Symposium on
       Trace Characterization - Chemical and Physical (NBS),
       Gaithersburg, Md, USA, Oct. 3-7 (1966).

10.60 C.C. Tsen and E.E. Martin, Cereal Chemistry, 48, 721
       (1971).

10.61 D.E. Wood, Isotopes and Radiation Technology, 9, 351
       (1972).

10.62 E. Vernin, et al., Proc. IAEA Symp in Bled, Yugoslavia
       10-14 April 1972, Nuclear Activation Techniques in the
       Life Sciences.

10.63 L. Kosta, et al., Proc. from IAEA/FAO Conf. in Röstånga
       Sweden, 17-21 June 1968, New Approaches to Breeding for
       Improved Plant.

10.64 D. Brune and A. Arroyo, Anal. Chim. Acta 56, 473 (1971).

10.65 D.M. Bibby and H.M. Champion, Radiochem. Radioanal. Let-
       ters, 18, 177 (1974).

10.66 M. Bormann et al., Handbook on Nuclear Activation Cross
       sections IAEA, Vienna Tr-no 156 (1974).

10.67 K.R. Blake and E.L. Hudspeth, Int. J. Appl. Rad. Isotopes,
       22, 233 (1971).

10.68   S. Semel et.al., Conf. Recent. Development in Neutron
         Activation Analysis 4-7 Aug. 1975, Churchill College
         Cambridge.

10.69   W.D. Ehmann and J.W. Morgan, Science 167, 528 (1970).

10.70   J.A. Czubek, Nuclear Techniques and Mineral Resources
         (symp. in Buenos Aires 5-9 Nov., 1968), IAEA Vienna 3(1969).

10.71   G.G. Santos et al., IBID, 463.

10.72   G. Gorski et al., Talanta 11, 1135 (1964).

10.73   D.F. Bespalov et al., Int. Conf. Peaceful Uses of Atomic
         Energy, Geneva 7 (1971).

10.74   J.R. Rhodes, et al., ORO-2980-18 (USAEC)(1968).

10.75   R. Van Grieken et al., J. Radioanal. Chem., 6,385 (1970).

10.76   S. Gangadharan et al., J. Radioanal. Chem., 24, 57 (1975).

10.77   R.E. Williams et al., J. Radioanal. Chem. 63, 187 (1981).

10.78   J. Perdijon, Anal. Chem., 39, 448 (1967).

10.79   M. Cuypers and J. Cuypers, J. Radioanal. Chem., 1, 243
         (1968).

10.80   S.S. Nargolwalla and E.P. Przybylowicz, Activation Ana-
         lysis with Neutron Generators, John Wiley and Sons (1973).

10.81   W.E. Jr. Kuykendall et al., J. Radioanal. Chem. 19, 351
         (1974).

10.82   M. Janssens et al., J. Radioanal. Chem. 26, 305 (1975).

10.83   I. Rhodes, Intern. Lab. July/Aug. (1973).

10.84   I. Kuleff et al., J. Radioanal. Chem., 62, 187 (1981).

10.85   J.A. Velandia and A.K. Perkons, J. Radioanal. Chem., 20,
         715 (1974).

10.86   P.J. Hewitt, J. Environ, Mgm. 3, 133 (1975).

10.87   R.A. Nadkarni and G.H. Morrison, Radiochem. Radioanal.
         Letters 24 (2), 103 (1976).

10.88   T.L. Isenhour and G.H. Morrison, Anal. Chem., 38, 162
         (1966).

10.89   D. Duffy et al., Nucl. Instr. Meth. 80, 1 (1970).

10.90   F.E. Senftle et al., Nucl. Instr. Meth. 93, 425 (1971).

10.91   W.G. Lussie and J.L. Brownlee Jr., Proc. from Int. Conf. Modern Trends in Activation Analysis, College Station, Texas, 194 (1965).

10.92   A. Bäcklin, et al., LFF-45, 1971, Forskningsrådens Lab., Studsvik, Sweden.

10.93   P.N. Tiwari et al., Int. J. Appl. Rad. Isotop. 22,587 (1971).

10.94   H. Zwittlinger, J. Radioanal. Chem. 14, 147 (1973).

10.95   R. Henkelmann and H.J. Born, J. Radioanal. Chem. 16, 473 (1973).

10.96   R.O. Bergman, Biological Effects of Neutron Irradiation, Proc. of Symp., Neuherberg, 22-26 oct., (1973), Vienna (1974) 83 (STI/PUB/352).

10.97   J.J.M. De Goeij et al., Anal. Chim. Acta 146, 161 (1983).

11 Charged Particle Activation Analysis

11.1 Introduction

In the literature, charged particle activation analysis is often abbreviated to CPAA. In the following, this term will be used for convenience.

CPAA is mainly used in the analysis of light elements such as carbon, oxygen, nitrogen and fluorine. The reason for this is the possibility of accomplishing highly sensitive analysis, due to the low Coulomb barrier in reactions involvning the light elements.

Further, from an analytical point of view there has been a lack of highly sensitive methods for determining light elements in various matrices.

However, CPAA is not necessarily restricted to light elemental analysis, as is exemplified later in this chapter.

A feature of charged particle behaviour, which is of special interest in connection with material surface research e.g. in diffusion studies, is their rapid deceleration on penetration into metals. This property renders them particularly suitable for the analysis of elements in surface layers formed by the interaction between metals and light elements, such as carbonized steel surfaces. CPAA can also be used for analysis of the depth distribution of light elements down to 10-15 µm beneath the surface (profile measurements) without altering the sample (non-destructive analysis).

Activation is promoted by irradiation with protons, deuterons, tritons, helium-3, or α-particles generated in an accelerator, for example a Van de Graaff accelerator. The prompt products of the nuclear reaction - protons, neutrons, α-particles or γ-quanta - together with the γ-radiation emitted by the nuclides thus formed can be registered to form a basis for characterizing specific elements.

11.1.1 DETECTION LIMITS

As pointed out above, the particular significance of
analyses of the light elements in a metal matrix is their
appreciably lower Coulomb barrier by comparison with that
of the heavier metallic elements involving higher reaction
rates. The Coulomb barrier (eq. 1.14) is proportional to the
atomic number Z of the bombarded atom. This factor influences
the degree of sensitivity to traces of light elements such as
carbon and oxygen in the metal surfaces. Table 11-1 below
gives some idea of the detection limits of the CPAA technique,
although it should be stressed that the figures refer to a
somewhat idealized case, namely that presented by pure foils[11.1].
In practice it can be difficult to attain these high levels of
sensitivity, partly because of interference from different elements which may form the same nuclide.

Table 11-1
Detection limits for various light elements[11.1].

	Energy (MeV)	Detection limit (in µg/g)			
		Boron	Carbon	Nitrogen	Oxygen
Protons	10-15	0.001	0.01	0.0005	0.01
Deuterons	10-15	0.0001	0.001	0.0002	-
^{3}He-particles	10-20	-	0.001	-	0.001
^{4}He-particles	40	-	0.01	-	0.001

11.1.2 NUCLEAR REACTIONS

We will now consider some nuclear reactions that are
used in CPAA work, either for surface studies or for depth
distribution evaluation, and describe the various possibilities for nuclear detection on which the quantitative chemical
analysis is based. In fig. 11-1 is given various nuclear
reactions used in CPAA of fluorine, carbon and oxygen. The
nuclear methods of identification of the elements are also
presented in the figure.

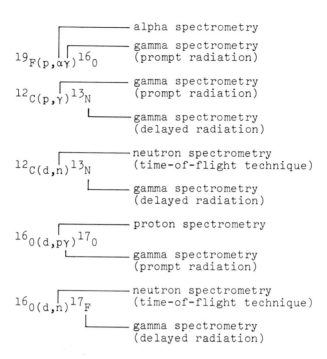

Fig. 11-1   Nuclear reactions used for CPAA of fluorine, carbon and oxygen[11.6, 11.10, 11.25].

11.1.3 THE ACTIVATION YIELD

The activity formed in a specific nuclear reaction is proportional to the amount of an element present in the sample as well as to the particle flux and the cross section. For non-resonant reactions the cross section often changes relatively slowly with energy. In this case, when the target thickness is small compared to the penetration depth of the incident charged particle, the yield, Y at an incident particle energy T is (see eq. 3.24):

$$Y = N\sigma(T)\Delta x \qquad \text{(thin target)} \qquad (11.1)$$

where Y is the number of reactions initiated per unit particle flux, N is the number of target atoms per cm^3, Δx is the target thickness in centimeters and $\sigma(T)$ is the reaction cross section in cm^2 at energy T. If the target is

thicker than the range of the incident particle the formula
becomes (eq. 3.87).

$$Y = N \int_{o}^{R} \sigma(x)dx \qquad \text{(thick target)} \qquad (11.2)$$

where R denotes the penetration depth of the incident particles in the target and $\sigma(x)$ is the reaction cross section
as a function of depth $\underline{x}$.

   In the literature, cross sections are usually given as a
function of particle energy in the form of so-called excitation functions. Such functions are represented either as
differential (cross section as a function of angle) or as
integral (cross section integrated over 4π geometry).

   The conversion of $\sigma(T)$ to $\sigma(x)$ can easily be performed
with a knowledge of range-energy relationships[11.2]. When
$\sigma(x)$ is known the above integral (eq. 11.2) can be evaluated
numerically and the quantity N can be determined (see sec.
3.9).

   Ricci and Hahn[11.3] have simplified such calculations
by introducing a thick target cross section, $\bar{\sigma}$, defined as:

$$\bar{\sigma} = \frac{1}{R} \int_{0}^{R} \sigma(x)dx \qquad (11.3)$$

A substitution for the integral in eq. 11.2 gives:

$$Y = N\bar{\sigma}R \qquad (11.4)$$

The significance of this expression is the possibility of
eliminating the parameter $\bar{\sigma}$ in comparison measurements.

11.1.4 HEAT AND DIFFUSION EFFECTS IN THE SAMPLE

   During the irradiation of a surface with charged particles considerable amounts of energy are dissipated into
rather small volumes, due to the slowing down of protons
or other particles. Experiments have revealed that when
thin iron foils are irradiated with protons of a few MeV at
currents of about 5 μA, they often melt in the zone exposed

to the particles. The heat evolved might further give rise
to diffusion effects, e.g. for carbon, resulting in carbon
losses from the volume analysed. In steel a transport effect
is already noticeable at a few hundred degrees C. At 500°C
carbon may move a distance of 5 μm in a few minutes[11.4,11.5].
In depth distribution analysis, layers are studied which have
a resolution of the order of magnitude of 1 μm. Consequently
it is necessary to take into account diffusion effects. For
this purpose the temperature rise in the irradiated volume
has to be known.

The heat flux incident on the irradiation spot is
assumed to be uniformly distributed over the area of the
spot; it is thus given by $P/\pi a^2$, where P is the beam power
(usually of the order of 1-50 W), and a is the radius of
the spot.

The volume in which the heat is generated takes the
form of a thin disk. The maximum temperature, t_{max}, is derived
from the following formula:

$$t_{max} = \frac{P}{\pi K a} \qquad\qquad (11.5)$$

where K denotes the thermal conductivity of the sample. In
connection with proton and deuteron irradiation the temperature rise in an iron specimen of thickness 5 mm and collimator hole 8 mm was measured at the front and at the rear face
at various beam powers[11.6]. The results are presented in
fig. 11-2.

From a consideration of the diffusion processes of
carbon in steel, the mass transport induced by a thermal
gradient, referred to as the Soret effect, is the temperature
limiting factor if analytical error is to be avoided. Thus
temperatures should not be allowed to exceed 300°C. Practically, this is achieved by using large spot radii (of about 4 mm)
with beam powers of the order of 10 W. In order to further
ensure the absence of thermal diffusion, forced cooling of
the specimen can be employed to promote heat drainage.

In practice the effect of heat dissipation due to proton
irradiation which results in carbon diffusion has been observed in connection with the evaluation of the carbon gradient

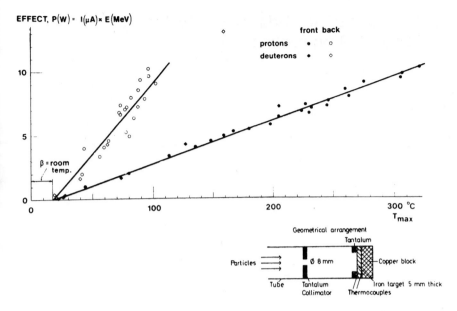

Fig. 11-2   Temperature distribution in an iron specimen
             during irradiation[11.6].

in steel samples (fig. 11-7). Originally, a sample contained
carbon distributed in accordance with the concentration gradient represented by Δ in fig. 11-7. After several runs this
gradient was radically altered so that a sample was obtained
that was homogeneous as regards the carbon content in the
surface, due to diffusion effects caused by heating during
irradiation.

11.1.5   GENERAL APPLICATIONS

     A low energy accelerator of the Van de Graaff type con-
stitutes a versatile instrument for chemical analysis when
the following techniques are combined, and it may then compete economically with other instruments[11.7].

-    Charged particle activation analysis,
-    Proton induced X-ray analysis,
-    Rutherford scattering,
-    Microprobe analysis,
-    Channelling studies.

365

11.2 Resonance methods for evaluation of concentration profiles

The nuclear reaction

$$^{19}F(p,\alpha\gamma)^{16}O$$

demonstrates a sharp resonance in the excitation function at a proton energy of 1 375 keV[11.8] (fig. 11-3). Such resonances can successfully be used for the determination of the depth distribution of an element below the surface. Amsel and Samuel[11.9] demonstrated the use of nuclear resonance reaction for depth profiling measurements in a study of anodic oxidation considering an ^{18}O-enriched oxide layer. Möller and Starfelt[11.10] used the resonance technique to evaluate the fluorine distribution below the surface of zircaloy samples.

11.2.1 PROTON-RESONANCE ACTIVATION OF FLUORINE

As an example of the resonance method we shall study in more detail the determination of the concentration of fluo-

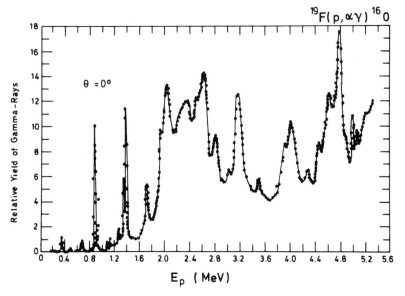

Fig. 11-3   Excitation function for the reaction $^{19}F(p,\alpha\gamma)^{16}O$[11.8].

rine in various layers beneath a zircaloy surface. For this
purpose the specimen is irradiated with a proton beam in a
Van de Graaff accelerator. Measurement of γ-rays is performed
on the prompt γ-radiation (6.1 and 7.1 MeV) emitted during
the course of the reaction at selected proton energies.

The process can be best understood by first considering
the effects of irradiating the metal with protons of various
selected energies. Thus if the protons in the beam have an
energy which coincides with the resonance energy (1 375 keV)
then the fluoride content at the surface will be revealed.
The protons penetrating below the surface will experience an
immediate change to lower energies, brought about by collision with the metal matrix, and will therefore be unable to
take part in reactions characterized by the 1 375 keV resonance energy. Other resonances in the excitation function
will, however, contribute to activity production. An increase
in the proton energy will shift the site of this reaction to
a layer beneath the surface at a depth determined by the mc-
derating action of the matrix. Thus it can be imagined that
for each metal there is a strict relationship between the
energy of the proton beam and the thickness of metal required
to reduce it to the level corresponding to the resonance
energy. The yields of prompt γ-radiation and the energy of
the proton beam can therefore be readily derived in terms
of the amount of fluorine at a given level below the metal
surface (see fig. 11-4 and sec. 3.9).

In the instance of fluorine in zircaloy the limit of
detection is of the order of 0.01 µg cm^{-2}, while the resolution of the separate layers is of the order of 0.1 µm close
to the surface. Owing, however, to the effect of straggling.
the layer resolution is reduced with increasing depth of penetration into the metal matrix.

According to the same principle Ahlberg et al.[11.11]
determined the distribution of fluorine and nitrogen in the
enamel of teeth of rats using a Van de Graaff accelerator. In
the case of fluorine the resonance in the excitation function
at 872 keV were considered, whereas for the proton induced
reaction in nitrogen, the resonance occurring at 897 keV was
used. Detection limits at the ppm level have been reported in

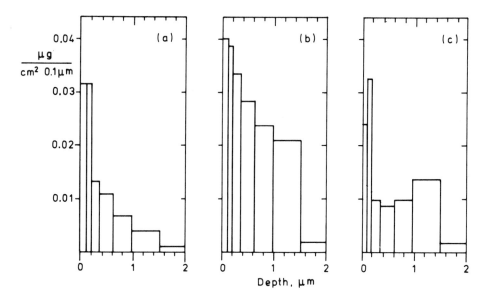

Fig. 11-4 Depth distributions of fluorine in different samples of zircaloy, which were ground, pickled or
electropolished and undergone autoclave treatments.

this investigation. Close to the surface the resolution of
the analyzed layers was of the order of 1 000 Å.

An interesting contribution to the profile measurement
technique has been given by Leich and Tombrello[11.12], who
have reversed the roles of target and beam in the reaction
$^{19}F(p,\alpha\gamma)^{16}O$ in a study concerning the evaluation of hydrogen
in lunar samples. Thus, the "inverse" reaction: $^{1}H(^{19}F,\alpha\gamma)^{16}O$
was initiated with fluorine ions in the energy region 16 -
- 18 MeV. They used the resonance occurring at 872 keV energy.
Fig. 11-5 below shows the hydrogen distribution in a lunar
fragment determined with this technique [11.12].

11.2.2 PROTON-RESONANCE ACTIVATION OF CARBON

We shall now study another example of proton resonance
activation for the purpose of profile evaluation. The determination of carbon by proton activation analysis can be carried out using the nuclear reaction:

$^{12}C(p,\gamma)^{13}N$

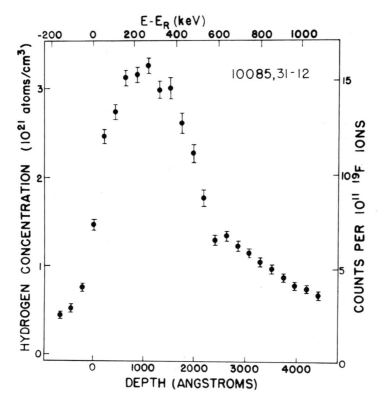

Fig. 11-5 Concentration profile for hydrogen in a lunar
           sample[11.12].

The excitation function of this reaction exhibits two resonances, namely one at 460 keV and one at 1 700 keV, in the
relevant energy region. The first of these resonances gives
rise to a transition in which γ-quanta of 2.36 MeV are emitted while the second provided 3.51 MeV quanta in an analogous
fashion. By making an appropriate choise of proton energy
these emissions can be used to determine the carbon concentration in surface layers of steel samples[11.13, 11.14], or
at different depths below the surface of the sample[11.6].
    Two measurement techniques for evaluating carbon can be
used:

a) the registration of the prompt γ-radiation which is emit-
    ted as a result of the reaction (2.36 and 3.51 MeV).

b)   the measurement of the activity associated with the reac-
     tion product ^{13}N, which is a positron emitter with a
     half-life of 10 min.

Of the above methods, a) is the simplest since information
regarding carbon concentration can be obtained by activity
measurements at each chosen proton energy level which corresponds to a specific depth in the sample. The method is limited, however, by its low sensitivity, estimated as being about
1% C in a metal like steel.

Method b) requires that the decay of ^{13}N can be followed
up after each irradiation in order to permit correction for
the contributions to the activity made by interfering nuclei.
The detection limit for carbon in metal surfaces is, in this
instance, about 0.05%. Depth resolution in the carbon containing layers calculated by eq. 11.7 falls from roughly 0.3 μm
for those near the surface to 2 μm at a depth of 15 μm (see
fig. 11-6).

Carbon profiles in carburized and decarburized samples
of steel have been measured with such a technique to a depth
of about 15 μm. Typical gradients thus obtained are shown in

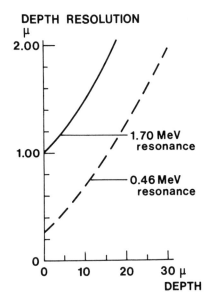

Fig. 11-6 Depth resolution versus depth in the steel matrix[11.6]

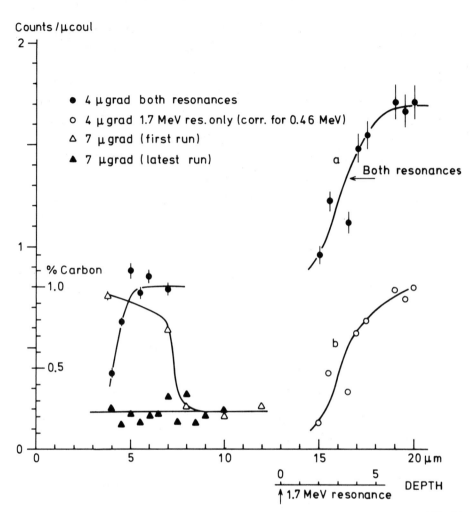

Fig. 11-7   Carbon concentration profiles in steel surfaces[11.6].

fig. 11-7. The method is, however, time consuming. (Concerning the anomalies associated with the gradient for 7 μgrad, see sec. 11.1.4 of this chapter dealing with heat and diffusion effects in the sample).

11.2.3 DEPTH-ENERGY RELATIONSHIP AND RESOLUTION

The relationship between the depth below the surface of an analyzed layer and the particle energy can easily be cal-

culated from an energy-range table[11.2], resonance energies
can be found by inspection of the excitation function[11.8].

In this connection it should be emphasized that the
depth of the analyzed layer, $\Delta R(T)$, is not equal to the total range of the particle at the initial energy, but rather
to the difference between the total range $R(T)$ and the range
of a particle with the initial energy T_r (resonance energy).
The following relationship is then valid for the resonance
method:

$$\Delta R\ (T) = R\ (T) - R\ (T_r) \qquad\qquad (11.6)$$

The thickness broadening of a layer at a defined depth
is described in terms of the following three parameters for
a proton induced resonance reaction:

Resonance energy width; the standard deviation of the resonance distribution σ_R. For a gaussian distribution

$$\sigma_R = \Gamma/2.355 \qquad\qquad (11.7a)$$

where Γ is the full width at half maximum.

Energy straggling; the standard deviation σ_T of the mean
residual energy after passing an absorber with thickness
ΔR is given by (eq.5.6)

$$\sigma_T^2 = 4\pi e^4 N Z z^2 \Delta R/(4\pi\varepsilon_o)^2 \qquad\qquad (11.7b)$$

where N is the atom density, Z is the atomic number of
the stopping atoms, and z is the charge number of the
incident particles.

Energy resolution in the beam σ_B.

The total energy broadening $2\sigma_D$ of the protons at a depth Δx
is now obtained from eq. 8.9. Usually the effect or energy resolution is so small that it can be neglected. The following
relationship is therefore applicable

$$2\sigma_D = 2(\sigma_R^2+\sigma_T^2+\sigma_B^2)^{\frac{1}{2}} \cong 2(\sigma_R^2+\sigma_T^2)^{\frac{1}{2}} \qquad\qquad (11.8)$$

This energy broadening of course corresponds to a thickness
broadening the size of which is obtained from stopping power
diagrams (fig. 5-2) and equations (eq. 5.3).

At the surface, the layer thickness for protons is de-

fined approximately by the resonance alone, but a deeper
levels straggling becomes the predominant factor. An estimate
of energy straggling can be obtained using eq. 5.6 deduced
by Bohr, which is valid under restricted conditions[11.15].

The effect of energy straggling has further been discussed in the comprehensive compilation of CPAA by Wolicki[11.16].

11.3 Non-resonance methods

The resonance technique applied to profile measurements
requires a successive variation of the energy of the incident
particle in order to determine the profile. The yield of
emitted γ-radiation or particles provides information about
the concentration of the element present in layers at various
depths. The resonance method can be used for a limited number
of charged particle induced reactions (see e.g. ref. 11.8).
Consequently, in concentration profile measurements with e.g.
deuterons, non-resonant methods have to be considered.

A charged particle continuously loses kinetic energy on
its way through matter. Consequently nuclear reactions induced
by monoenergetic charged particles will give rise to product
particles with various energies, depending on where in the
target the nuclear encounter has taken place. Thus, the energy
of the emitted particles can give information about the depth
profile of the element, whereas their yield is proportional to
the concentration of the element.

Methods based on registration of emitted particles have
obviously the advantage that depth profiles can be obtained
in a single run. The result is dependent on the energy of the
incident particle to the extent that the experimental conditions are optimized as regards desired range and resolution.

A variety of charged particle induced techniques for the
study of depth profiles are described in the review of
Wolicki[11.16].

11.3.1 DEUTERON ACTIVATION

A method was previously outlined for the evaluation of
carbon concentration in steel using proton activation. Using
a deuteron activation technique, the analysis of carbon can

be accomplished faster and also with lower detection limits. Let us first consider the reaction:

$$^{12}C(d,p)^{13}C$$

The excitation function for this reaction is rather non-resonant and is found e.g. in ref. 11.8.

Pierce used this reaction to evaluate carbon concentration profiles in steel for routine work[11.17]. He measured the emitted protons with a surface barrier detector using the proton spectroscopic method. In a section of a steel sample, profiling up to 400 μm was carried out in less than 2 hours through scanning with deuteron microprobe techniques (microprobe analysis will be discussed later). Evaluation of carbon concentration profiles at depths up to several μm without cutting the sample can be achieved by considering the energy retardations of both deuterons and protons in the sample.

The experimental set-up for use with surface barrier techniques is given schematically in fig. 11-8. Further, the

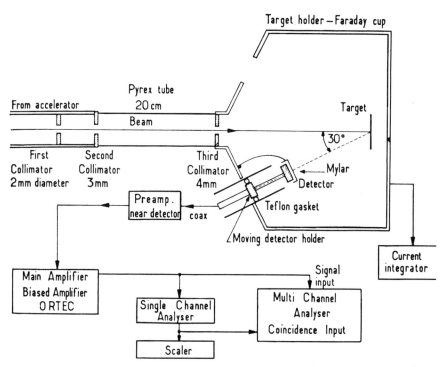

Fig. 11-8   Arrangement for detection of charged particles[11.18].

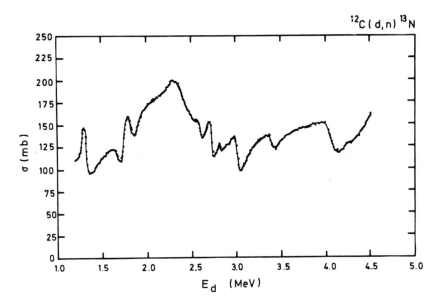

Fig. 11-9   Excitation function for the reaction
^{12}C(d,n)^{13}N$^{(11.8)}$.

following non-resonant (d,n) reaction can be used for carbon
analysis:

$$^{12}C(d,n)^{13}N$$

The excitation function for this reaction is given in fig. 11-9.
    The measurement of carbon can then be done using either
of the following methods.
a)   Measurement of the induced ^{13}N activity (annihilation
    radiation).
This technique permits the detection of carbon contents down
to the 0.01% level in steel surfaces. It is not, however,
useful for studies of depth distribution[11.13].
b)   Measurement of the energy of emitted neutrons.
Using the neutron spectroscopic method which will be described
in more detail below, the carbon concentration profile can be
obtained by determining the energy spectrum of the neutrons
that are produced in the nuclear reaction[11.16, 11.19].

11.3.2 NEUTRON SPECTROMETRY

Complex problems are associated with the evaluation of the retardation of both incident and emitted charged particles for profile measurements; these problems can be avoided by **substituting protons for neutrons and measuring the energy** spectrum of the emitted neutrons. Thus microanalysis has been carried out using (d,n) reactions in combination with a neutron time-of-flight technique (sec. 9.5.1) on light elements in metal surfaces and in gases[11.20, 11.21].

The irradiation of carbon, nitrogen and oxygen present in metal surfaces with deuteron beams produces neutrons through (d,n) reactions occurring with different Q values. Since neutrons are neutral particles they are scarcely affected by the metal surface. Accordingly the neutron energy subsequent to the nuclear reaction will constitute a measure of the deuteron energy at a given depth under the surface. Measurement of the intensity of neutrons of a given energy thus provides a measure of the extent of the nuclear reaction occurring at the depth corresponding to this energy. By scanning the intensities at a range of neutron energies it now becomes possible to form a picture of the concentration profile for the specified light element using a single, though suitable chosen, deuteron bombarding energy.

Surface studies of oxygen, carbon and nitrogen in metals using a time-of-flight measurement technique have previously been accomplished by Möller et al[11.20]. However, in a study by Lorenzen[11.19] this technique has been improved so that it allows the performance of fast and simple routine analysis of carbon in steel surfaces. The main advantage of his technique rests on the use of a "standard" which is homogeneous with respect to the element to be analyzed in the surface.

Thus, the time-of-flight spectrum for the sample with the unknown carbon gradient is measured together with that of the standard sample, which is homogeneous with respect to carbon. The two spectra are then divided channel by channel whereby the carbon gradient of the unknown sample is obtained after normalization and subtraction of the background.

Fig. 11-10 presents examples of such spectra revealing the carbon concentration profile. It should be mentioned that

376

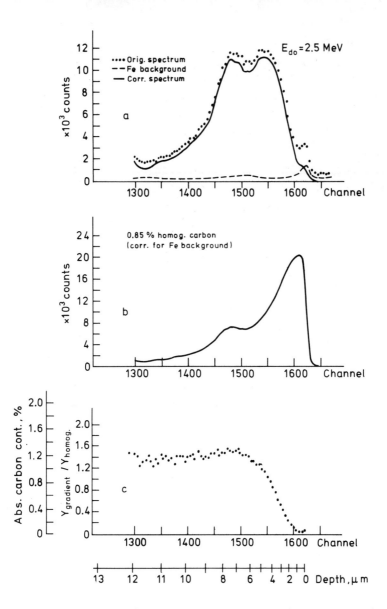

Fig. 11-10   Time-of-flight spectra produced in the reaction
$^{12}C(d,n)^{13}N$ (11.19).

    a) spectrum for the sample with unknown carbon gradient

    b) spectrum of standard sample with homogeneous carbon constant

    c) the two spectra divided channel by channel.

these measurements make use of a liquid scintillator detector for neutron registration. The time-of-flight technique has also been applied to the analysis of oxygen and nitrogen concentration profiles in steel surfaces using (d,n) reactions.

The depth resolution afforded by the neutron-spectrometric method is mainly dependent on the time resolution of the spectrometer, which in combination with the stopping power of the matrix provides an instrumental resolution.

Thus the total energy broadening, $2\sigma_D$, is composed of the instrumental resolution part, σ_I, and the energy straggling parameter, σ_T:

$$2\sigma_D = 2(\sigma_I^2 + \sigma_T^2)^{\frac{1}{2}} \qquad (11.9)$$

In the $^{12}C(d,n)^{13}N$ reaction, the depth resolution corresponding to the energy broadening of analyzed steel layers amounts to the order of 1 μm up to 15 μm depth[11.19].

11.3.3 NITROGEN DETERMINATION IN SINGLE SEEDS

Using (d,p) and (d,α) reactions in ^{14}N, Sundqvist et al.[11.22] have developed a method for the analysis of the nitrogen content of single seeds.

They irradiated their samples in a tandem Van de Graaff accelerator (sec. 4.3.1), and measured the emitted proton and α-particles with a surface barrier detector.

The method has been reported to allow determination of the depth distribution of nitrogen at points below the surface up to 0.3 mm.

11.3.4 ^{3}He-ACTIVATION

The application of ^{3}He-activation in CPAA was introduced by Markowitz and Mahoney[11.23] in a study of oxygen determination. The following reaction was used:

$$^{16}O(^3He,p)^{18}F$$

The positron emitter, ^{18}F, is registered. Detection limits of the order of part per billion (ppb) i.e. ($\mu g \cdot kg^{-1}$) where estimated in an aluminium matrix. In the case of practical analysis it should, however, be stressed that various interfering reactions would reduce the detection limits in real samples. Thus, for oxygen analysis based on ^{18}F-detection, the same nuclide is produced from fluorine through the following reaction (see fig. 11-11):

$$^{19}\text{F}(^3\text{He},\alpha)^{18}\text{F}$$

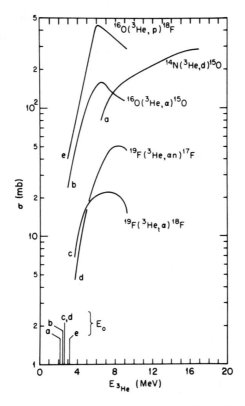

Fig. 11-11   Excitation functions for helium induced reactions in oxygen and fluorine[11.24].

379

11.4 Surface and depth analysis

11.4.1 SEMI-BULK ANALYSIS

The above account shows that two main features of CPAA should be emphazised:

- Accomplishment of highly sensitive analysis of light elements in surfaces
- Accomplishment of depth distribution evaluation below the surface.

With a low energy accelerator of the Van de Graaff type, surface and depth distribution evaluation is carried out at particle energies of a few MeV. Surface layers of fractions of a μm can then be studied by the resonance method.

However, at particle energies exceeding, let us say, 10 MeV the concept of surface analysis is no longer valid. Thus the range of a 5 MeV proton in iron exceeds 100 μm. The concept of semibulk analysis seems more adequate in such cases.

With increasing particle energy the "surface depth" analysed is increased. In fig. 11-12 below this effect is demonstrated for oxygen in a zircaloy matrix using (d,p) and (d,n) reactions[11.25].

In the analysis of light elements in metal surfaces the specimen is irradiated with charged particles produced in various types of accelerators at energies up to about 50 MeV[11.26-11.32].

In a study by Barrandon et al.[11.32] the analysis of moderately heavy elements such as Ti, V, Cr, Fe, Ni, Cu and Zn using protons with energies up to 20 MeV has been accomplished. Several matrices were investigated. In aluminium and tantalum detection limits of the elements were reported to be lower than one part per million (ppm).

With a proton activation technique using protons in the range of 11-15 MeV, Krivan[11.33] determined chromium, nickel and copper by means of the following reactions:

$$^{52}Cr(p,n)^{52m}Mn, \quad ^{60}Ni(p,n)^{60}Cu \text{ and } ^{63}Cu(p,n)^{63}Zn$$

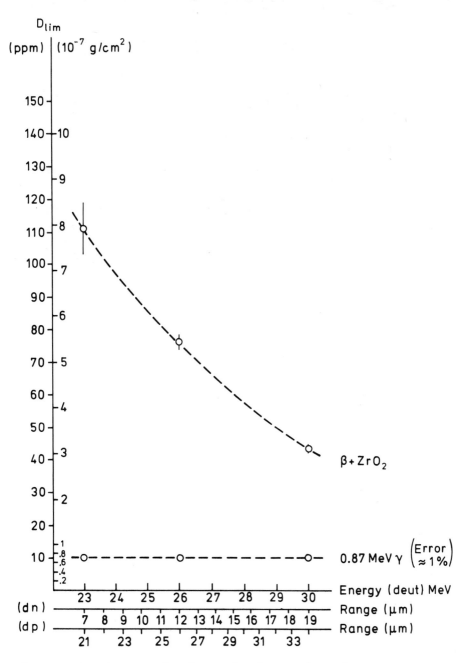

Fig. 11-12   Yield of (d,n) and (d,p) reactions in oxygen
versus deuteron energy corresponding to various
depths (range)[11.25]. The detection limit of
oxygen using the (d,n) reaction decreases with
increasing deuteron energy.

Detection limits for chromium, nickel and copper in a pure cobalt matrix were reported to be 0.4, 3.9 and 12.8 ppm respectively.

Detection limits in the range of $\mu g \cdot g^{-1}$ - $\mu g \cdot kg^{-1}$ have been reported using CPAA techniques. However, in practical analysis it might be difficult to attain these levels due to interfering reactions.

Moreover, proton activation analysis has been applied to determine lead concentrations in airborne particulate matter collected on filter papers according to Desaedeleer et al[11.34]. The measurements were based on identification of the radionuclide ^{204}Bi produced through 40 - 50 MeV proton interaction with lead through the following reactions:

$$^{206}Pb(p,3n)^{204}Bi, \quad ^{207}Pb(p,4n)^{204}Bi \text{ and } ^{208}Pb(p,5n)^{204}Bi$$

Up to 40 samples were bombarded simultaneously for the period of half an hour. Detection limit below the ng level of lead was reported.

11.4.2 THE PROTON MICROPROBE

The electron microprobe permits micro-analyses of metal surface specimens with a resolution of the order of 1 μm. The basis of the method is excitation by electrons and subsequent measurement of the characteristic X-rays produced from the atoms of the elements present. This technique, which is widely used in the metallurgical industry, is not, however, well suited for quantitative measurements of light elements like carbon.

An extension of this technique is represented by SIMS (Secondary Ion Microprobe Spectroscopy) which allows determination of the spatial distribution of most elements in various surfaces. The system is based on the effect of sputtering with heavy ions and analysis with mass spectrometry, e.g.[11.35]

As a third alternative, microprobe analysis of metals can be accomplished by means of proton microbeam activation techniques. The proton microprobe technique became possible when Cookson and his collaborators at Harwell succeeded in producing a proton beam of a diameter of about 4 μm, using collimators and focusing with the aid of quadrupole lenses (11.36). By scanning the proton beam over the sample surface or by moving the sample so that the beam strikes the exposed surface in a regular pattern, it becomes possible to build up a picture of the constitution of the metal surface.

The proton microbeam can be employed for analysis in different ways:

1 Analysis of light elements is based on charged particle activation analysis according to the principles previously outlined.

2 For the excitation of atoms with the subsequent registration of the emitted X-rays, i.e. the so-called PIXE technique, introduced by Johansson, Akselsson and Johansson(11.37). This method is suitable for the characterization of medium and heavy weight elements (see sec. 11.5).

With proton-induced X-ray techniques and thin target arrangements extremely low detection limits can be obtained for several elements(11.37).

However, in studies aimed at determining the distribution of a certain element in a metal surface using proton microprobe and X-ray techniques, detection limits are considerably increased due to the high background levels that are introduced when normally thick targets are exposed to the proton beam. Detection limits at levels in the range of 10 - 1 000 ppm are achieved under conditions affording the simultaneous measurement of several elements in steel samples with this technique(11.38). The general principle of a nuclear microprobe employing focusing technique is presented in fig. 11-13. The development of microprobes since the advent of the Harwell Microprobe has followed two main routes, viz. those cases where one has stuck to the original format with four quadrupoles arranged to form a Russian quadruplet and those where another system solution has been chosen. Among those operated with a Russian quadruplet are the Melbourne Proton Microprobe(11.40), the Studsvik Nuclear Microprobe(11.41) and the Namur Microprobe(11.42). The Heid-

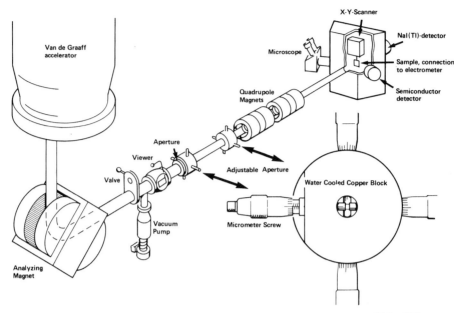

Fig. 11-13 General principle of a proton-microprobe[11.39].

elberg Microprobe[11.43] using two quadrupoles giving an asymmetric picture has been model for e.g. the Karlsruhe, Bochum, Amsterdam and GSI (Gesellschaft für Schwerionen-forschung, Darmstadt) microprobes. Typical beam sizes in both groups are 2-5 μm either circular or rectangular[11.44]. The state of the art is represented by the Oxford 1 μm Micro-probe[11.45].

The field of application of nuclear microprobes is vast and includes biology, medicine, metallurgy, mineralogy and en-vironmental studies. The Studsvik microprobe is mainly devo-ted to life science studies and allows a resolution of the beam of about 5 μm[11.41]. It has been used to determine the depth distribution of fluorine and, simultaneously, to char-acterize the superficial distribution of elements in both healthy enamel and in regions adjacent to amalgam restora-tions according to Lindh and Tveit[11.46]. Also the distri-bution of lead within structural units (osteons) of bone tissue from industrially exposed workers could be assessed with such techniques[11.47]. Microprobe assessment of concentration profiles within osteons of human femur is outlined in fig. 11-14.

Elemental mapping has been accomplished by Legge and Mazzolini with a scanning proton microprobe[11.40] and an example of such mapping of various elements in an epidermal strip of a wheat leaf is presented in fig. 11-15. Microanalysis of cells and tissues could advantageously be carried out with the mapping technique.

11.4.3 ACTIVATION WITH HIGH ENERGY PROTONS

Consider further the application of high energy particles such as protons of about 200 MeV produced in cyclotrons (e.g. the 185 MeV Uppsala synchrocyclotron). For CPAA studies

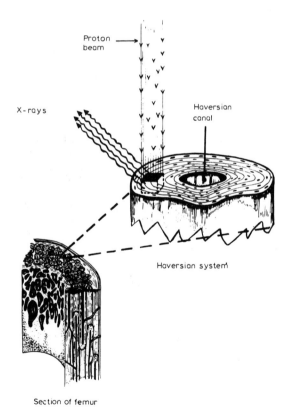

Fig. 11-14 Microprobe analysis of a section of human bone (femur)[11.47].

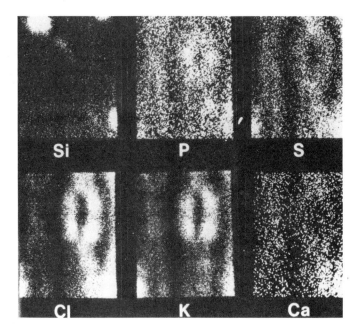

Fig. 11-15 Elemental mapping with a scanning proton microprobe[11.40].

this type of technique seems to be suitable as regards bulk
analysis, since the retardation effects in this energy region,
which affect the activation rate, can be neglected in small
samples. The range of 185 MeV protons in aluminium exceeds
1 cm. Nondestructive analysis of larger pieces can also be
achieved with this technique by sweeping the focused proton
beam in a recticulate pattern. On the other hand, high energy
proton irradiation gives rise to the formation of several
radionuclides in the matrix through spallation processes,
which considerably reduces the detection limits of various
elements[11.48].

Let us consider the analysis of carbon and oxygen in
aluminium matrices using high energy proton activation. Thus,
185 MeV proton irradiation of carbon and oxygen produces the
positron emitters ^{11}C and ^{15}O by means of the following reactions:

$$^{12}C(p,pn)^{11}C$$

$$^{16}O(p,pn)^{15}O$$

$$^{16}O(p,3p,3n)^{11}C$$

In the aluminium matrix the same nuclides are also formed through spallation. Consequently it is unlikely that detection limits of orders better than 10^2 ppm - 10^3 ppm oxygen or carbon in aluminium samples will be obtained[11.49]

11.4.4 CHANNELLING

About two decades ago the discovery was made that the penetration of energetic ions in a crystalline material exhibits a strong directional dependence, i.e. penetrating ions can be steered through the lattice by correlated collisions (see sec. 5.3). In this process, the ions suffer a lower rate of energy loss and are thus able to penetrate deeper into the material. In amorphous solids, the penetration profile shows an approximately gaussian form, whereas polycrystalline samples exhibit a skewness with a tail of deep penetration, as observed by Davies et al.[11.50] in range studies of 40 keV ions in polycrystalline aluminium.

The effect of channelling has to some extent found applications in CPAA studies where the localization of various impurity atoms in single crystals is desired. Thus, quantitative information about the fraction of impurity atoms located on different lattice rows can be obtained by bombarding the crystal with the particles in certain selected directions[11.51].

For analytical purposes channelling is often combined with back-scattering measurements. Thus, substitutional or interstitial impurity atoms in crystals can be distinguished by their different back-scattering properties[11.51].

Matzke et al.[11.52] combined channelling with back-scattering measurements and CPAA in order to localize excess oxygen in various crystal structures of uranium oxide. Thus, uranium was detected by deuterium and helium back-scattering and oxygen through the nuclear reaction $^{16}O(d,p)^{17}O$ (proton spectroscopy). This channelling study revealed that the uranium lattice was only slightly disturbed by excess oxygen.

11.4.5 ELASTIC SCATTERING FOR THE ANALYSIS OF HEAVY ELEMENTS IN SURFACES

Scattering of charged particles by heavy atoms in a surface provides a valuable complement to the surface analysis described above for light elements, and it is therefore deserving of mention although nuclear reactions are not involved.

When a beam of monoenergetic particles strikes a metal surface, a fraction of the particles experience elastic collisions. Departure from the incident path, scattering, and loss of energy to an extent depending upon the relative masses of the colliding atoms, are the natural consequences of this process. The energy of the scattered particles is also dependent upon the angle of scatter. The total number of scattered particles is a function of the number of atoms in the surface which produces the scattering effect.

Thus, by measuring the energy and counting the number of the scattered particles at defined angles, the chemical identity and number of atoms in the solid can be determined. Consider scattering for surface atoms; the energy of the scatte-

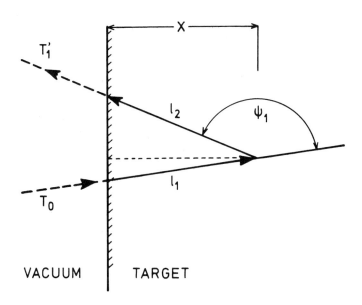

Fig. 11-16 Elastic scattering below surface[11.53].

red particle T_1' is related to the incident energy T_1 by the equation (conservation of energy and momentum), see eq. 3.4d:

$$T_1' \simeq \left[\frac{(m_1 \cos\psi_1 \pm m_2)}{m_1+m_2} \right]^2 T_1 \qquad (11.10)$$

if $\quad \dfrac{m_1 \sin\psi_1}{m_2} \ll 1$

where m_1 is the mass of the incident ion, m_2 is the mass of the struck atom, and ψ_1 is the laboratory scattering angle.

If the process takes place beneath the surface the energy losses of both the incident and the emerging particles at paths ℓ_1 and ℓ_2 must be taken into account (see fig. 11-16). This has been treated in detail by Machintosh and Davies (11.53). A practical analytical case of elastic scattering

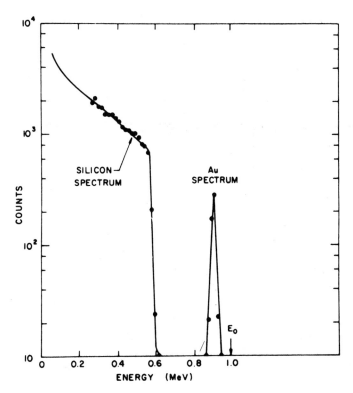

Fig. 11-17 Spectrum of helium ions scattered from a silicon surface coated with gold(11.53).

is demonstrated in fig. 11-17. This figure shows the energy spectrum of helium ions scattered from a silicon target whose surface has been covered with a monolayer of gold. The gold peak is found at a higher energy than the spectral part produced by silicon, which is consistent with eq. 11.10. Erikson et al.[11.54] used the elastic scattering technique to obtain information about depth distribution profiles of various components between coatings and paper substrates.

Abel et al.[11.55] report detection limits of various heavy elements such as Au, Hg and Pb at the level of 10^{10} atoms$\cdot$cm^{-2}. Scattering experiments were achieved by means of a 2 MeV Van de Graaff accelerator using singly charged ^{12}C and ^{16}O beams with currents up to 0.5 µA. Surfaces of iron and silicon were examined.

The scattering technique further allows the analysis of various light elements. Thus, with 16 MeV protons incident on an air particulate specimen Akselsson et al.[11.56] analyzed C, N, O, Na, Al, Si and S with the scattering method.

Moreover, Rutherford back-scattering in crystal analysis has been reviewed by Götz and Schwabe[11.57].

11.5 Particle-induced X-ray emission (PIXE)

The advantages using charged particles in the analysis of constituents in surfaces have previously been pointed out in this chapter dealing with charged particle activation technique. Analysis of light elements is thereby suitably accomplished through measurement of reaction products like charged particles, neutrons and γ-radiation arising from various nuclear reactions. In such works a Van de Graaff accelerator is often used as irradiation facility. Such accelerators can also suitably be used as devices for producing X-rays through proton impact. Nuclear reactions can consequently be combined with the non-nuclear X-ray emission technique in chemical analysis. It should further be stressed that the back-scattering technique is suitably accomplished with the Van de Graaff accelerator as previously pointed out.

A comprehensive review of theory as well as of analytical

applications of particle induced X-ray emission (PIXE) has
been given by Johansson and Johansson[11.58].

11.5.1 PRODUCTION OF X-RADIATION

Production of X-rays is conveniently accomplished by
means of electrons. They are used for analytical purposes
mainly in connection with electron microscopy. The electron
beam when transversing the sample produces X-rays which are
recorded by means of a crystal spectrometer or a semiconductor
detector, thereby giving information about the composition of
the sample.

The X-ray production cross section for electrons in the
10 - 100 keV range is about the same as for protons in the
MeV range. In the electron case the background is dominated
by the direct bremsstrahlung produced by the beam, which is
much greater than the proton bremsstrahlung because of the
smaller mass of the electron. In the proton case secondary
electron bremsstrahlung contributes heavily to the background
at lower energies (see fig. 11-20). Still electron beams give
an increased background compared with proton beams of the
order of 3 to 4 magnitudes. Hence, the limit of detection is
rather high, about 0.1%. An electron microprobe can consequently not assay the trace elements but only the main constituents of a sample. Because of the thin samples and the fine
focus of the beam, extremely small amounts of various elements
can be detected, i.e. about 10^{-16} g.

The interaction between accelerated, heavy charged particles and target atoms may lead to the emission of characteristic X-rays. This phenomenon, involving the removal of at
least one inner-shell electron, has been studied since its
discovery by Chadwick in 1921. In theoretical descriptions
the process is visualized as coulombic and its major features
are well understood. It is useful to make a distinction between point charge particles such as protons and α-particles
on one hand and heavier ions on the other, since the principles of interaction between incident particles and target atoms
are different in these cases. For analytical purposes, only
protons and α-particles are of immediate interest. Heavier
ions, however, show many interesting properties and may be

used for special analytical tasks, although they involve less wellknown problems such as X-ray energy shift and changing fluorescence yields due to multiple ionization. For surface studies they are attractive in some cases where the particle penetration depth is critical. For low energy applications, selective ionization occurs but this has not yet been utilized for analytical purposes. The yield of X-ray production for proton impact as a function of proton energy is given by

$$\sigma_p = \sigma_i \; \omega \; k \qquad\qquad\qquad (11.11)$$

where σ_i is the ionization cross section, ω is the fluorescence yield and k the relative line intensity of possible transitions to fill an inner-shell vacancy.

In fig. 11-18 the K and L ionization cross sections are obtained from the y-axis. The product of the electron binding energy squared, u_i^2, and the ionization cross section, σ_i, is plotted versus the expression $T/\lambda u_i$, where T is the proton energy and λ the ratio of the proton mass to the electron mass.

Bambynek et al.[11.63] have reviewed X-ray fluorescence yields and Freund[11.64] has collected experimental values for K-shell fluorescence yields.

The yield of X-ray production can be increased using heavy ion impact. With equal velocity ions the cross section for X-ray production is proportional to Z^2, where Z denotes the atomic number.

11.5.2 BASIC PRINCIPLES OF THE METHOD

The principle experimental arrangement of the PIXE technique can be described as follows:
Protons, α-particles or heavy ions pass through an irradiation chamber. The intensity of the beam is often made uniform by means of a diffuser foil. The beam is then defined by a series of collimators. The target is typically a thin foil of carbon or plastic upon which the sample to be analysed has been deposited. Thick targets such as sections of organic tissue or powder compressed to a pellet may also be used. The beam is dumped in a Faraday cup connected to a beam integra-

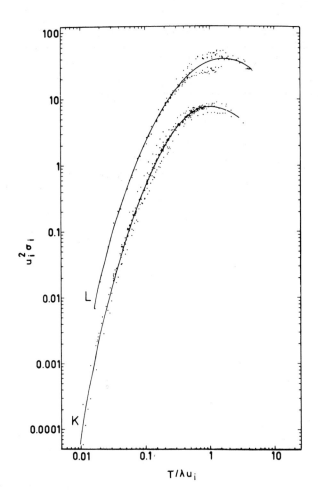

Fig. 11-18   A plot of the K and L ionization cross sections
in proton impact. On the y-axis is plotted $u_i^2 \cdot \sigma_i$
where u_i is the electron binding energy and σ_i the
ionization cross section. On the x-axis is plotted
$T/\lambda u_i$ where T is the proton energy and λ the ratio
of the proton mass to the electron mass.

tor. X-rays emitted by the sample pass through a thin window
in the chamber and are detected by a silicon detector[11.58].
The pulses from the detector are analysed in a multi-channel
analyser (fig. 11-19).

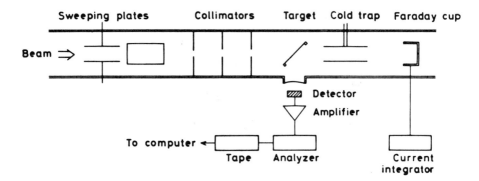

Fig. 11-19 A schematic diagram of the PIXE experimental
           arrangement[11.58].

11.5.3 BACKGROUND

In PIXE analysis the trace elements to be determined are
always mixed with the bulk of the material in the sample, the
so-called matrix. Examples are the organic tissue in biological samples or a carbon backing on which an aerosol sample
has been deposited. The X-rays registered by the silicon detector are not only characteristic X-rays from the trace element but also background radiation from the matrix. As an
illustration of this, fig. 11-20 shows a spectrum of a thin
carbon foil bombarded with 1.5 MeV protons. It consists of a
continuous distribution peaked at rather low energy and having a high energy tail. The shape at low energies is influ-
enced by the cut-off due to absorption in the windows of the
experimental set-up.

Several processes can contribute to the background, e.g.
bremsstrahlung from secondary electrons and from the incident
particles. The contribution to the background arising from
the latter process is indicated by a line in fig. 11-20.

11.5.4 SENSITIVITY

Let us calculate the number of X-ray pulses registered
by a silicon detector in a normal geometry when a small amount
of a trace element is bombarded with protons in the MeV range.
We will find that a sufficient number for a convenient regi-

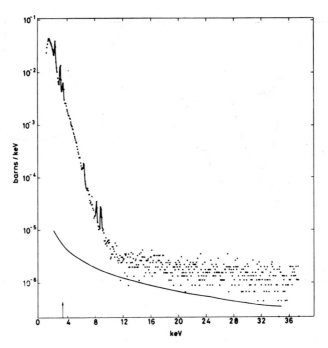

Fig. 11-20   Background radiation from carbon foil bombarded
              with 1.5 MeV protons.

stration is obtained even with extremely small amounts of
matter, of the order of 10^{-16} g.

   This is, however, not a representative value for PIXE
analysis. The trace elements to be measured are, always contained in a matrix, which could be organic tissue or a thin
foil of carbon or plastic used for collecting aerosols. Background radiation arises inevitably in the interaction of the
incident particles with the matrix atoms. This background sets
a limit to the sensitivity which can be obtained since, in
order for a characteristic X-ray peak to be discerned, it
must rise above the background in a statistically significant
manner.

   Concentrations of the order of 1 ppm is conventionally
obtained. The detection limits expressed in terms of absolute
mass are very low, however, typically less than 1 ng - and are
several orders of magnitude lower still for microbeams (with
beam diameters in the range of 1 to 10 μm).

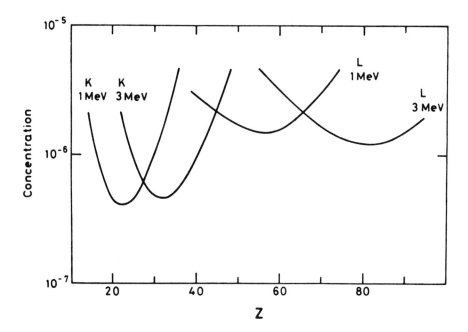

Fig. 11-21   Minimum detectable concentrations versus atomic
              number[11.58]. The following experimental para-
              meters were used: detector resolution 165 eV,
              solid angle $0.003 \cdot 4\pi$, collected charge 10 μC and
              target thickness 0.1 mg/cm^2.

Various detection limits are presented in figs. 11-21 and
11-22. Due to the slowing down of the particles in the sample
and to absorption of X-rays in the sample itself, only a thin
layer of the order of 0.1 to 5 mg·cm^{-2} is normally analyzed.
If thick targets are considered the background level increases
and also the limits giving a lower sensitivity as discussed by
Ahlberg et al[11.59].

11.5.5 EXPERIMENTAL ARRANGEMENTS
    A typical experimental arrangement for the PIXE techni-
que is demonstrated in fig. 11-23.

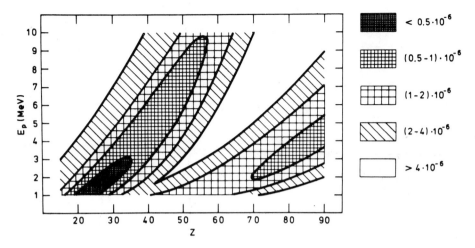

Fig. 11-22   Minimum detectable concentrations as a function
              of atomic number and bombarding energy[11.58].
              The experimental parameters are the same as in
              fig. 11-21.

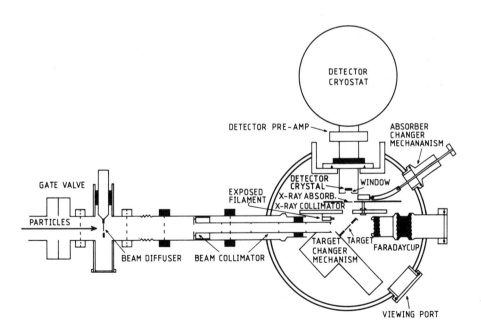

Fig. 11-23   Experimental arrangement for the accomplishment
              of PIXE analysis[11.58].

11.5.6 INTERFERENCES

The large number of X-ray lines and the limited resolution power of Si(Li)-detectors makes line interference unavoidable. Fortunately, in most analytical cases there are methods to handle these problems. In the computer code for spectrum evaluation by Kaufmann, Akselsson and Courtney[11.60], interferences are handled intrinsically in the code, a nice consequence of their approach. For other spectrum evaluation methods, line-interference corrections must be made after peak locations and areas have been determined.

A frequently encountered interference problem is that caused by K_α of element Z and K_β of element Z-1. This interference occurs in the Cr-Cu region where the detector does not resolve the two K_α and K_β lines. For higher Z-elements, the energy differences permit the detectors to separate these interfering lines. From Z values around 35 and up, the interferences between L-lines from these elements and K-lines from lighter elements have to be considered. For elements in the lead region interference between M-lines from these elements and K-lines from elements in the sulphur region occurs. In typical applications, not more than 10-15 elements are detected so interference problems are rather limited and methods to cope with them exist in most cases.

11.5.7 BACKING MATERIAL

Several materials have been used for backing. The material ought to possess a low Z value in order to minimize peak interferences from higher Z elements which shall be analyzed in the specimen prepared on the backing. Also the material should be thin in order to reduce X-ray background and thus increase the sensitivity. A good strength of the material is further of utmost importance.

Characteristic properties of commonly used backing materials are given in table 11-2. The compilation is taken from ref. 11.58.

Table 11-2

Various characteristic properties of various backing materials[11.58].

	Thickness $\mu g \cdot cm^{-2}$	Performance in beam	Impurities	Mechanical strength	Comments
Carbon	20	μA,hours	Fe,Ni,Cu Zn(Cl,Ca) ($\approx 0.1\text{-}1$ $ng \cdot cm^{-2}$)	fragile	
Collodion	<1 μm	$5\ nA \cdot cm^{-2}$ $30\ nA \cdot cm^{-2}$ 30 min if sandwiched		strong	Easy to handle: the low beam capacity has been improved by evaporating carbon or aluminium onto the film
Formvar	10 ≈ 50 100 aluminium	200 nA, 3MeV,300 nA($\approx ng \cdot cm^{-2}$)	Fe,Cu,Zn	fragile	
Kapton	20	>300 nA	K,Ca,Ti,Cr Mn,Fe,Ni,Cu Zn($\approx ng \cdot cm^{-2}$)	fragile	Preparation similar to polystyrene. Difficult to float off and handle. Impurity levels probably not inherent in material and may be reduced
Millipore	≈ 5000	$300\ nA \cdot cm^{-2}$	-	strong	
Mylar	500			very strong	
Nuclepore	≈ 1000	$300\ nA \cdot cm^{-2}$	Mn,Fe,Ni,Cu Zn($\approx ng \cdot cm^{-2}$)	strong	
Polystyrene	40	100 nA	Ca,S,Mn,Fe, Zn($\approx ng \cdot cm^{-2}$)	strong	

11.5.8 APPLICATIONS

The PIXE technique has found a wide range of application. An ideal situation for the application of PIXE is the investigation of many small samples, preferably as thin foils, containing 10-15 elements in amounts of 0.01-10 ng. Aerosol

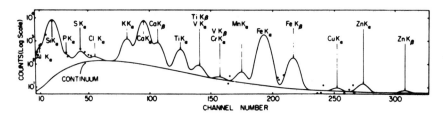

Fig. 11-24   X-ray spectrum of an aerosol sample[11.59].
            Points deviating more than two standard devia-
            tions are plotted.

samples present such a case. Already in the first publication
on this subject[11.37], it was indicated that PIXE can be
used for aerosol studies.

    Samples are collected on filters or by cascade impactors.
Both Millipore and Nuclepore filters have been used in the
analysis. Aerosol size fractionation is obtained by cascade
impactors, where the aerosol to be sampled is drawn through
jets of decreasing diameters. The aerosol is directed towards
a plate covered with a thin plastic material which will catch
the material to be analysed. Studies of this type are described in the compilation of Johansson and Johansson[11.58]. A
typical X-ray spectrum of an aerosol specimen is presented in
fig. 11-24. Further in fig. 11-25 is given a spectrum from
PIXE analysis of the human kidney cortex.

    It should further be mentioned that PIXE technique has
been suitably used in the analysis of various elements in
trace quantities of human tooth enamel and dentine according
to a study of Ahlberg and Akselsson[11.62]. Results of their
investigation are presented in table 11-3.

11.5.9 ADVANTAGES AND DISADVANTAGES
    According to Johansson and Johansson[11.58] the follow-
ing advantages and disadvantages characterise the PIXE technique.

Advantages:
1)   PIXE is multielemental. Up to 20 elements can be deter-
     mined simultaneously.

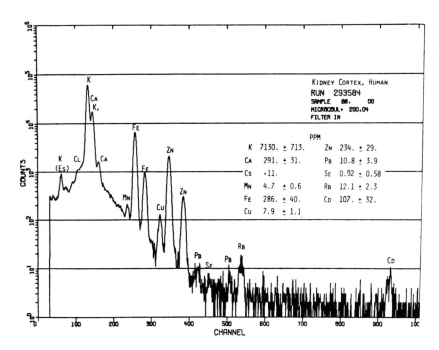

Fig. 11-25   X-ray spectrum of various elements from the
human kidney cortex[11.61].

2)   Practically the whole periodic table can be covered in
one single run and the sensitivity is fairly constant
over this region. For elements with Z>12, it does not
deviate more than a factor of 3 from the mean value.

3)   The sensitivity is very high. This means for microbeams
minimum detectable amounts as low as 10^{-16} g.

4)   PIXE is fast. For most samples a running time of 2-5 min
is sufficient. If the resulting spectra are fed into a
computer they can be analysed and a print-out be available a few minutes after a run.

5)   Very small samples can be analysed with full sensitivity.
Using the microprobe technique, even microstructures can
be analysed.

6)   PIXE is non-destructive.

7)   Good economy. Taking into account the fact that information for all elements above a certain, quite low, con-

401

Table 11-3

Concentration of various elements in human enamel and dentine[11.62]

Element	Enamel	Dentine	This work			
			External enamel	Internal enamel	Dentine	Pulpal wall of dentine
P%	18.3±2.2	13.5±2.8	25	22	19	19 -
S%	-	-	0.42	<0.09	0.12	0.27
Cl%	0.65±0.30	0.39±0.11	1.1	0.56	2.4	3.4
K%	0.05-0.3	-	0.22	0.07	0.06	0.16
Ca%	37.4±1.0	28.2±1.2	35	30	27	25
Mn ppm	0.54±0.08	0.19±0.06	10	<5	<5	<5
Fe ppm	338±109	110±22	48	9.0	5.5	14
Ni ppm	-	-	4.5	<2.5	<2.5	4.0
Cu ppm	0.26±0.11	0.21±0.10	18	<2.2	<2.2	13
Zn ppm	276±106	199±78	1200	40	150	510
Br ppm	4.6±1.1	4.0±2.0	2.5	<2.2	6.0	7.9
Sr ppm	94±22	70±18	120	170	200	190
Pb ppm	-	-	140	50	8.9	29

centration are obtained in a single run of short duration, it is cheaper than most other methods.

Disadvantages:

1) Interferences occur between K and L X-rays from light and heavy elements, respectively, and between K_α and K_β peaks in neighbouring elements. This has a negative effect on the sensitivity. Increased detector resolution could give great improvements. Another possibility is to analyse the X-rays by means of a crystal spectrometer. Then, however, the multielemental character of the method is lost.

2) PIXE is best suited for thin samples (thickness less than 1 mg/cm^2). Thick samples can also be used but the analysis of the resulting spectra is more complicated and the accuracy is reduced. One possibility to improve this situation is to use internal standards.

11.6 References

11.1   E. Ricci, Energia Nuclear 14, 5 (1970).

11.2   C.F. Williamsson, et al., Tables of range and stopping power of chemical elements for charged particles of energy 0.5 to 500 MeV. (CEA-R-3042),(1966).

11.3   E. Ricci and R.L. Hahn, Anal. Chem. 40, 54 (1968).

11.4   P. Shewmon, Acta Met. 8, 605 (1960).

11.5   L.S. Darken and R.A. Oriani, Acta Met. 2, 841 (1954).

11.6   D. Brune, et al., AE-451 (1972).

11.7   G. Deconninck, Panel meeting on the utilization of low energy accelerators, IAEA, Zagreb, 30 Sept-4 Oct (1974).

11.8   Handbook on Nuclear Activation Cross-Sections, IAEA, Vienna TR-no 156 (1974).

11.9   G. Amsel and D. Samuel, J. Phys. Chem. Solids 23, 1707 (1962).

11.10   E. Möller and N. Starfelt, Nucl. Instr. Meth. 50, 225 (1967).

11.11   M. Ahlberg et al., Odont. Rev. 26, 267 (1975).

11.12   D.A. Leich and T.A. Trombello, Nucl. Instr. Meth. 108 67 (1973).

11.13   J.J. Point, Radioisotopes in Scientific Research, Proc. of the first (UNESCO) intern conf., Paris September 1957 (ed. R.C. Extermann), Pergamon Press, Vol II, 180 (1958).

11.14   T.B. Pierce et al., Nature 204, 571 (1964).

11.15   N. Bohr, Kongl. Danske Vidensk Selsk - Mat-fys Medd 18, 8 (1948).

11.16   E.A. Wolicki, in New Uses of Ion Accelerators (ed. J.F. Ziegler), Plenum Publ. Corp. (1975).

11.17   T.B. Pierce, Harwell, ENGLAND (private communication).

11.18   G. Amsel and D. Samuel, Anal. Chem. 14, 1689 (1967).

11.19   J. Lorenzen, AE-502 (1975).

11.20   E. Möller et al. Nucl. Instr. Meth. 50, 230 (1967).

11.21   W.J. Nande et al., J. Radioanal. Chem. 1,231 (1968).

11.22   B. Sundqvist et al., Int. J. Appl. Rad., Isotopes 27 273 (1976).

11.23   S.S. Markowitz and J.D. Mahony, Anal. Chem. 34, 329 (1962).

11.24   E. Ricci and R.L. Hahn, Anal. Chem. 40, 54 (1968).

11.25   J. Lorenzen and D. Brune, Nucl. Instr. Meth. 123, 379 (1975).

11.26   Ch. Engelmann et al., Proc. conf. on Modern Trends in Activation Analysis, Oct 7-11, 1968, Gaithersburg, Maryland. NBS Special Publ. 312, Vol. II, 819 (1969).

11.27   T. Nozaki, Ibid, 842.

11.28   Minh Due Tram et al., Ibid, 811.

11.29   M. Peisach and R. Pretorius, Ibid, 802.

11.30   J.M. Butler and E.A. Wolicki, Ibid, 791.

11.31   J.N. Barrandon and Ph. Albert, Ibid, 794.

11.32   J.N. Barrandon et al., Nucl. Instr. Meth. 127, 269 (1975).

11.33   V. Krivan, J. Radioanal. Chem. 26, 151 (1975).

11.34   Desaedeleer et al., Anal. Chem. 48, 572 (1976).

11.35   Modern Physical Techniques in Material Technology (eds. T. Mulvey and R.K. Webster). Oxford University Press, London, 172(1974).

11.36   J.A. Cookson et al., J. Radioanal. Chem. 12, 39 (1972).

11.37   T.B. Johansson, R. Akselsson and S.A.E. Johansson, Nucl. Instr. Meth. 84, 141 (1970).

11.38   M. Ahlberg et al., Nucl. Instr. Meth. 123, 385 (1975).

11.39   D. Brune et al., Nucl. Instr. Meth. 142, 51 (1977).

11.40   G.J.F. Legge and A.P. Mazzolini, Nucl. Instr. Meth. 168, 563 (1980).

11.41   U. Lindh, Nucl. Instr. Meth., 197, 185 (1982).

11.42   G. Demortier and T. Hackens, Nucl. Instr. Meth., 197, 223 (1982).

11.43   R. Nobiling et al., Nucl. Instr. and Meth., 142, 49 (1977).

11.44   D. Heck, Nucl. Instr. Meth., 197, 91 (1982).

11.45   F. Watt et al., Nucl. Instr. Meth., 197, 65 (1982).

11.46   U. Lindh and A.B. Tveit, J. Radioanal. Chem., 59, 167 (1980).

11.47   U. Lindh et al., Sci. Total Environ, 16, 109 (1980).

11.48   E. Bruninx, Cern reports No 61-1, 64-17, 62-9.

11.49   D. Brune, Kem. Tidskr. 11, 50 (1971).

11.50   J.A. Davies et al., Can. J. Chem. 38,1, 526 (1960).

11.51   J.A. Davies, J. Vacuum Sci. Technol. 8,487 (1971).

11.52   H.J. Matzke et al., Can. J. Phys. 49, 2, 215 (1971).

11.53   W.D. Mackintosh and J.A. Davies, Anal. Chem. 41, 27A (1969).

11.54   L. Erikson et al., J. Radioanal. Chem. 12, 287 (1972).

11.55   F. Abel et al., J. Radioanal. Chem. 16, 587 (1973).

11.56   R. Akselsson et al., Bulletin American Physical Society Ser. II, 20, 155 (1975).

11.57   G. Götz and F. Schwabe, J. Radioanal. Chem. Vol. 28, 19 (1975).

11.58   S.A.E. Johansson and T.B. Johansson, Nucl. Instr. Meth. 137, 473 (1976).

11.59   M. Ahlberg and R. Akselsson, Nucl. Instr. Meth. 123, 385 (1975).

11.60   H.C. Kaufmann et al., Nucl. Instr. Meth., 142, 251 (1977).

11.61   N.F. Mangelson et al., Anal. Chem. 51, 1187 (1979).

11.62   M. Ahlberg and R. Akselsson, Int. J. Appl. Rad. Isotop 27, 279 (1976).

11.63   W. Bambynek et al., Rev. Mod. Phys. 44, 716 (1972).

11.64   H.U. Freund, X-ray Spectrom. 4, 90 (1975).

12 Photon Activation Analysis

Just like charged particle reactions, photonuclear reactions offer a complementary technique to neutron activation analysis by making available a different range of nuclides for analytical purposes. Thus, a number of elements may be determined more easily, or with increased sensitivity, following irradiation with photons. The technique may be used in connection with chemical separation procedures to obtain optimum sensitivity and nuclide selection. Alternatively, in many cases analysis may be accomplished without destruction of the sample.

Photon activation analysis has not been used extensively in the past, partly because sources of high energy electrons have not been widely available and partly because one usually produces far less activity per unit irradiation time with bremsstrahlung than with moderate-flux reactors. The new generation of high-intensity easily-handled electron accelerators has partly changed this situation.

Nowadays, analysis by photonuclear reactions is employed in almost all fields of science and technology. It is used to determine minute traces of light or other elements in very pure metals or semiconductors, to measure trace elements in biological media, to analyse geological samples or to monitor atmospheric pollution. However, serious complication exists in photonuclear reaction analysis caused by the energy distribution of the photons in the bremsstrahlung beam giving problems with interfering reactions and yield determinations.

We will thus first discuss how to evaluate photonuclear yield data before we describe the three different methods of analysis used, which are prompt analysis of the photonuclear reaction products, post-irradiation measurements of activities induced and studies of resonance scattered photons. As a com-

plement we at last discuss photon induced X-ray fluorescence
for analysis.

12.1 Photonuclear yield curves

Fig. 12-1 shows diagrammatically the energy distribution
of bremsstrahlung photons and the cross section curve for a
typical photonuclear reaction. By irradiating with electron
energies greater than the threshold energy (E_1) for a particular reaction, this reaction will take place and the yield will
increase as the electron energy is raised, since a greater proportion of the photons will have energies above the reaction
threshold. The bremsstrahlung activation yield per energy content of the beam can be written as (eq. 3.102):

$$y(E_0) = \frac{\int_0^{E_0} \sigma(E)\ n(E,E_0)dE}{U} = \frac{\sigma_q}{E_0} \qquad (12.1)$$

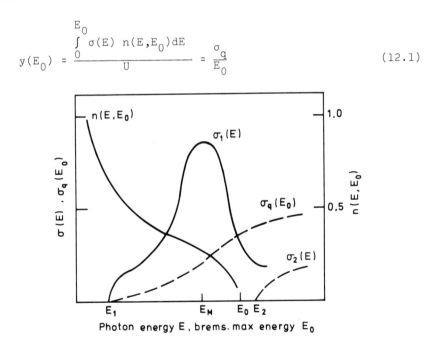

Fig. 12-1.   Photon energy E, bremsstrahlung max energy E_0.
Variation with energy of cross section for ty-
pical photonuclear reaction and bremsstrahlung
γ-source intensity.

where the laboratory factors $f(E)$ and $R(E_0)$ has been neglected.

When the electron energies exceed E_2 a second reaction may occur, which may be an interfering reaction producing the same nuclide as that used for measurement, or an interfering activity from the matrix.

If, e.g., E_1 is 10.5 MeV, the threshold for the $^{14}N(\gamma,n)^{13}N$ reaction and E_2 is 18.7 MeV, the threshold for the $^{12}C(\gamma,n)^{11}C$ reaction, nitrogen can be determined non-destructively in diamond using the 10 min β^+ activity of ^{13}N without interference from the 20 min β^+ activity of ^{11}C, provided the electron energy is kept below 18 MeV. This is a good example of how judicious choice of γ-irradiation energy can increase the selectivity of non-destructive photon activation analysis.

It is possible to calculate the induced activities of photonuclear reactions in an approximate way, replacing the energy integral in eq. 12.1 by a simple summation:

$$y(E_0) \simeq \frac{\sum\limits_i \sigma(E_i) \, n(E_i,E_0) \, \Delta E_i}{U} \qquad (12.2)$$

Experimentally determined values of the reaction cross sections can be obtained from the Photonuclear Data Center at NBS[12.1] which offers a digital data library of cross section data. Data in graphical form are also available in two other compilations[12.2]. Bremsstrahlung spectrum values are obtained in table 3-3 and the energy content of the beam in table 3-2.

Calculations of this kind of activities of photon-induced reactions have been performed by Lutz[12.3] who has estimated the sensitivities in photon activation analysis in (γ,n) and (γ,p) reactions using a 100 µA electron beam of 25, 30 and 35 MeV current and exposure times of 10 minutes and 4 hours.

12.2 Prompt analysis of photonuclear reactions

Prompt analysis is scarcely used, except to measure deuterium and beryllium concentrations by the photonuclear reactions

$$^2\text{H}(\gamma,n)^1\text{H} \qquad\qquad E_t = 2.23 \text{ MeV}$$

$$^9\text{Be}(\gamma,n)^8\text{Be} \qquad\qquad E_t = 1.67 \text{ MeV}$$

In these two reactions, the threshold energies are low enough to permit the use of γ-rays from radionuclides, e.g. ^{24}Na, ^{124}Sb (see subsec. 4.2.1.2) as photon source but bremsstrahlung gives a useful improvement in sensitivity as is shown by Guinn and Lukens[12.4] who detected the neutrons emitted by the reaction $^{55}\text{Mn}(n,\gamma)^{56}\text{Mn}$.

Frequently the photoneutrons are recorded in conventional ways (BF$_3$ counter, induced activity etc. see chap. 9). For both reactions, the detection limit is some µg, which can be compared with the limits for some other reactions (table 12-1). A lower detection limit is obtained for the ^{3}He induced reaction but, in this case, only surface layers can be analyzed.

The method of direct observation with radioelement γ-sources has been used to study deuterium in biological media by George and Kramer[12.5] and beryllium concentration in metals, alloys, compounds and solutions by Kienberger[12.6].

It is obvious in view of the threshold energies that it is always possible to determine beryllium in the presence of deuterium by using a photon source of energy below 2.2 MeV, such as ^{124}Sb, whereas deuterium can only be measured in the absence of beryllium.

In discussing prompt analysis, we have also to consider the special problem of testing nuclear fuels by γ-irradiation and detecting prompt neutrons from (γ,n) reactions and

Table 12-1

The detection limit (µg) for beryllium by using various nuclear reactions.

Reaction	Bombarding energy	Intensity	Detection limit µg
$^9\text{Be}(\gamma,n)^8\text{Be}$	2.15 MeV	1 mA	24
$^9\text{Be}(^3\text{He},n)^{11}\text{C}$	18 MeV	10 µA	0.002
$^9\text{Be}(n,\alpha)^6\text{He}$	14 MeV	$10^{13}\text{n/m}^2\text{s}$	110

photofission. These methods are useful for the measurement of the fissile elements ^{235}U and ^{239}Pu, fertile elements like ^{232}Th and ^{238}U and for the determination of the ^{238}U/^{235}U and ^{239}Pu/^{235}U concentrations in fuels. The bremsstrahlung beams used are produced by small electron linacs of energy between 5 and 10 MeV[12.7].

12.3 Activation analysis methods using photonuclear reactions

The activation analysis method is the most widely used of the photon methods discussed. In many cases, this method can be superior to thermal neutron activation analysis, but it must be regarded only as a complementary and not as a competing method. The reactions (γ,n), (γ,2n), (γ,p), (γ,2p) or more generally (γ,xnyp) of the target nuclides often lead to products other than those resulting from neutron irradiation, increasing the probability of forming products with half-lives and γ-ray energies convenient for analysis.

The yields of the different photonuclear reactions depend strongly on the atomic number of the target nucleus and the complexity of the reaction. The yields are also very dependent on the exposure energy. In fig. 12-2 different yields are shown when nuclei are exposed to 30 MeV bremsstrahlung. A similar compilation is also made at 60 MeV bremsstrahlung[12.8].

In comparison with thermal neutron activation analysis, the photon method has the advantage of giving high sensitivity for the determination of the light elements C, N and O. According to Engelmann[12.9], photon activation analysis is also useful for many other elements in the following circumstances:

a) When neutron activation leads to the production of nuclides of inconveniently short half-lifes (for example, ^{20}F, 11.0 s).
b) When neutron activation produces nuclides with specific activity too low for the estimation of small traces (as with Ti, Ni, Zr, Pb).
c) When neutron activation yields nuclides with unsuitable properties (for example, soft β-radiation or very low energy γ-radiation).

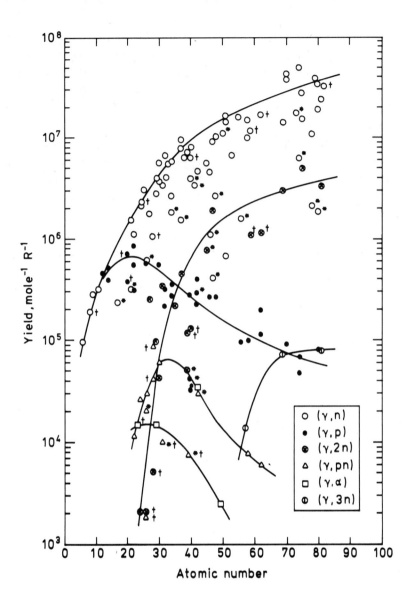

Fig. 12-2.   The yields of photonuclear reactions as a function
             of atomic number (30 MeV bremsstrahlung). *-the
             yields of metastable isomers, †-the yields for the
             reactions on target nuclides with "magic" neutron
             or proton numbers[12.8].

411

d) For the estimation of the minor constituents in a material
    (such as Li, B, Cd, Eu, Dy) characterized by heavy absorp-
    tion of thermal neutrons and consequent flux depression.
e) For the determination of trace constituents in materials
    (such as Na, Mn, rare earths, Hf) which themselves become
    inconveniently active under thermal neutron bombardment.
f) In trace element determinations which are difficult or im-
    possible with neutrons because of interference from secon-
    dary nuclear reactions. Suppose, for example, that it is
    desired to estimate nickel or zinc in a sample of cop-
    per. Thermal neutron irradiation of copper gives ^{64}Cu and
    ^{66}Cu. ^{64}Cu decays in a variety of ways with emission of
    electrons, positrons and γ-rays as well as by K-capture,
    leading to ^{64}Ni and ^{64}Zn (fig.2-21). - These products are
    identical with naturally-occuring stable nuclides of the
    two elements. They are immediately activated themselves to
    produce ^{65}Ni and ^{65}Zn - the very nuclides used for the es-
    timation of nickel and zinc. Consequently it is not possib-
    le by thermal neutron activation analysis to estimate nickel
    or zinc in copper or its alloys.

    Although reactions such as (n,2n) and (n,np) induced by
14 MeV neutrons yield a similar set of products as in the photon case, one can obtain much greater fluxes of photons from
electron accelerators than of high energy neutrons from neutron
generators.

    In comparison with charged particle activation analysis -
which is also highly specific for the determination of light
elements - photon activation is sometimes preferable because
a) it always allows the possibility of avoiding interference
from neighbouring elements and b) the activation of the sample
is relatively uniform and not confined to a layer of thickness
determined by the penetrating power of the activating particles.

    Of the photon reactions, the (γ,n) reactions is the one
most commonly used in activation analysis and allows the estimation of a large number of elements, sometimes by non-destructive technique and, in other instances, by post-irradiation
chemical separation.

Another kind of useful reaction is the (γ, γ')-reaction
where a radioactive isomer of the element under investigation
is produced. This method is non-destructive, since the nuclear
isomer decays to the same stable element from which it was formed.

For some light nuclei (B, O, F), delayed neutron emitters
may be produced by photonuclear reactions giving a sensitive
and fast method of detection, provided that no fission of heavy
elements occurs simultaneously, in which delayed neutron emitters
are also produced.

12.3.1 ANALYSIS OF LIGHT ELEMENTS C, N, O AND F BY (γ, n) REACTIONS

The photon activation method is more suited than thermal
neutron activation for the determination of the light elements
C, N, O and F. In all cases except F, which is mono-isotopic,
neutron activation can only induce radioactive species from
low abundance nuclides and these species are unsuited for analysis, having either too long or too short half-lives. The
photon activation products, on the other hand, are produced
from high abundance nuclides and have much more amenable half-
lives as shown in table 12-2.

The half-lives of the activities produced are long enough

Table 12-2

Photonuclear reactions suitable for the determination of C, N,
O and F and principal interfering reactions attributable to
neighbouring elements[12.9].

Element to be determined	Appropriate nuclear reaction	Threshold MeV	Principal interfering reactions	Threshold MeV	Half-life of activity produced
C	$^{12}C(\gamma,n)^{11}C$	18.7	$^{14}N(\gamma,t)^{11}C$	22.7	20.3 m
			$^{16}O(\gamma,\alpha n)^{11}C$	25.9	
N	$^{14}N(\gamma,n)^{13}N$	10.6	$^{16}O(\gamma,t)^{13}N$	25.0	9.96 m
			$^{19}F(\gamma,\alpha 2n)^{13}N$	25.4	
O	$^{16}O(\gamma,n)^{15}O$	15.7	$^{19}F(\gamma,tn)^{15}O$	27.4	2.03 m
			$^{20}Ne(\gamma,\alpha n)^{15}O$	20.4	
F	$^{19}F(\gamma,n)^{18}F$	10.4	$^{20}Ne(\gamma,d)^{18}F$	21.1	109.7 m
			$^{23}Na(\gamma,\alpha n)^{18}F$	20.9	

to allow chemical etching of large samples before counting or chemical separation. By etching the sample, it is possible to remove the superficial layer of impurities formed before or during irradiation. This facility is very important, especially in the estimation of elements such as C, N and O in high-purity materials. The possibility of surface etching after irradiation is one of the major advantages of photon activation analysis in comparison with other methods of trace estimation such as mass spectrometry, atomic absorption and charged particle induced activation. Etching eliminates not only the reagent blank errors but also avoids the problem of contamination in the course of analysis.

In table 12-2 a number of interfering reactions is also mentioned as are the threshold energies at which these occur. As can be seen, three of the four possible interfering reactions may be avoided as long as activation is confined to energies below 30 MeV. The energy dependence of interference is shown in fig. 12-3.

The approximate limits of detection are given in table 12-3 where a comparison also is made with data from proton and fast neutron analysis.

Table 12-3

Detection limits for C, N, O and $F^{(12.10)}$.

Element	Sensitivity of detection μg		
	$\gamma^{x)}$	$p^{xx)}$	$n^{xxx)}$
C	0.02	0.002	–
N	0.1	0.0001	90
O	0.04	0.002	30
F	0.02	0.0002	24

x) γ: E_{max} = 40 MeV $\qquad$ $\bar{I}_e$ = 30 μA

xx) p: E_p = 10-15 MeV $\qquad$ $\bar{I}_p$ = 10 μA

xxx) n_{14}: E_n = 14 MeV $\qquad$ $\varphi = 10^{13}$ $n \cdot m^{-2} s^{-1}$

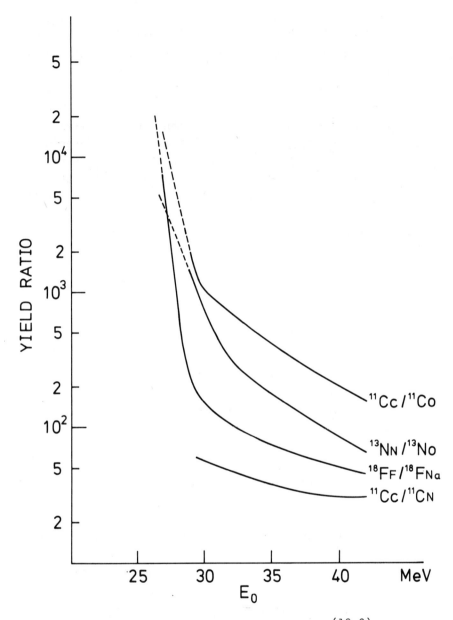

Fig. 12-3. Energy dependence of interference[12.9] (top and
bottom curves) - from nitrogen and oxygen in estimation of carbon, (second-curve)-from oxygen in
estimation of nitrogen, (third curve)-from sodium
in estimation of fluorine.

As rather large samples can be handled in photon exposures, very small concentrations of light nuclei can be detected. For example, carbon and fluorine concentrations have been measured down to 1 ng·g^{-1}.

Often, it is not applicable to measure the light nuclei discussed by non-destructive methods since the products in all cases are pure β^+-emitters with no element-specific γ-spectrum. In view of the activities induced in other elements present in the sample, the radioelements or elements concerned thus must be isolated by chemical or physico-chemical separation methods before counting. Counting is carried out by detecting the coincident annihilation radiation in two antiparallel NaI(Tl)-crystals.

For this reason, fast chemical separation methods have been developed to isolate ^{11}C, ^{13}N and ^{15}O. In fig. 12-4, a method[12.9] is shown, which isolates ^{15}O simply and quickly by reduction melting in a graphite crucible. The carbon monooxide formed is swept away by a stream of helium or argon. After passing through a filter which traps solid particles, the monooxide is oxidized into dioxide. A new filter traps away oxides of nitrogen and halogens and then an ascarite trap collects the dioxide. This trap is placed between two NaI(Tl)- detectors which measure the annihilation radiation from the β^+-decay of ^{15}O.

Photoactivation measurements of light nuclei have been used to determine trace quantities in very pure semiconducting elements (Si, Ge), in metals (Be, Al, Ti, Cr, Fe, Ni, Cu, Zr, Nb, Mo, Ta, W, Pb, Bi) and in alkalimetals (Na, Cs). Impurities of light elements like carbon disturb properties of metals even at low concentrations. A sensitive method of detection is needed. Carbon can be determined in metals and alloys by putting the sample in a nickel crucible containing a fused alkaline bath. After acidification the released CO_2 was swept by air and absorbed in KOH absorbers. With this method a detection limit of 20 ng·g^{-1} has been obtained[12.11].

The presence of non-metallic impurities especially carbon, nitrogen and oxygen is an important factor with respect to corrosion of materials by liquid sodium - a problem of great significance in fast neutron reactor technology. A method has thus

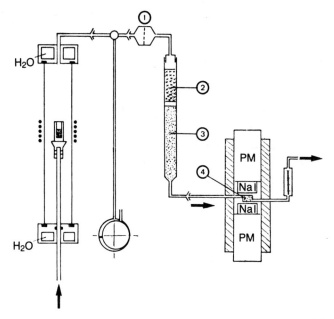

Fig. 12-4. Diagram of apparatus for separation of ^{15}O by reduction fusion[12.9].
1 = silicon wool filter and porous glass filter
2 = Schutze's reagent
3 = MnO_2
4 = tube of diameter 50 mm containing ascarite.

been developed for simultaneous chemical separation of ^{11}C and
^{13}N which is based on combustion of the sample, after irradiation and etching, in a mixture of oxidizing acid fused salts
(B_2O_3 and Pb_3O_4) containing a dispersing agent and the respective carriers. Each analysis takes about 40 min altogether. Carbon and nitrogen contents of the level of 10 ng·g^{-1} has been
measured[12.12].

Nitrogen analysis is a tool in the measurements of the
protein content in different vegetable samples. Small amounts
of fluorine can easily be determined in both biological and
geological materials.[12.13]

12.3.2. ANALYSIS OF HEAVY ELEMENTS BY (γ,n) REACTIONS

Many heavy elements can be detected, too, by photon-induced reactions. Large Ge(Li) detectors are often used to measure the emitted γ-rays and, in many cases, analysis can be performed without chemical separation. In table 12-4, those elements are given which can be determined with rather high sensitivity by photon activation. In the table the detection limits, when thermal neutrons are used, are also given. In most cases, the neutron activation method is more sensitive. For some elements, however, better results are obtained with photons (Ti, Ni, Zn, Sr, Zr, Tl, Pb).

Table 12-4

Limits of detection (ng)$^{(12.10)}$.

Element	(γ,n)	n_{th}
Sc	20	0.1
Ti	2	10
Ni	10	40
Cu	1	0.4
Zn	5	10
As	10	0.4
Ga	5	0.4
Sr	0.8	4
Zr	10	100
Ag	10	0.04
Sb	20	1
I	10	0.4
Pr	20	4
Ta	10	10
Au	1	0.04
Tl	10	–
Pb	100	–

(γ,n): E_{max} = 30-40 MeV $\bar{I}_e$ = 30 μA

n_{th}: $\varphi = 10^{17}$ $n \cdot m^{-2} \cdot s^{-1}$

The photon activation method is advantageous when biological materials are to be analyzed. These materials contain much sodium, which is strongly activated by thermal neutrons, giving ^{24}Na. Photon activation produces another isotope ^{22}Na with much longer half-life which thus interferes much less. Many elements have been studied. Here will be mentioned some measurements on Sr and Pb which have been reviewed by Hislop. These elements have γ-activation products which possess more suitable nuclear properties for analytical purposes than the corresponding thermal neutron activation products[12.14].

Two main nuclides ^{87m}Sr and ^{85m}Sr, may be produced by γ-irradiation of strontium. During the investigation on the non-destructive analysis of teeth using 12 MeV bremsstrahlung, Andersson et al.[12.15] found they were able to measure simultaneously the strontium and the fluorine content. The levels for strontium reported are in the range of 35-350 ppm. Strontium has also been determined in freeze-dried sea water by Gordon and Larson[12.16], who measured the ^{87m}Sr activity after radiochemical separation. They report an average concentration of approximately 8 ppm.

The increased awareness of the danger of lead pollution has increased the demand for analysis of this element in biological and environmental materials. Photon activation is particularly suited to the determination of lead since two nuclides may be produced, ^{203}Pb, by the reaction ^{204}Pb(γ,n) and ^{204m}Pb produced from both ^{204}Pb(γ,γ') and ^{206}Pb(γ,2n) reactions. Both product nuclides decay with convenient half-lives (52 h and 68 min respectively) and γ-emission. Hislop and Williams[12.17] studied the behaviour of lead and a number of other elements present in the bone during dry ashing at a series of temperatures following 35-40 MeV γ-irradiation. Using the radiochemical separation of ^{203}Pb, they also determined lead in human bone ash and standard reference plant materials in the concentration range 2-50 ppm. The accuracy of the technique, as assessed by comparison with results obtained by a variety of independent methods, was approximately ± 5% and the limit of detection was approximately 0.1 μg[12.18].

Lutz[12.19] has also reported the destructive and non-destructive analysis of lead in environmental samples using bremsstrahlung produced from electrons of 20-45 MeV. The samples studied included plant material, soil and coal and the results obtained have been compared with those from independent methods. The limit of detection for non-destructive determination using a high resolution γ-spectrometer is about 0.5 ppm, although this limit will be critically dependent on the matrix analysed. With a rigorous chemical separation of ^{203}Pb about 0.01 µg can be determined. Soils have also been analysed non-destructively for lead by Chattopadhyay and Jervis[12.20] using 35 MeV bremsstrahlung, followed by measurement of the ^{204m}Pb activity using a Ge(Li) detector. The method has been applied to garden soils having lead contents of 7-22 ppm. Samples containing lead concentrations less than 1-5 ppm required radiochemical separation of ^{203}Pb. These authors claim a detection limit, using chemically separated ^{203}Pb, of 0.001 µg lead.

An extemely small value for the lead content of milk was measured by Dutilh and Das[12.21]. Using 40 MeV bremsstrahlung and radiochemically separated ^{203}Pb, the sensitivity observed was 0.5 µg. However, the use of a 25 g powdered milk sample, divided into four portions which were irradiated simultaneously, enabled a concentration of < 0.005 ppm to be obtained.

In addition to the determination of a specific element in biological and environmental materials, there is an increasing interest in the use of photon activation, followed by γ-ray spectrometry, for the non-destructive, simultaneous determination of a wide range of elements in these materials. Such measurements have been reported on air particulates, blood, bone, hair, kale, soil, tobacco, tree bark and urine[12.14].

To facilitate investigations of this kind, several compilations of photonuclear reaction products and associated γ-rays exist[12.22-12.26]. Photoactivation analysis is of particular interest in bulk studies of non-destructible samples. The disadvantage is the small yields obtained. For thin samples and surface determinations, multi-element analysis can with advantage be performed using particle-induced X-ray analysis discussed in sec. 11.5.

12.3.3 ANALYSIS BY COUNTING DELAYED NEUTRONS

Among the lightest nuclides, some decay by emission of neutrons, so-called delayed neutrons (see sec. 2.5). These are $^9Li(0.17$ s$)$, $^{16}C(0.74$ s$)$ and $^{17}N(4.2$ s$)$. As there does not normally exist any neutron background, these nuclides can be identified very clearly. The half-lives are too short for chemical separation to be implemented for distinguishing among them and identification has thus to be done by measuring their decay-rates. To detect the neutrons, a battery of BF_3 counters surrounded by paraffin wax in order to thermalize the neutrons (fig. 12-5) is used.

The three nuclides can be used to determine the amount of B, O and F in a sample. Interfering reactions exist but, at exposure energies below 23,5 MeV, these are limited to the interference of oxygen in the estimation of fluorine (table 12-5).

The sensitivities with the delayed neutron method are one or two orders of magnitude lower than those of (γ,n) activation, but the short analysing times make the method very suitable for routine analysis. An interesting case is the possibility

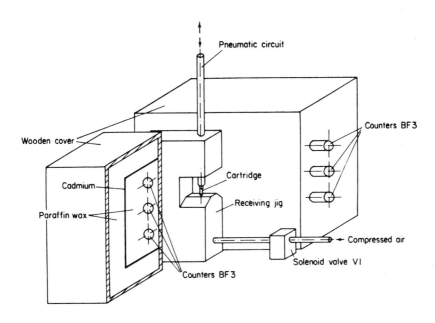

Table 12-5

Principal photonuclear reactions on light elements yielding ^9Li, ^{16}C and ^{12}N (12.9).

Table 12-5. Principal photonuclear reactions on light elements yielding ^{9}Li, ^{16}C and ^{12}N (12.9).

Element to be determined	Nuclear reaction	Threshold, MeV	Principal possible interfering reactions	Threshold, MeV
B	^{11}B$(\gamma,2p)^9$Li	30.9	^{12}C$(\gamma,3p)^9$Li	46.9
O	^{18}O$(\gamma,2p)^{16}$C	29.1	^{19}F$(\gamma,3p)^{16}$C	37.1
	^{18}O$(\gamma,p)^{17}$N	15.9	^{19}F$(\gamma,2p)^{17}$N	23.9
			^{20}Ne$(\gamma,3p)^{17}$N	36.7
			^{23}Na$(\gamma,\alpha2p)^{17}$N	34.4
			^{23}Na$(\gamma,^6$Be$)^{17}$N	35.8
			^{24}Mg$(\gamma,\alpha3p)^{17}$N	46.1
			^{27}Al$(\gamma,2\alpha2p)^{17}$N	44.5
			^{27}Al$(\gamma,^{10}$C$)^{17}$N	40.8
F	^{19}F$(\gamma,2p)^{17}$N	23.9	^{18}O$(\gamma,p)^{17}$N	15.9
			^{20}Ne$(\gamma,3p)^{17}$N	36.8
			^{23}Na$(\gamma,\alpha2p)^{17}$N	34.4
			^{23}Na$(\gamma,^6$Be$)^{17}$N	35.8
			^{24}Mg$(\gamma,\alpha3p)^{17}$N	46.1
			^{27}Al$(\gamma,2\alpha2p)^{17}$N	44.5
			^{27}Al$(\gamma,^{10}$C$)^{17}$N	40.8

to use the ^{16}O$(\gamma,p)^{15}$N reaction determining the ^{16}O content and the ^{18}O$(\gamma,p)^{17}$N method for the ^{18}O content and in this way measure the isotope-ratio ^{18}O/^{16}O in samples.

12.3.4 ANALYSIS BY DETECTION OF HIGH-ENERGY BETA RAYS

Another technique, also non-destructive and highly specific for certain elements, is based on the detection of high-energy β-rays (> 3 MeV) emitted by some short-lived radioelements. This method which has been developed by Engelmann[12.10] is particularly suitable for the measurement of lithium, beryllium and sulphur at concentrations above 10^{-2}% in 200 to 500 mg samp-

les and, like the previous method, is ideal for industrial analyses on very large batches of samples. A measurement can be
carried out in less than a minute and its price is therefore
highly competitive.

The nuclear reactions used for these analyses are as follows

$$^7Li(\gamma,p)^6He \qquad t_{1/2} = 0.81 \text{ s} \qquad T_{\beta^-,o} = 3.5 \text{ MeV}$$

$$^9Be(\gamma,p)^8Li \qquad t_{1/2} = 0.84 \text{ s} \qquad T_{\beta^-,o} = 16.0 \text{ MeV}$$

$$^{32}S(\gamma,n)^{31}S \qquad t_{1/2} = 2.6 \text{ s} \qquad T_{\beta^+,o} = 4.4 \text{ MeV}$$

This technique, based on the detection of short-lived high-energy β-emitters, can be used to determine many other light or
medium-weight elements.

Morover, this field of application can be greatly extended
by combining γ-detectors with the β-counter.

12.3.5 ANALYSIS BASED ON METASTABLE ISOMER PRODUCTION

There are about 40 stable nuclides which have metastable
isomers with half-lives greater than 0.5 seconds. Many of these
isomers can be produced from the corresponding stable state by
excitation with photons of energies sufficiently low that there
will be no photonuclear disintegration reactions produced in
any stable nuclides other than beryllium and deuterium.

The mechanism of isomer production involves the absorption
of a photon by the nucleus. The nucleus is excited to a high-
energetic level but γ-decays very rapidly ($<10^{-10}$ s) to the
metastabel level. The transition from the metastable state to
the ground state then occurs with a characteristic half-live
and with the emission of characteristic γ-rays (sec. 2.4).

The cross section-energy relationship for these reactions,
frequently refered to as (γ,γ') reactions, shows two peaks: one
just below the threshold energy for nucleon emission and the
other at the giant resonance. The peak cross section is usually less than a few millibarn. Although the sensitivity is not
particularly favourable, the specificity is very good. If the

423

irradiations are conducted at a suitably low energy (<7 MeV), the only activities produced will be due to the metastable isomers. No interfering reaction occurs.

This method is ideal for systematic routine analysis of very large batches of samples. Table 12-6 lists orders of magnitude of the detection limits for some elements measurable in this way, according to Engelmann[12.10]. It is assumed that the samples analysed weigh more than 10 g and are irradiated with bremsstrahlung produced by a 7 to 8 MeV electron beam of average intensity 100 μA.

Veres[12.27] has shown that many photoexcitable elements can be activated by radioactive sources (table 12-7). The (γ,γ') method has numerous applications. The silver contents of coins have been determined by isomer activation of $^{107m, 109m}$Ag. Another example may be the determination of Br, Se and Er. The corresponding isomers have half-lives of 4.8, 18 and 25 s, respectively, thus rapid investigation is possible. In this way

Table 12-6

Some orders of magnitude of (γ,γ') activation analysis detection limits[12.10].

Element	Detection limits $\mu g \cdot g^{-1}$
Se	10
Br	5
Sr	100
Y	50
Ag	10
Cd	5
In	5
Ba	100
Hf	1
W	200
Ir	20
Pt	50
Au	1
Hg	200

Table 12-7

Overall cross sections and other data of nuclear isomers excited by different radioactive sources[12.27].

Nucleus	Activation state (MeV)	Isomeric state		Cross section (nb = 10^{-33} cm^2)				
		E_i (keV)	Half-life	^{46}Sc	^{60}Co	^{142}Pr	^{116m}In	^{24}Na
^{73}Ge	-	66	0.53 s	-	-	-	-	-
^{77}Se	1.12	160	17.5s	14	95	285	330	490
^{79}Br	1.12	208	4.9s	6	10.7	57	-	65
^{83}Kr	-	43	1.9h	-	-	-	-	-
^{87}Sr	1.12	388	2.8h	3.2	9-22	-	15	78
^{89}Y	1.33	913	16.5s	-	-	-	-	-
^{90}Zr	-	2315	0.8s	-	0.02	-	-	-
^{93}Nb	-	30	12 y	-	-	-	-	-
^{99}Te	-	140	6 h	-	-	-	-	-
^{103}Rh	1.33	40	57 m	-	25	-	5?	-
^{105}Pd	-	200	23 s	-	-	-	-	-
^{107}Ag	1.325	93	44 s	+	4-6	-	3.6	41.5
^{109}Ag	1.210	88	40 s	+	4-6	-	3.3	41.5
^{111}Cd	1.3	396	49 m	+	13	26	6.9	45
^{113}Cd	1.33	270	14 y	-	12	-	-	-
^{113}In	1.132	393	1.7h	-	12	-	46	-
^{115}In	1.078	335	4.5h	7	20-90	97	150	135
^{117}Sn	-	320	14 d	-	-	-	+	56
^{119}Sn	-	89	279 d	-	-	-	+	175
^{123}Te	1.33	248	104 d	-	1	-	-	520
^{125}Te	1.33	1451	58 d	-	-	-	-	870
^{129}Xe	-	236	8.9d	-	-	-	-	-
^{131}Xe	-	163	12 d	-	-	-	-	-
^{135}Ba	1.33	268	28.7h	-	0.5	-	+	-
^{137}Ba	1.33	661	2.6m	-	10^{-3}	-	+	-
^{167}Er	1.33	208	2.5s	+	4	2050	750	214000
^{176}Lu	1.33	180	3.7h	-	1000-2400 / 23-47	-	+	-
^{178}Hf	-	1149	3 s	-	-	-	+	-
^{179}Hf	-	376	19 s	-	-	-	2	-
^{180}Hf	-	1143	5.5h	-	0.01	-	+	-
^{180}Ta	-	-	8.15h	-	-	-	-	-
^{183}W	-	290	5.5s	-	+	+	+	340
^{187}Os	-	-	39 h	-	+	-	+	-
^{188}Os	-	1600	180 d	-	+	-	-	-
^{189}Os	-	30	5.7h	-	-	-	+	-
^{190}Os	-	1700	10 m	-	0.2	-	+	-
^{191}Ir	1.33	170	4.5s	-	56	-	65	173
^{193}Ir	-	80	11.9d	-	-	-	+	830
^{195}Pt	1.18	260	4.1d	-	4.2	-	0.43	328
^{197}Au	1.33	407	7.2s	-	0.7	87.5	2.6	9800
^{199}Hg	1.33	526	44 m	-	0.05	-	-	42
^{204}Pb	-	2186	67.5m	-	-	-	-	-
^{207}Pb	-	1633	0.8s	-	0.08	-	+	-
^{235}U	-	0.07	26.5m	-	-	-	-	+

+ Isomeric activity not detected

the bromine content in bromine tablets has been studied[12.28].
Veres[12.29] has shown that the erbium content in earth rock
samples (3 $\mu g \cdot g^{-1}$) and meteorites (0.2 $\mu g \cdot g^{-1}$) can be determined. He also reports that the selenium content in food stuff,
which is of importance to know, can routinely be determined[12.30].
The (γ, γ') method, indeed, has numerous applications. Its simplicity, specificity and accuracy are well worth mentioning.

12.4 Gamma ray resonance scattering

A method worth noting which is as yet not very much used
and developed is that of γ-ray resonance scattering[12.31, 12.32].

In atomic physics, the resonant fluorescence process is
an important and easily observed phenomenon. However, many
of the early attempts to detect resonance radiation from nuclei failed because of the extreme narrowness of the absorp-
thion lines in relation to the recoil energy losses when a
γ-ray is emitted and absorbed by a nucleus. The recoil energy loss for emission is equal to $E_\gamma^2/2mc^2$ where E_γ is the
γ-ray energy, $\underline{m}$ the nuclear mass and $\underline{c}$ the velocity of
light. The same amount of energy is transferred to a nucleus
when a γ-ray is absorbed. Thus, for resonant excitation
of a nucleus with γ-rays from a similar nucleus, the energy is off resonance by E_γ^2/mc^2. For a typical resonance scattering experiment involving a $\simeq$ 1 MeV transition in a medium-
weight nucleus, the natural width of the level $\Gamma \simeq 10^{-3}$ eV,
the thermal Doppler width $\Delta \simeq$ 1 eV caused by the thermal motion
of the nucleus in the sample and the recoil energy loss is of
the order of 10 eV. Different methods exist to compensate the
recoil energy loss of the γ-source in order to get resonant
scattering. In all cases, motion is giving to the emitting nuclei either mechanically, thermally or by utilizing the nuclear recoil following radioactive decay.

(a) The source velocity, required to Doppler shift mechanically the incident γ-ray energy by an amount equal to the
recoil energy loss E_γ^2/mc^2, is $v = E_\gamma/mc$. For low energy γ-rays
in heavy nuclei this amounts to velocities in the range 100 to

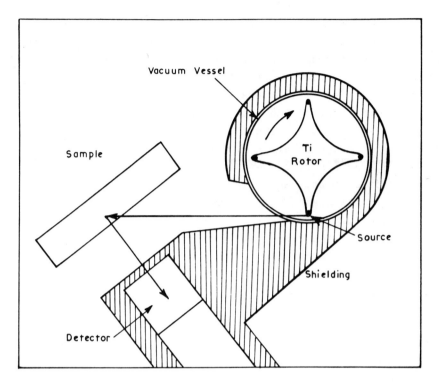

Fig. 12-6. Schematic diagram of an experimental arrangement
which has been used to measure resonance scattering with the centrifuge technique[12.33].

1000 ms^{-1}. Such velocities have been achieved using high speed
rotors (fig. 12-6). The tensile strength of the best rotor materials limits peripheral velocities to less than 1300 ms^{-1}
which in turn limits the range of possible applications for
elemental analysis. The most favourable elements are Tl and Hg
using sources of ^{203}Hg and ^{198}Au respectively.

(b) The thermal method is based on thermal broadening of
the emission and absorption lines with increased temperature.
Compared to the centrifuge method discussed above, the ratio
of resonance to background counts is small. However, the thermal method is simple with few demands on source quality.

(c) The motion of the emitting nucleus due to previous radioactive decay offers another method of achieving resonance
overlap. Recoil velocities at the time of γ-ray emission can

427

be used to compensate for the recoil energy loss. The energy
of nuclear recoils following radioactive β, γ or electron-capture decay is typically of the order of 10 eV. The slowing
down time of the source needs to be greater than the lifetime
of the γ-emitting level. Thus gaseous sources, with typical
slowing down times of the order of nanoseconds, are more favourable than solid and liquid sources with corresponding
times of the order of 0.1 - 0.01 picoseconds.

(d) If, in a nuclear reaction, the product nucleus is left
in an excited state, the de-excitation can take place by the
emission of a γ-quantum to the ground state. Such γ-radiation
can be used for resonance scattering. The recoil being compensated for by the velocity which the emitting nucleus has
attained in the reaction. Sources emitting α-particles have been
used to initiate such nuclear reactions. Even the lifetime of the
γ-emitter must be shorter than the slowing down time which restricts the method to lifetimes less than some picoseconds. The
Coulomb barrier restricts α-induced reactions to light nuclei
($Z < 18$) and furthermore the efficiency of γ-ray production is
low (around $1.6 \cdot 10^{-5}$ γ-rays/α-particle).

The use of accelerators to produce reactions suitable for
resonance scattering analysis theoretically seems to have many
advantages as a greater flexibility is obtained and the γ-ray
energy is dependent on the direction of emission. However in
practice intensities often would be too low to be useful in
resonance scatter logging. The narrowness of nuclear resonance absorption lines makes it difficult to use continuous bremsstrahlung beam for nuclear fluorescence studies. The signal to
background ratio is low and more than one nuclide can be excited. Even when photon-monochromaters are used only average
elastic photon scattering cross sections are observed[12.34].

The use of neutron capture γ-rays as photon sources
is more promising. Many instances of nuclear resonance scattering have been observed when various elements are irradiated with such monoenergetic photons[12.35]. Resonance scattering occurs because there is a chance overlap between the
energy of the γ-ray and an individual level in a target nucleus. Most resonances have been observed in the 6-9 MeV range
in heavy nuclei. Elements and capture γ-ray sources for which

Table 12-8

Elements which can be analysed by the resonance scattering of neutron capture γ-rays. Only those elements for which the effective cross section is greater than 100 mb are listed[12.35].

Scatterer	γ-ray energy (MeV)	Neutron capture material	Effective cross section (mb)
^{208}Pb	7.28	Fe	4100
Zr	8.50	Se	3500
^{209}Bi	7.15	Ti	2000
Hg	4.90	Co	385
Tl	7.64	Fe	370
Tl	7.16	Cu	120
Sr	7.01	Cu	110
Cd	7.47	Co	110
^{141}Pr	6.12	Cl	110
^{144}Sm	9.00	Ni	100
^{209}Bi	7.42	Se	100

the effective cross section exceeds 100 mb are listed in table 12-8. As the probability of complete overlap is very small, very few cross sections exceeding about 1 mb have been observed. All the experiments performed using this technique have employed neutrons from nuclear reactors.

In table 12-9 a comparison of the various method of γ-ray resonance fluorescence with special attention to borehole analysis is given. Of the methods discussed, the one using a gaseous radionuclide source seems to be the most promising one.

In table 12-10 a list of the elements most suited to analysis is given. Source strengths of about 20 Ci can be used to obtain resonance count rates of about 10 count $\cdot$ s^{-1} per % wanted element in boreholes of diameter > 15 cm. The most favourable elements are Cu, Ni, As and Cr.

Table 12-9

Comparison of the various methods of γ-ray resonance fluorescence for borehole analysis[12.32] (1 Ci = $3.7 \cdot 10^{10}$ Bq).

Method	Most favourable elements	Borehole applications		
		Source strength[a]	Potential[b]	Major limitations
Radiosotope source				
(a) Mechanical motion	Tl,Hg	≥ 1 Ci	F	high M, low E
(b) Thermal agitation	Tl,Hg,Pt,Cs	≥ 10 Ci	F	high M, low E
(c) Preceding radiation, solid or liquid source	Li,Ce,Sm,Sr	≥ 1 Ci	F	very restricted range of elements
(d) Preceding radiation, gaseous source	Cu,Ni,As,Li,Cr, W,Hg	~ 1 Ci	G	borehole diameter ≥ 15 cm
Nuclear reactions				
(a) α-induced	C,N,Na,Al,Si,S	$\approx 6 \cdot 10^5$ Ci ^{238}Pu	P	very high source activity
(b) Accelerator-induced	wide range	$\sim 10 \mu A$ of MeV particles	P	MeV particle accelerator
Bremsstrahlung	Li,B,N,Al	~ 10 mA of MeV electrons	P	betatron,Ge(Li)
Compton backscatter	As,Ti,V,Ga	$\sim 10^4$ Ci	P	high source strength, difficult shielding
Neutron capture γ-rays				
(a) Resonant scattering in sample	Pb,Zr,Bi,Hg	10^3 Ci ^{124}Sb (for Pb)	P	high source strength,background problems
(b) Neutron capture in sample	Fe,Se,Ti,Co	10^{10} neutrons/ sec(for Fe)	P	borehole diameter ≥ 28 cm,background problems

a) Approximate source strength required for resonance photopeak count rate of 1 count/s per % wanted element in a borehole geometry.

b) G = good, F = fair, P = poor.

Vartsky et al.[12.36] have developed a technique for determination of elements in human body in-vivo, utilizing nuclear resonant scattering of γ-rays. 847 keV photons emitted from a gaseous ^{56}MnCl$_2$ source are resonantly scattered from ^{56}Fe present in the body. The detection of these γ-rays is used to estimate the iron content of the liver or heart of patients. Elements of interest which may be detected at physiological concentrations by the same technique include Mg (using ^{24}Na vapour source) ^{28}Si (using ^{28}Mg vapour source) and ^{65}Cu (using ^{65}ZnI$_2$ vapour source).

Table 12-10

Elements most suited to analysis by γ-ray resonance scattering
using a gaseous radionuclide source[12.32].

Element	Radioactive source	Half-life	Source details		Relative resonance cross section[a]
			γ-ray energy (MeV)	Lifetime of level(ps)	
Li	^{7}Be	53d	0.48	0.07	1980[b]
Ti	^{46}Sc	84d	0.89	4	4.2
Ti	^{48}V	16d	0.98	3	39.4
V	^{51}Cr	28d	0.32	200	0.3[b]
Cr	^{52}Mn	5.7d	1.43	0.7	63.0
Ni	^{60}Co	5.3y	1.33	0.7	19.4
Cu	^{65}Zn	245d	1.12	0.4	24.1[b]
Ge	^{74}As	18d	0.60	12	25.6
As	^{75}Se	120d	0.27	12	6.3
Cd	^{111}Ag	7.5d	0.34	27	10.6
Cs	^{133}Ba	10.7y	0.38	10	2.1[b]
W	^{184}Re	38d	0.90	1.5	15.9[b]
Hg	^{198}Au	2.7d	0.41	20	21.4
Tl	^{203}Hg	47d	0.28	280	5.8

a) Product of the resonance cross section (mb) and the natural abundance of the stable isotope
which participates in the resonance scattering process.
b) Electron capture decays for which the resonance cross section was calculated assuming a 1%
overlap per eV of the emission and absorption lines.

12.5 Photon-induced X-ray fluorescence analysis

12.5.1 PRINCIPLE

The basic principle for energy-dispersive X-ray fluorescence spectrometry is to produce vacancies in the inner shells of the atoms of interest by irradiating the sample with ionizing radiation. When the vacancies are filled by transitions of electrons from higher levels, emission of characteristic X-rays takes place. The measurement of the energies and intensities of these X-rays is the basis of X-ray fluorescence spectroscopy analysis. The production of secondary photons by a monoenergetic photon beam is schematically shown in fig. 12-7.

Photons with energy E_0 are incident as a colinear beam on a surface with an incident angle θ. The number of photons transmitted through a thickness z of the sample depends on the average mass attenuation coefficient of the sample at the energy E_0 (eq.5.27).

$$\mu_{m0} = \frac{\mu_{\ell 0}}{\rho}$$

431

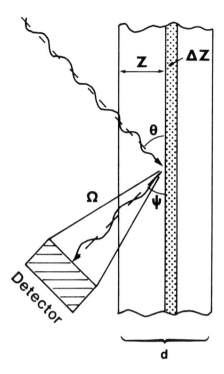

Figure 12-7. The irradiation geometry for secondary excitation
at the depth 'z' in a sample of thickness 'd'.
The incident angle for the photons is θ and the
secondary photons are emitted at an angle ψ in
direction to the detector which cover a solid
angle Ω.

The fraction of primary photons penetrating to depth z is:

$$\frac{n}{n_0} = e^{-\frac{\mu_{m0}\rho z}{\cos\theta}} \qquad (12.4)$$

The photons at this depth interact in the layer of thickness
dz with the sample nuclei to produce vacancies in the K- or L-
shells of the element 'k', followed by X-ray emission.

As the linear attenuation coefficient μ_ℓ also is the macroscopic cross section (sec. 5.5) we define the macroscopic cross
section of photoelectric absorption in element 'k' as

432

$$\mu_{\ell k}^{pe} = \Sigma_{pe,k} = \rho_k \frac{N_A}{A_k} \sigma_{pe,k} \qquad (12.5)$$

The number of vacancies produced in layer dz now can be written (eq. 5.25).

$$dn = n \; \Sigma_{pe,k} \; (\cos\theta)^{-1} dz \qquad (12.6)$$

The number of X-rays produced is

$$dn_x = n\omega_k \; \Sigma_{pe,k} \; (\cos\theta)^{-1} dz \qquad (12.7)$$

where ω_k is the fluorescence yield[12.37-39].

The intensity of fluorescence X-rays produced at depth z escaping the sample at angle ψ in a relative solid angle $g = \Omega/4\pi$ can thus be written

$$dI = I_0 \; g \; \omega_k \; \Sigma_{pe,k} \; e^{-(\frac{\mu_{m0}}{\cos\theta} + \frac{\mu_{m1}}{\cos\psi})\rho z} \; (\cos\theta)^{-1} dz \qquad (12.8)$$

where μ_{m1} is the mass attenuation coefficient of the sample at the energy E_1 (the X-ray energy from element 'k').

If the detector efficiency of X-rays with energy E_1 is i_1 and if we intergrate over the sample thickness d, which is assumed to be modest, we get the following expression for the counting rate of fluorescence radiation in the detector.

$$c_k = I_0 g \; i_1 \; \omega_k \; \Sigma_{pe,k} \; (\frac{\mu_{m0}}{\cos\theta} + \frac{\mu_{m1}}{\cos\psi})^{-1} \rho^{-1} [1 - e^{-(\frac{\mu_{m0}}{\cos\theta} + \frac{\mu_{m1}}{\cos\psi})\rho d}] (\cos\theta)^{-1}$$

$$(12.9)$$

For very thin samples eq. 12.9 reduces to

$$c_{k,thin} = I_0 \; g \; i_1 \; \omega_k \; \Sigma_{pe,k} \; (\cos\theta)^{-1} d \qquad (12.10)$$

Here we observe that all information about the element studied is given by the product $\omega_k \; \Sigma_{pe,k}$ which of course is proportional to the density of atoms of the element in the sample.

For very thick samples in which $\rho d \rightarrow \infty$ eq. 12.9 reduces

$$c_{k,thick} = I_0 \; g \; i_1 \; \omega_k \; \Sigma_{pe,k} \; (\frac{\mu_{m0}}{\cos\theta} + \frac{\mu_{m1}}{\cos\psi})^{-1} \rho^{-1} (\cos\theta)^{-1}$$

$$(12.11)$$

This is the normal case in trace element analysis where the matrix is known. The density of atoms of the trace element does not significantly alter the matrix absorption (μ_{m0} and μ_{m1}).

These formulas from monoenergetic excitation can easily by extended to excitation with broad-energy distributed beams by summation or integration over measured X-ray distributions.

Photon-excited X-ray fluorescence analysis has been extensively used and appears to be a most useful method for general analytical application. The characteristics of photon-induced fluorescence and detector limits are best discussed in term of a hypothetical fluorescence spectrum, as shown in fig. 12-8.

There are three important modes of interaction for photons in the energy range of interest (sec. 5.5.1):

- Rayleigh scattering
- Compton scattering
- Photoelectric interaction

In Rayleigh scattering, an incident photon interacts with a tightly bound atomic electron and is thereby elastically scattered with no loss of energy. In Compton scattering, the inter-

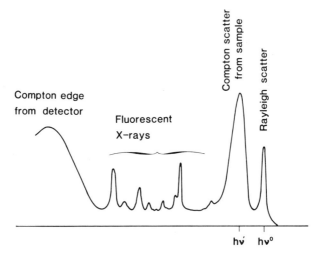

Fig. 12-8.   Idealized hypothetical X-ray fluorescence spectrum resulting from excitation of trace elements in a low atomic-number matrix using monoenergetic photons.

action takes place with a loosely bound electron. In the photoelectric interaction, the energy of the incident photon is
totally absorbed by an electron.

The Compton process and the photoelectric interaction are
described in detail in sec. 5.5.

The photons scattered by either Rayleigh or Compton processes can reach the detector and contribute with undesirable
bumps in the spectrum. The fluorescence radiation of the element in question and its relative cross section for K-or L-shell
ionization causes the spectrum peaks. The production is related
to the photoelectric absorption. An idealized X-ray fluorecence
spectrum resulting from excitation of trace elements in a low
atomic-number (Z) matrix using monoenergetic photons (E_0) is
shown in fig. 12-8.

The highest energy peak in the spectrum occurs at the energy corresponding to the incident photons, and is due to photons
that Rayleigh-scatter from the sample and are totally absorbed
in the detector. The broad peak at slighly lower energy is due
to the inelastic Compton scattering (E_γ') in the sample. The width
of this peak is determined by the width of the scattering angles of the photons who are recorded according to the Compton
relationship (eq. 5.33). The series of small peaks is due to
fluorescence radiation from the trace elements, which are totally absorbed in the detector.

The continuous distribution beneath the peaks is due to multiple scattering and bremsstrahlung from the sample, or poor charge collection in the detector. The low-energy continuous distribution is due to Compton scattering in the detector in connection with escape of the scattered photons.

The low-Z elements in the sample contribute to most of the
scattered X-rays, and the high Z-trace elements produce the
fluorescence peaks.

In a sample of high-Z material such as minerals or alloys
the magnitude of the Rayleigh peak increases in relation to the
Compton peak, because of the approximate Z^2-dependence of the
Rayleigh scattering cross section compared to the Z-dependence
of Compton scattering cross section.

The spectra of the fluorescent X-rays from high-Z samples
are often very complex with several overlapping peaks.

The relationship between trace element concentration and the area under the corresponding X-ray peaks are often nonlinear due to absorption and enhancement effects in high-Z samples.

12.5.2 EXCITATION SOURCES

In actual practice it is very difficult to provide a perfectly monochromatic photon source for excitation. Characteristic X-rays generated directly by an X-ray tube or by a radioactive source are used. A secondary fluorescence system in which the X-rays from the primary source strike a secondary pure element target is also in use.

In some cases, however, it is more convenient to use a continuous bremsstrahlung spectrum, which produces a wide-band excitation and thus gives a more uniform sensitivity over a wide range of elements. In this case, however, the background below the fluorescence peaks is higher than with excitation by monoenergetic photons, because the scattered radiation extends over the whole spectrum.

The geometries used for photon excitation are illustrated in fig. 12-9 and fig. 12-10. Radionuclide excitation is caused by one of the radioactive sources listed in table 12-11. These sources are commercially available ready-mounted for either direct

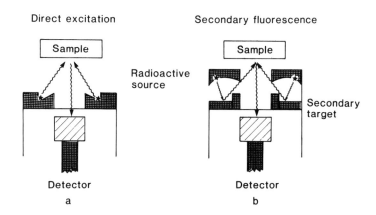

Fig. 12-9.   Schematic geometries for excitation by using a radioactive source
a) Geometry for direct excitation, b) Geometry for secondary fluorescence excitation.

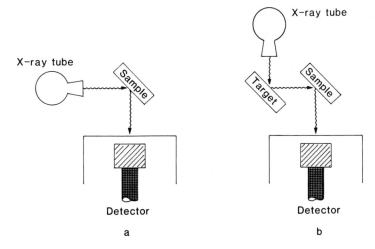

Fig. 12-10.   Schematics of typical geometries for excitation
              by using an X-ray tube as radiation source.
              a) Geometry for direct excitation
              b) Geometry for secondary fluorescence or polari-
              zed X-ray excitation.

or secondary fluorescence excitation[(12.40)]. In the case of direct excitation, the source is mounted in a ring collimator of
very pure heavy metal above the detector, as shown in fig. 12-9a.
A secondary target can easily be installed in the system, as
shown in fig. 12-9b.

   X-ray tubes provide a well controlled photon source, but
require high-voltage power supply and associated items, and
are therefore more expensive than a radioactive source.

   The output from a typical X-ray tube consists of the elec-
tron-excited characteristic X-rays of the anode material together with bremsstrahlung spectrum, as shown in fig. 12-11.

   The output can be filtered with an absorber of the same
material as the anode to provide a source predominately composed of the characteristic X-rays of the anode material. The
most common anode materials in X-ray tubes are tungsten and
molybdenum, which give characteristic X-rays of about 18 keV
and 60 keV, respectively.

   A better method of producing monoenergetic radiation with

Table 12-11

Radionuclide sources used for photon excitation in X-ray fluorescence analysis.

RADIONUCLIDE	HALF-LIFE	USEFUL RADIATION		TYPICAL ACTIVITY
		Type	Energy keV	
^{3}H + Ti-target	12.3a	Ti KX-rays	4-5	200 GBq
		continuum	3-10	
^{3}H + Zr-target	12.3a	continuum	2-12	200 GBq
^{55}Fe	2.7a	Mn KX-rays	5.9	800 MBq
^{57}Co	270.0d	Fe KX-rays	6.4	100 GBq
		γ	14	
		γ	122	
		γ	136	
^{109}Cd	1.3a	Ag KX-rays	22	120 MBq
		γ	88	
^{125}I	60 d	Te KX-rays	27	
		γ	35	
^{147}Pm + Al-target	2.6d	continuum	12-45	100 GBq
^{153}Gd	236d	Eu KX-rays	42	40 MBq
		γ	97	
		γ	103	
^{210}Pb	22a	Bi LX-rays	11	40 MBq
		γ	47	
^{238}Pu	89.6a	U LX-rays	15-17	
^{241}Am	470a	NP LX-rays	11-22	
		γ	26	
		γ	59.6	

X-ray tubes employs the secondary target system, as shown in
fig. 12-10b.

The use of monochromatic X-ray excitation results in particularly high sensitivity for detection of the element with an
absorption edge slightly lower in energy than the exciting
radiation. This selective method is the opposite of the method
using continous radiation for excitation, where the detection
sensitivity is lower, but a coverage of a wide range of elements
is achieved in one exposure.

A recent development using polarized X-ray beams gives
certain advantages, as no scattering occurs parallel to the
direction of polarization of the exciting radiation, which results in a drastically reduced background[12.41-12.43].

The current techniques for generating polarized X-ray beams
involve multiple scattering, which results in a considerable

438

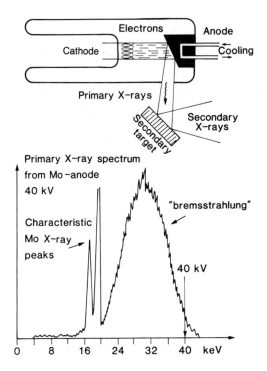

Fig. 12-11.   Schematics of an X-ray tube and the output spectrum.

reduction of intensity. The synchrotron radiation produced by
very high-energy electron storage rings has high intensity
and will be a future alternative to polarized photon sources.

12.5.3 SAMPLE PREPARATION AND ANALYSIS

X-ray fluorescence analysis is a non-destructive method,
but the measurement of trace levels requires that the sample
is prepared in some standardized fashion. This implies some
simple sample preparation, such as polishing to obtain a plane
surface on solid samples, or producing pellets of powder samples. Microanalysis can require some concentration process, such
as collecting small particulate samples on a filter. Low concentrations of soluble elements can be concentrated on ionex-
change resin. It is desirable that the substrate (filter or
ion-exchange resin) on which the sample is collected (or mounted) is of low-Z-material, (as pure as possible).

Table 12-12

Detection limits for vaious elements in different types of sample matrix in parts per million (ppm).

Element	Iron and steel W-target 2240W,10 min	Fe and Ni base alloys W-target 2025W,100s	Mineral oil Cr-target 1000W, 10 min	Mineral oil W-target 1000W, 10 min	Whole blood
Mg					
Al					
Si	170	4			
P		35	5		
S		8	2		
Cl			0.7		
Ca			0.15		
Ti	1				
V	2			0.2	
Mn	1.4	5			
Ni	5.4			0.1	
Cr	4.0	1			2.6
Fe					
Cu	8.5	12			
Zn				0.1	
As	6.8				1.7
Zr	4.6				
Br					

Single-element detection limits obtained in a series of different types of samples using Mo K X-ray for excitation are in the range of 1-10 ppm for most elements at a counting time of 1000 s or 17 min. Similar detection limits can be achieved using a radionuclide source for excitation, but with a counting time in the order of 3-24h[12.44,12.45].

A summary of the detection limits for X-ray fluorescence trace-èlement analysis using an X-ray tube for excitement is given in table 12-12[12.46-12.48].

The method of photon-induced X-ray fluorescence analysis has a wide range of applications in metallurgy, geology, mining and for processing in archeology and cosmology.

12.6 References

12.1   E.G. Fuller et al., NBS Special Publication 380 (1973) Suppl. 1(1978).

12.2   B.L. Berman, Lawrence Livermore Laboratory Preprint, UCRL-78482 (1976).
       B. Bülow, B. Forkman, Tech. Report Series 156 IAEA 1974.

12.3   G.J. Lutz, Anal. Chem. 41, 424 (1969).

12.4   V.P. Guinn, H.R. Lukens, ANS Trans. 9, 106 (1969).

12.5   K.D. George, H.H. Kramer, Nucl. Appl. Tech. 7, 385 (1969).

12.6   C.A. Kienberger, TID - 4500 (1970).

12.7   D.E. Rundquist, Proc. Int. Conf. Photonucl. Reac. and Appl. Asilomar, 619 (1973).

12.8   T. Kato, Y. Oka, Talanta 19, 515 (1972).
       T. Kato et al., Jour Radioanal. Chem. 32, 51 (1976).

12.9   Ch. Engelmann, Advances in Activation Analysis 2 (1972) Academic Press.

12.10 Ch. Engelmann, Proc. Int. Conf. Photonucl. Reac. and Appl. Asilomar, 1137 (1973).

12.11 M. Fedoroff et al., Jour. Radioanal. Chem. 38, 101 (1977).

12.12 P. Bock et al., Jour. Radioanal. Chem. 38, 97 (1977).

12.13 D. Brune, Anal. Chim. Acta 44, 9 (1969).

12.14 J.S. Hislop, Proc. Int. Conf. Photonucl. Reac. and Appl. Asilomar 1159 (1973).

12.15 G.H. Andersson et al., Nuclear Techniques in the Life Sciences, Proc. of a Symposium Amsterdam IAEA, 99 (1967).

12.16 C.M. Gordon, R.E. Larson, Radiochem. Radianal. Letters 5, 369 (1970).

12.17 J.S. Hislop, D.R. Williams, Nuclear Activation Techniques in the Life Sciences, Proc. of a Symposium, Bled IAEA, 51 (1972).

12.18 J.S. Hislop, D.R. Williams, Analyst 97, 78 (1972).

12.19   G.J. Lutz, Jour. Radioanal. Chem. 19, 239 (1974).

12.20   A. Chattopadhyay, R.E. Jervis, Radiochem. Radioanal.
        Letters 11, 331 (1972), Anal. Chem. 46, 1630 (1974).

12.21   C.E. Dutilh, H.A. Das, Radiochem. Radioanal. Letters 6,
        195 (1971).

12.22   M.G. Davydov et al., Atomic Energy Jour. 39 nov. INDC
        (CCP)-104/LN IAEA (1975).

12.23   V. Galatanu, M. Grecescu, Jour. Radioanal. Chem. 10,
        315 (1972).

12.24   P. Andersson et al., AERE - R 7823.

12.25   M.E. Toms, Jour. Radioanal. Chem. 20, 177 (1974).

12.26   G.J. Lutz, Ch. Segebade, Jour. Radioanal. Chem. 33,
        303 (1976).

12.27   A. Veres, K. Yoshihara, N. Ikeda, Radioisotopes 27, 43
        (1977).

12.28   I.A. Abrams, L.L. Pelekis, Latv. PSR Zinatnu. Akad.
        Vestus un Techn. Zin. ser. No 5, 45 (1967).

12.29   A. Veres, private communication.

12.30   A. Veres, J. Radioanal. Chem. 38, 155 (1977).

12.31   K.G. Malmfors, in Alpha-, Beta- and Gamma-Ray Spectro-
        scopy (ed. K. Siegbahn; North-Holland (1965)), 1281.

12.32   B.D. Sowerby, Nucl. Instr. and Meth. 108, 317 (1973).

12.33   H. Langhoff Phys. Rev. 135, B 1 (1964).

12.34   R.M. Laszewski, Thesis Urbana Illinois 1975.

12.35   G. Ben-David et al., Phys. Rev. 146, 852 (1966).
        B. Arad, G. Ben-David, Ann. Rev. Nucl. Sci. 24, 35 (1974).

12.36   D. Vartsky et al., Nucl. Instr. and Meth. 193, 359 (1982).

12.37   W.J. Gallagher and S.J. Cipolla, Nucl. Instr. Meth. 122,
        405 (1974).

12.38   W. Bambynek et al., Rev. Mod. Phys. 44, 716 (1972).

12.39   J.S. Hansen, H.V. Freund and R.W. Fink, Nucl. Phys. A142,
        604 (1970).

12.40   J.R. Rhodes, Design and Application of X-Ray Emission
        Analyzers Using Radioisotope X-Ray or Gamma Ray Sources,
        in: Energy Despersion X-Ray Analysis: X-Ray and Electron
        Probe Analysis. ASTM Special Technical Publ., 485, 243
        (1970).

12.41   T.G. Drubay, B.V. Jarrett and J.M. Jaklevic, Nucl. Instr.
        Meth. 115, 297 (1974).

12.42   L. Kaufman and D.C. Camp, Advan. X-Ray Anal. 18, 2
        (1975).

12.43   R.H. Howell, W.L. Pickles and J.L. Cate, Jr. Advan.
        X-Ray Anal. 18, 265 (1975).

12.44   R.D. Giuque, F.S. Goulding, J.M. Jaklevic and R.H. Pehl
        Anal. Chem. 45, 671 (1973).

12.45   J.C. Russ, A.O. Sandborg, M.W. Barnhart, C.E. Söderquist,
        R.W. Lichtinger and C.J. Walsh, Advan. X-Ray Anal. 16,
        284 (1973).

12.46   W.J. Campbell and J.W. Thatcher, Bureau of Mines Report
        of Investigations 5966 (1962).

12.47   R. Louis, Z. Anal. Chem. 208, 34 (1965).

12.48   M.F. Lubozynski, R.J. Baglan, G.R. Dyer and A.B. Brill
        Int. J. Appl. Rad. Isotopes 23, 487 (1972).

Part D.
Applications

13 Sample Preparation and Chemical Separation

13.1 Pretreatment of sample before irradiation

Usually samples to be analyzed with NAA technique consist of specimens weighing from a few milligrams up to some grams. Solid pieces such as metals are, for example wrapped in pure aluminium foil and then placed in aluminium cans, which are introduced in a desired reactor position pneumatically. Water samples and biological specimens are usually dried (freeze dried or dried at $105^{\circ}C$) prior to irradiation. In well thermalized reactor positions with fluxes of about 10^{12} n cm^{-2}s^{-1} irradiations can usually be accomplished without any serious complications. However, in medium high flux reactor positions with thermal fluxes of about 10^{13} n cm^{-2}s^{-1} containing a high proportion of fast neutrons as well as a high γ-flux, the effect of radiolysis in water or biological speciment will complicate the irradiation procedure. In water samples fast neutron and γ-interactions lead to the formation of various radicals and atomic species such as $OH^{\cdot}$, $H^{\cdot}$, e^{-}_{aq} (solvated electron), H_2, O_2 and H_2O_2.

Thus, in sealed quartz containers of such samples irradiated under undried conditions in gram quantities, high gas pressures are built up in the ampoules. Due to this pressure, the container may break during irradiation, or the opening of the ampoule after irradiation may be complicated inasmuch as the sample may disperse in fragments when it is broken.

The pressures generated by gaseous decomposition products of biological material such as freeze-dried orchard leaves has been measured by Becker and La Fleur[13.1]. According to their experiments 1/2 gram of freeze-dried orchard leaves irradiated for 6 hours in a neutron flux of $5 \cdot 10^{13}$ n cm^{-2}s^{-1}

produced about 1 cm^3 of gaseous decomposition products, corresponding to approximately 1 atm excess pressure in the ampoule used. It should, however, be pointed out that several
irradiation positions in various research reactors, yielding
a thermal neutron flux similar to that one used in this experiment, will cause a considerably higher radiolytic pressure
due to higher γ-radiation levels.

Because of these effects, the water content of the sample
is usually eliminated by drying procedures prior to irradiation. In dried biological specimens radiolysis processes still
occur to some extent with the matrix elements. The pressure
in the sealed ampoules can, however, be strongly reduced by
cooling the ampoules in liquid nitrogen, prior to opening.

In the drying procedure the utmost care must be taken
with volatile elements such as Hg, As, Sb and Se. According
to studies by Westermark and Sjöstrand[13.2] and Gorsuch[13.3]
serious losses of these elements may occur during drying procedures. The studies by Gorsuch[13.3] refer to dry ashing at
a temperature of 600° C. Losses are very dependent on the
composition of the matrix. Freeze drying procedures seem so
far, to be a reliable drying method, though it has been reported that significant losses of various elements may even
occur with this method[13.4].

According to Parr[13.5], considerable amounts of chromium are lost from various biological specimens during drying
procedures, even at low temperatures.

Considerable volume reduction occur during drying. Dry
ashing of biological material such as tissues reduces the
amount to the percentage level. Drying at 105° C or freeze-
drying procedures result in a sample volume reduction by a
factor of about 3 to 5.

13.2 Contamination of the sample with impurities from the container suface

There are several reports in the literature of methods

that yield detection limits lower than 10^{-10} g for various elements. However, such methods are valid only under ideal conditions.

All analytical techniques involve various inherent sources of contamination, e.g. in the sampling procedure, or through contact with chemicals, reagents and containers in the analytical procedure.

Thus in trace element analysis at the ppm ($\mu g \cdot g^{-1}$) level or lower the application of non-nuclear techniques imposes great demands on the purity of the reagents employed during the course of analysis. Thus it is possible that the element which is to be assayed is found in the reagents, even in those of pro-analysis quality, at the level of 1/100 to 1 ppm. Such contamination effects are easily eliminated in activation analysis since the samples are only chemically treated after the activation procedure.

However, impurities in the container material may give rise to a contamination of the sample in NAA procedures, as explained below.

Nuclides from the container material may strike the sample directly owing to the recoil mechanism in the neutron capture process. The contamination caused by this effect is, however, almost negligible as a nuclide in the mass region of 50 would have range of magnitude of less than 100 Å (10 nm) in common container materials. However, the containers are often rinsed in order to obtain a quantitative removal of the sample from the container and during this operation the magnitude of the contamination might be considerably increased, due to the dissolution of activities from a certain depth of the material.

The pre-concentration of an aqueous sample by evaporation to dryness before irradiation is accompanied by contamination risks that arise during the cleaning process when the substance is separated from the container. In such cases, the container is often washed with acid so as to ensure the quantitative removal of the sample, and the nuclides in the surface of the container consequently become associated with the sample. When the container is rinsed, the contamination

Table 13-1

Amounts of Cu, Mn, Na and Sb extracted through washing of polyethylene and quartz containers in HCl, expressed in µg per
cm^2 container area[13.6]. Single, double or triple measurements
indicated.

Element	Polyethylene container			Quartz container	
Cu	0.004	0.007	0.008	0.001	0.001
Mn	0.00006,	0.00008,	0.00008	0.00002,	0.00003
Na	0.002			0.03	
Sb	0.00008			0.0001,	0.0001

is increased owing to the extraction of activities from depths
of up to 10^{-2} and 10^{-3} mm in containers made of polyethylene
and quartz, respectively. Owing to the operation of the Szilard-Chalmers process, the active nuclides can easily be extracted from the surface[13.6].

Leaching experiments with pure containers revealed that
when the polyethylene and quartz containers were rinsed by
thorough shaking with hydrochloric acid, the nuclides ^{64}Cu,
^{56}Mn, ^{24}Na and ^{122}Sb were identified in the extracts. The
amounts of the elements corresponding to these activities
are given in table 13-1.

What then is the implication of such interference effects for practical analysis? Let us take as an example the
analysis of the copper content of drinking water, assumed
to be 1 ppb or 1 µg·l^{-1}. A water sample of 1 ml volume is irradiated in a "clean" polyethylene vial with 5 cm^2 surface area
containing 0.1 µg·g^{-1} copper as an impurity. From this container copper can be leached out down to a depth of the order of
0.01 mm, thus yielding an analytical result which is about
50% of the "real" copper in the aqueous sample. Similarly,
biopsy or autopsy specimens may be contaminated by the metals
of which surgical instruments are made to yield analytical
errors of one order of magnitude in unfavourable cases.

In trace element analysis of sea-water samples, diver-
gent results have been obtained with reference to absorption

and contamination effects with polyethylene containers used in sampling. Thus, it has been reported that mercury in dilute solutions may be lost through adsorption mechanisms on container walls of polyethylene. However, it has also been observed that acidified sea-water samples can be contaminated with mercury due to leaching processes from the container walls or due to diffusion from the ambient air. Volatilization losses of mercury can be strongly reduced adding various compounds to sample or standard[13.7].

13.3 Contamination during sampling

Detailed studies of contamination associated with the collection of liver biopsies and human blood samples have been carried out by Versieck and Speecke[13.8]. They simulated the contamination process occuring when sampling the specimens, using medical equipment that was made radioactive by irradiation prior to cutting.

Thus, needles for veinpuncture, surgical blades and Menghini needles for percutaneous liver biopsy were irradiated in a reactor to obtain a high specific activity of the various elements in the instruments. Conditions for collecting human blood and liver biopsies were simulated in vitro. Blood samples of about 20 ml were collected by puncturing a plastic tube filled with blood, using the irradiated needles. Liver biopsies were taken with the radioactive surgical blades or with Menghini needles. Under these experimental conditions the radioactivity transferred into the various biological samples represents possible contaminations. The values, converted into ppm, were compared with data given in the literature for the concentrations of trace elements in normal human serum and liver tissue. During the collection of blood, contaminations of manganese, copper, chromium, iron, nickel, cobalt, zinc and tantalum occur. The contaminations were largest in the first 20 ml flowing through the needle. For manganese, contaminations of 0.5 to 0.7 ppb were obtained in serum, the concentration of manganese in normal

serum being approximately 0.6 ppb. Iron contamination reached about 5 to 10% of the normal Fe concentration in serum.

In liver biopsies taken with irradiated surgical blades
contaminations of manganese, copper, scandium, chromium, iron
nickel, cobalt, zinc, silver, antimony and gold were observed. The calculated concentrations usually remained low by
comparison with the concentrations in normal human liver tissue.

Contaminations induced by taking liver biopsies with a
Menghini needle were large: they could exceed the concentrations in normal human liver tissue. For cobalt, contaminations of up to 0.24 ppm were found, while in the literature
values of approximately 0.06 to 0.20 ppm (wet tissue) are
given as a normal concentration. For chromium, contaminations up to about 4 ppm were found. Thus, the use of this
sampling method is unreliable for determinations of certain
trace elements. Wester[13.9] has demonstrated that these
contamination effects may be avoided by using quartz knives
when cutting autopsy specimens.

Contamination in the NAA of specimens of various kinds
can be avoided to a large extent by applying a technique that
is illustrated in the following two examples.

Consider the bulk analysis of a piece of metal. The specimen is first contaminated in the sampling step by contact
with tools. The sample is then exposed to the irradiation
procedure and appropriate amounts of activities of various
nuclides are induced in the sample. However, after completion
of the irradiation procedure, the contaminated surface layer
can be removed from the bulk of the sample through chemical
cleaning or mechanical polishing. In this way, a "pure" sample
is obtained ready for subsequent activity measurements, which
are carried out either directly on the "pure" material or after
appropriate chemical separation of the desired nuclides. The
addition of chemicals and reagents, and contact with containers
after irradiation cannot introduce any interference in the
analysis.

In the trace element analysis of aqueous samples, food
sources and biological specimens, the technique used is in
principle the same. Hence the irradiation of the sample can

be carried out while the sample is kept frozen, and the contaminated surface layer may be removed after irradiation.

Thus, the irradiation of normally liquid samples is
carried out in the frozen state. The surface of the frozen
aqueous sample, which has been in contact with the container
is then allowed to melt and is discarded prior to analysis.
For a frozen biological specimen, the contaminated surface
can be cut off using a microtome technique. In this case,
both contaminations that arises in the sampling step in
connection with medical tools and that from container effects
can be prevented[13.10].

13.4 Technique of cold irradiation

From a practical point of view, the radiolytic pressure
which arises in liquids during irradiation complicates the
irradiation when the samples are exposed to higher radiation
doses, as the pressure built up in closed containers can reach
several atmospheres and cause explotions. In frozen samples,
however, the radiolytic pressure is strongly suppressed. Those
samples which are liquid at room temperature can therefore be
irradiated in the solid state in nuclear reactors at approp-
riately low temperature for prolonged periods, if such a procedure is preferred instead of irradiating the dried specimen
at ambient reactor temperature.

Moreover, the technique of cold irradiation facilitates
the opening of containers that have held biological specimens
or other normally liquid samples, because the pressure is reduced. Accordingly, the loss of volatile gases that are produced during irradiation can be avoided when the ampoule is
opened.

As previously pointed out (see sec. 13.1), volatile elements and their compounds may be lost during a drying procedure of an aqueous or biological specimen. Such losses can be
avoided by irradiating the sample while frozen.

Various cooling devices for the neutron irradiation of
samples at low temperatures are currently in use in connec-

tion with experiments in solid-state physics or physical metallurgy.

Liquid nitrogen cooling: The sample is inserted in pure
liquid nitrogen in the reactor. The cooling medium is kept at
a lower temperature through circulation in a heat exchanger
installed outside the reactor core. Commercial liquid nitrogen used for cooling is applied in the heat exchanger. Within
the reactor core, pure liquid nitrogen containing no oxygen
has to be used in order to avoid explosions. In this way,
samples can be irradiated at temperatures down to 77 K[13.11].

Liquid helium cooling: With a liquid helium facility inserted in the reactor core, it is possible to perform the neutron irradiation procedure at 5 K. Such facilities are mainly
used in "solid-state" experiments[13.12].

Cold nitrogen gas cooling: With such devices, neutron
activation analysis of liquid specimens can easily be performed for routine purposes at low expense. The refrigerator
system is capable of cooling several samples simultaneously
at about 250 K for long irradiation periods[13.13].

Dry-ice cooling: In reactor positions with low γ-heat
i.e. less than 0.05 Wg^{-1}, it is possible to irradiate normally
liquid specimens in the frozen state by using dry-ice cooling
in a double-walled aluminium container. For short irradiation
periods, this system seems favourable for neutron activation
analysis applications. The sample temperature is about 230 K
[13.14].

The neutron activation of liquids or biological specimens has previously been carried out by irradiating the samples in the frozen state, using a dry-ice cooling technique,
or by employing cold nitrogen gas or helium cryostat for cooling purposes. This technique is used mainly in neutron activation analysis and in hot atom chemistry[13.10].

13.5 Destruction of the irradiated sample

The irradiated sample, e.g. a metal fragment or a dried
biological specimen, is preferentially dissolved in a hot

mineral acid, or in a mixture of hot acids after the addition
of carriers. Various mixtures are used in the wet combustion
such as H_2SO_4 + H_2O_2, H_2SO_4 + $HClO_4$, or HNO_3 [13.15, 13.16].
A rapid destruction can be accomplished by fusing the samples
in mixtures of Na_2O_2 and NaOH [13.17].

Dry ashing at a low temperature is suitably performed
through combustion in oxygen streams containing excited oxygen atoms produced in discharge processes. Types of apparatus suitable for such procedures are available on the commercial market.

When selecting a suitable destruction method, however,
the effect of losses of volatile elements or their compounds
from a given matrix should always be kept in mind. The necessary precautions should be taken, for example the use of closed systems during destruction, or the use of reflux conden-
sor techniques.

13.6 Radiochemical separations

13.6.1 ADDITION OF CARRIERS

According to an investigation by Lavrukhina et al. [13.18]
elements present in solution at trace levels are found in
chemical states similar to those in macro element chemistry,
i.e. in the ionic state, as molecules, as colloides etc.

However, at low concentration levels anomalies can be
observed regarding the chemical behaviour of trace elements
in experimental studies. Often such phenomena can be ascribed
to adsorption effects on container walls or adsorption on
particles present in solution. To avoid these effects so called carriers, i.e. small amounts of the element to be determined, of the order of e.g. mg quantities, are added to the
sample.

During the production of the radionuclide (e.g. in a
(n, γ) process) the nuclide formed is "hot", and will often
attain a valency state differing from that of its presursor
atom. However, through repeated oxidation and reduction processes after the addition of carriers a desired valency state

can be obtained according to classical chemical rules. Due to
radiation effects in solutions (so called radiolysis) which
produce solvated electrons and oxygen radicals together with
several other species, the valency states of radionuclide and
tracer element may change with time in solutions of strong
specific activity as a result of oxidation or reduction mechanisms.

13.6.2 CHEMICAL SEPARATION METHODS

   The current literature gives details of several hundred
alternative separation methods. The majority of these methods,
however, are scarcely original and depart only slightly from
methods already familiar in conventional chemical analysis
(see e.g. bibliography in ref. 13.19).

   Thus precipitation, distillation, extraction, electro-
lysis, ion exchange, isotopic exchange, paper chromatography
and other methods are widely used in activation analysis for
the separation of individual nuclides or groups of nuclides
(e.g. 13.20).

   The simultaneous determination of a large number of elements has been found to be of great interest in metallurgy,
biology, geochemistry, environmental control and medicine.
Group separation systems based on the above mentioned chemical techniques have been developed and applied by several investigators (e.g. 13.21-13.26).

   When biological materials are irradiated with thermal
neutrons, the activities induced are strongly dominated by a
few nuclides, mainly ^{38}Cl, ^{24}Na, ^{42}K, ^{82}Br and ^{32}P (13.27),
this makes it essential to apply a chemical separation tech-
nique for the satisfactory determination of most of the other
elements in such materials (see fig. 13-1).

   Successful separations of negatively charged metal complexes have been performed with anion exchangers. An extensive study of the adsorption of a large number of elements
on anion-exchange resins in a hydrochloric acid medium has
been carried out in the classical study by Kraus and Nelson
(13.28). Since the complex equilibrium is displaced when the
chloride concentration is changed, desired adsorptions and

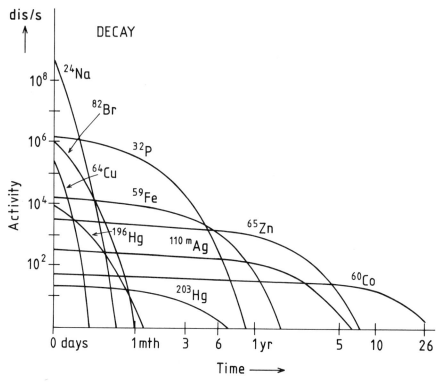

Fig. 13-1. Growth and decay of induced activities in various elements in 1 cm³ of normal blood according to Comar[13.27].

elutions are easily performed by appropriate changes of the hydrochloric acid concentration.

This separation technique has been applied successfully to neutron activation analysis in connection with group separations. Separations of this type have been used in the early days of NAA by Blanchard and Leddicotte[13.22] in the determination of various elements in water and by Samsahl[13.23] when separating a large number of elements in various materials. The technique has also been widely applied in trace element analysis of biological specimens like blood, human and animal heart materials, and liver tissues in normal and diseased cases[e.g. 13.29, 13.30].

13.6.3 REMOVAL OF SODIUM

After destruction of the irradiated sample a selective chemical separation method is applied to isolate a desired nuclide before activity measurement. Alternatively, optimized separation schemes are used in group separations. These methods only differ from those used for inactive elements in that a small quantity of radioactive material is also present.

In connection with the problems associated with the dominance of the activity of ^{24}Na in biological, geological and several other matrixes, special methods for the removal of this nuclide have been developed. According to Girardi and Sabbioni ^{24}Na is retained on a column of hydrated antimony pentoxide (HAP). The approach seems to constitute a reliable basic tool for activation analysis applications in the biomedical field and in studies of water pollution. ^{24}Na has thus been reported to be quantitatively retained on HAP with a capacity of 30 mg Na per g HAP. Retention studies of about 60 ions in hydrochloric acid medium on HAP have been done. Only fluorine was reported to be partly retained, while tantalum was quantitatively adsorbed on this column. The other ions passed through the column[13.31].

Using another system reported by Menon and Wainerdy[13.32] ^{24}Na can be eliminated from solutions by precipitation as NaCl in a mixture of butanol and hydrochloric acid, where the solubility of NaCl is low.

Further, ^{24}Na can be separated by precipitation with organic reagents like 5-benzamino-anthraquinon-2-sulfonic acid, according to Bock-Werthman and Schulze[13.33]. A column of tindioxide can selectively retain arsenic while sodium pass the column[13.34].

13.6.4 GROUP SEPARATION

The two past decades have seen the growth in popularity of the so called group separation methods. There are now a large number of such systems based on combination of methods whereby, using defined conditions, between 10 and 40 elements can be determined in a matrix. For this purpose the treatment provides a separation of the different radionuclides in groups

457

suitable for γ-ray spectrometry without mutual interference.

One of the inventors of the group separation approach was Samsahl[e.g.13.15]. His system achieved a grouping of the radionuclides by the successive application of ion exchange, distillation, extraction chromatography, selective sorption etc. In the automated device, specimens are prepared which permit the simultaneous determination of up to 40 elements in a wide range of matrixes, for example biological or geological samples. The technique is especially suited to application in medical screening tests and for monitoring the pollution of waterways. According to this system bromides and oxides of volatile elements such as As, Hg, Sb and Se are distilled from the irradiated biological specimens after destruction with sulphuric acid and hydrogen peroxide. After distillation the elements are subdivided into various groups.

Time consuming separation steps are avoided in Samsahl's[13.15] automatic group separation system. The procedure involves the combined use of chelating resins, anion exhange resins, inorganic exchangers and partition chromatographic columns. Selective sorption is performed on columns in series. Before entering the column the solution is adjusted for selective sorption of the specific group by mixing the effluent of the preceeding column with a suitable solution. With this technique the adsorption-elution cycle, usually applied in ion exchange separations, is reduced to that of the first adsorption-washing step, consequently separation times are reduced. The whole procedure can be completed within interval of less than 2 hours. The principle of the separation scheme is outlined in fig.13-2.

In another development of the group separation approach to neutron activation analysis, Morrison and Nadkarni have developed a system capable of dividing more than 40 elements into six groups. This system has been applied with success to the analysis of lunar samples. Since each group contains several nuclides, measurement with high resolution Ge(Li) spectrometry is necessary[13.35].

13.6.5 SOLVENT EXTRACTION

Solvent extraction procedures are widely used in various radiochemical operations, mostly in the separation of fission

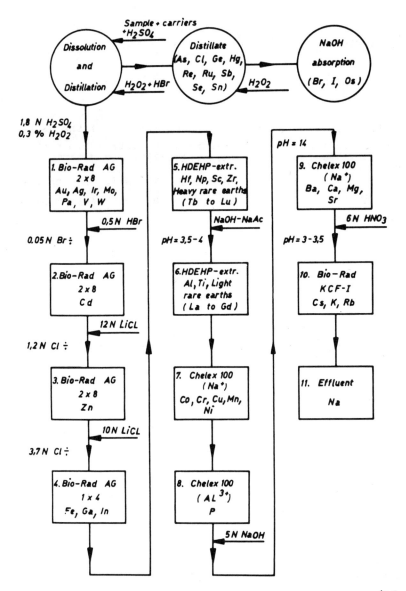

Fig. 13-2. Group separation scheme according to Samsahl[13.15].

products. The technique has further found applications in activation analysis.

Consider a solute to be distributed between two phases, i.e. two immiscible solvents at a defined temperature. At

equilibrium conditions the distribution coefficient, K_D, is
simply defined by the ratio:

$$K_D = \frac{[x_1]}{[x_2]} \tag{13.1}$$

where $[x_1]$ and $[x_2]$ denote the concentrations of the solute
in the two solvents. In literature such distributions are described by various factors[13.36, 13.37].

Now, let us consider a metal ion M^{N+} which reacts with
an organic reagent HA, forming an uncharged chelate MA_N, distributed between two phases:

$$M^{N+} + N(HA)_{org} \rightleftharpoons (MA_N)_{org} + NH^+ \tag{13.2}$$

The equilibrium constant, K, of the above reaction is:

$$K = \frac{[MA_N]_{org}\,[H^+]^N}{[M^{N+}][HA]_{org}^N} \tag{13.3}$$

K depends on the temperature and on the ionic strength of the
aqueous phase.

The practical performance of a solvent extraction process
is achieved by bringing the two solvents into intimate contact
e.g. by vigorous stirring whereby the solute is extracted in
the desired solvent in one or repeated operations. The two
solvents are then separated preferentially in a funnel.

The general theory and technique of extraction is described e.g. by Stary[13.36] or Marcus and Kertes[13.37].

As pointed out above, solvent extraction can be performed
by bringing two solvents in intimate contact with each other
by stirring. Solvent extraction may also be carried out on a
column. In this case the extractant is absorbed on a bed of
inert hydrophobic particles. The solvent extraction process
is then carried out by flowing the aqueous solution through
this bed. This technique is called reversed-phase partition
chromatography, and was introduced by Siekierski and Kotlinska[13.38]. The technique has been successfully utilized in

the separation of transuranium elements. In this variant of
the solvent extraction procedure, the usual solvent extraction processes - extraction, scrubbing etc - are carried out
on the column[13.39].

A comprehensive study of the extraction of a large number of elements from mineral acids by means of the reagent
Di-(2-ethylhexyl) orthophosphoric acid (abbreviated HDEHP)
has been performed by Quershi et al.[13.40] for NAA purposes.
A group separation scheme for the analysis of biological materials based on solvent extraction of As, Cd, Co, Cu, Fe, Hg,
Mo, Sb and Zn with zincdiethyldithiocarbamate and N-benzol-N-
phenylhydroxylamine from a strong acid and neutral media has
been developed by Kučera[13.16]. Further, Goode et al.[13.41]
have developed an automated radiochemical group separation
method based on solvent extraction. This separation scheme
comprises the fractionation of a large number of radionuclides
in six groups with extraction in various organic solvents,
e.g. tributyl phosphate. The separation scheme is outlined
in fig. 13-3.

13.6.6 ION EXCHANGE SEPARATIONS

Ion exchange separations of both cations and anions may
be effected by utilizing the exchange properties of groups
of synthetic resins or of natural products such as zeolites.

In nature ion exchange processes occur with e.g. clays
and minerals. Among the minerals the zeolites are of importance for ion exchange processes. These consist of aluminium
silicates of a complicated composition. Zeolites are also produced synthetically and are mainly used for softening water.
The softening proceeds by the passage of water containing
calcium ions through a bed of cation exchanger to which sodium ions are attached. Calcium ions in the water are then
exchanged for sodium ions according to the reaction mechanism:

$$Ca^{2+} + 2Na^{+}(Z) \underset{\text{regeneration}}{\overset{\text{sorption}}{\rightleftharpoons}} 2Na^{+} + Ca^{2+}(Z) \qquad (13.4)$$

Z is the cation exchanger matrix.

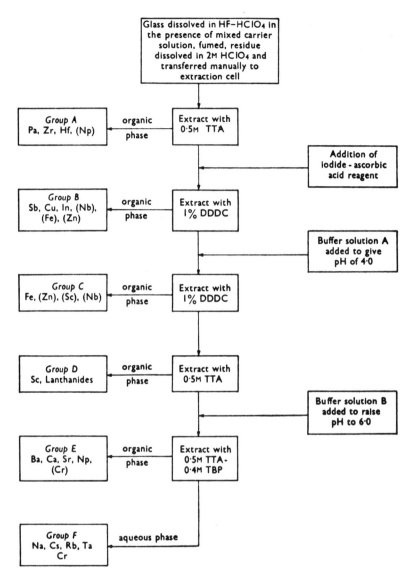

Fig. 13-3. Group separation scheme according to
Goode et al.[13.41]

When the sodium ions of the bed are exhausted, the bed
is regenerated by treatment with a strong solution of sodium
chloride. In the regeneration step the calcium ions are displaced and the exchanger is reconverted into the sodium form.

Otherwise, ion exchangers consisting of organic materials are widely used. Anion exchangers often contains a basic group e.g. an amino group, bound to the insoluble resin. The amino groups have the ability to form salts with common acids. The anions bound to the resin in this way may then be exchanged for other anions. Thus nitrate ions in a solution may be exchanged with chloride ions attached to the amino group of the resin R_S:

$$NO_3^- + Cl^- \ H_3N^+ \ (R_S) \ \xrightleftharpoons[\text{regeneration}]{\text{sorption}} \ NO_3^- \ H_3 \ ^+N \ (R_S) + Cl^-$$

$$(13.5)$$

Let us consider the desalination of water by means of ion exchange techniques. The process is described schematically in fig. 13-4.

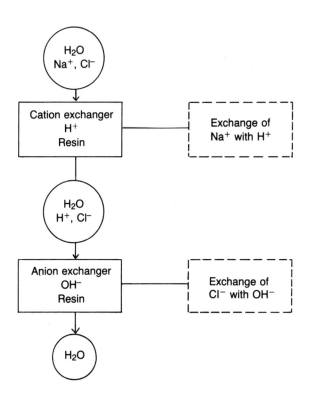

Fig. 13-4. Desalination of water.

As a rule of thumb, the capacity of an ion exchange resin which depend strongly on the type is often of the order of 1 milli equivalent per ml of the resin.

Successful separation of negatively charged metal complexes have been performed with anion exchangers. An extensive study of the adsorption of a large number of elements on anion exchange resins in a mineral acid medium has been carried out by Kraus and Nelson[13.28], as previously mentioned. Their technique has been successfully applied in a variety of NAA separations.

The ion exchange separation technique is described in e.g. the textbook by Rieman and Walton[13.42].

13.6.7 ISOTOPE AND AMALGAM EXCHANGE

Radiochemical separation by means of amalgam exchange techniques has found several applications in NAA studies. Rapid and selective radiochemical procedures have been developed by De Voe et al.[13.43] for the elements cadmium, thallium, lead and bismuth.

Since the amalgam exchange process is rapid, the technique is useful for the analysis of short-lived nuclides. The separation of a radionuclide is accomplished through the exchange of an element present in a dilute amalgam and its ion in the solution. Amalgams of bismuth, cadmium, gallium, indium, lead, tin, thallium and zinc have been prepared by combining the metals with mercury. The amalgams contained 2% by weight of the element to be exchanged.

The principle of the amalgam separation process is as follows.

Consider a radioactive monovalent metal ion $*M^+$ and the liquid amalgam M(Hg) in intimate contact; the following equilibrium conditions are attained:

$$*M^+ + M(Hg) \rightleftharpoons *M(Hg) + M^+ \qquad (13.6)$$

In order to obtain a quantative separation of $*M^+$ in the amalgam phase, there must be a large quantity of the metal M in the amalgam compared to the amount of $*M^+$.

In this case the radioactive ion $^*M^+$ in solution exchanges with a metal atom in the amalgam phase. The $^*M^+$ ion is reduced and passes into the amalgam system, while an inactive M^+ ion is dissolved.

The equilibrium is usually obtained within a few minutes provided good contact is obtained within the phases formed (e.g. through electrovibration).

As an example, we may take the separation of mercury from a nitric acid solution in droplets of mercury, described by Kim and Silverman[13.44] in a study relating to determination of mercury in seeds. The asterisk denotes the radioactive nuclide.

We then have the following equilibria:

$$^*Hg_2^{2+} + 2\ Hg(solid) \rightleftharpoons 2\ ^*Hg(solid) + Hg_2^{2+} \qquad (13.7)$$

$$^*Hg^{2+} + Hg(solid) \rightleftharpoons\ ^*Hg(solid) + Hg^{2+} \qquad (13.8)$$

This separation technique has also been used with success in the separation of mercury from fish tissues. With such technique, the exchange time increases with increasing amount of solution. The correlation yield expressing the degree of separation for various sample volumes versus exchange time in a nitric acid solution is illustrated in fig. 13-5[13.45].

In this figure the mercury exchange curves obtained for aqueous solutions with volumes of 20, 100 and 200 ml respectively are shown. From these curves it is seen that exchange times of only a few minutes are needed to obtain yields exceeding 95 per cent for a solution with volumes of 20 ml. For larger volumes the exchange time would have to be increased if higher yields were desired. Thus for a 500 ml volume the exchange time needed to attain a yield of 95 per cent exceeds one hour. It should be mentioned that the present results were obtained with vigorous stirring of the solutions, and that the curves were found to be less steep with less vigorous stirring.

Monnier and Loepfe[13.46] separated silver, gold and

465

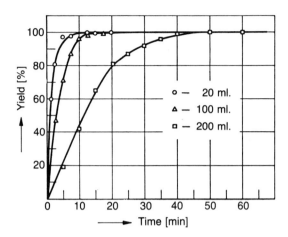

Fig. 13-5. Yield of mercury amalgam exchange verus
exchange time[13.45].

platinum in solution by reduction through contact with mercury and then amalgamation. The amalgam separation technique
has been found to be selective and yields for silver, gold
and platinum exceeding 99% have been obtained in time periods
of 0.5 - 4 min. The following reactions have been reported to
take place:

$$8 \; Hg + 2 \; ^{*}AuX_4^{-} \rightarrow 2 \; (Hg) \; ^{*}Au + 3 \; Hg_2X_2 + 2 \; X^{-} \qquad (13.9)$$

$$5 \; Hg + {}^{*}PtX_6^{2-} \rightarrow (Hg) \; ^{*}Pt + 2 \; Hg_2X_2 + 2 \; X^{-} \qquad (13.10)$$

X is a monovalent complexing ion, e.g. Cl^{-}.

A rapid isotope exchange technique has been developed by
Braun and Farag[13.47], comprising the separation of radioiodide and radiosilver respectively on plasticized iodide and
silver dithizonate foams with a column technique. The reaction of radioiodide and the immobilized iodine in the foam is
as follows:

$$2 \; ^{*}I^{-} \; (aq.) + I_2(foam) \rightleftharpoons 2 \; I^{-} \; (aq.) + {}^{*}I_2(foam)$$
$$(13.11)$$

The asterisk denotes the radioisotope of iodine.

Analogously, the isotope exchange reaction of silver on the primary silver dithizonate foams is presented by the following equation:

$$*Ag^+(aq.) + AgHDz(foam) \rightleftharpoons Ag^+ (aq.) + *AgHDz(foam)$$

$$(13.12)$$

The asterisk denotes the radioisotope of silver.

Chemical separations have further been accomplished directly on polyurethane foams according to Bowen[13.48].

13.6.8 ELECTROPLATING

During about one decade considerable attention has been paid to the analysis of mercury in connection with pollution studies. Several methods are given in the literature for performing NAA of this element in various matrices[13.49-13.52]. According to a method utilized by Ljunggren et al.[13.49], the irradiated sample comprising e.g. fish tissue is digested in a mixture of nitric and sulphuric acid. Mercury is distilled as the chloride and then electroplated as metal on a gold foil. Detection limits down to 0.3 ngg^{-1} sample have been reported for this technique.

Electrodeposition is also needed for preparation of samples for α-spectrometry. This is particularly important when analyzing the actinide elements. Methods have been developed for separation of actinides from interfering bulk elements. The samples are ashed and dissolved in 1M HNO_3 - 93% methanol solution. When this solution passes through an anion exchange column the actinides are absorbed while interfering elements pass through. The actinides can then be eluted with 0.5 M HNO_3 and electroplated from $(NH_4)_2 SO_4$ medium on stainless steel discs. The α-particles are measured with silicon surface barrier detectors according to Holm and Persson[13.53]. This method can also be used for sequential elution of uranium, thorium, plutonium, americium and curium[13.54].

13.6.9 SUBSTOICHIOMETRIC SEPARATIONS

To a certain extent, so called substoichiometry has been applied to chemical separations pertinent to NAA. The technique, which is in principle similar to isotope dilution, ensures a simpler chemical procedure, comprising analytical steps such as precipitation, solvent extraction, ion exchange chromatography etc. The assessment of a complete chemical recovery is not required with this technique. Work in the field has been introduced by Ruzicka and Stary[13.55].

The method implies 1) dissolution of the irradiated specimen, 2) addition of a known amount of carrier of the element to be determined, X. The specific chemical separation step for the element is then carried out; this does not need to be complete. The activity of the desired nuclide in the separated fraction is denoted $^*r$.

The total activity of the nuclide in the original specimen, r, is:

$$r = {}^*r \, \frac{X}{M} \qquad\qquad (13.13)$$

where M is the recovered fraction.

Treating the standard sample in the same way, we have the similar relationship for the standard, denoted by s:

$$r_s = {}^*r_s \, \frac{X_s}{M_s} \qquad\qquad (13.14)$$

We denote the amount of the element sought in the sample Y and the amount of the standard Y_s. Then, if the same amount of the element is added to sample and standard, i.e. $X = X_s$, we obtain the following simple expression for calculating the amount of the element present in the original sample assuming $M = M_s$;

$$Y = Y_s \, \frac{{}^*r}{{}^*r_s} \qquad\qquad (13.15)$$

The substoichiometric separation technique has been used in e.g. the analysis of trace amounts of mercury in biological specimens[13.56].

A comprehensive review of substoichiometry in trace element analysis, covering the principle of the method as well as practical applications, has been given by Kudo and Suzuki [13.57]. Further, the limits of detections in substoichiometric isotope dilution has been surveyed by Das et al [13.58]. Substoichiometric separation of chromium has advantageously been accomplished by Shigematsu and Kudo [13.59].

13.6.10 THE SZILARD-CHALMERS EFFECT

The Szilard-Chalmers effect is widely applied in radiochemical separations, as is well known. This process is based on the change in chemical valence which occurs when an atom is subjected to a nuclear process, e.g. an (n,γ), (n,p) or (n,α) reaction, and leaves its original position due to the recoil effect. The recoil energy is obtained from the following expression:

$$\bar{T}_R = \frac{\Sigma I_\gamma E_\gamma^2}{1862\ M \Sigma I_\gamma} \qquad (13.16)$$

Where $\bar{T}_R$ denotes the weighted mean value of the recoil energy, M the atomic mass of the product nucleus and E_γ the energy of the quantum emitted with intensity I_γ. For atoms with masses of between 20 and 200 the recoil energy would vary between about 1000 eV and 100 eV, assuming a single prompt γ-ray emission of 6 MeV; this might lead to the rupture of most of the chemical bonds, which usually have energies of some eV. The radioactive nuclides produced may then exist in chemical states that are different from the inactive nuclides allowing chemical separation. In an ideal case the isolated active nuclides are carrier-free.

The recoil atoms, "hot" atoms, resulting from the neutron capture reactions travel a certain distance in the material before being thermalized. Thus, in haemoglobin molecules of unit-cell dimensions 109·63·53 Å, the recoil length of an ^{59}Fe particle might be about 30 Å, assuming the slowing down processes to be essentially due to elastic encounters with carbon and oxygen.

The question is then how the Szilard-Chalmers effect may

be utilized in NAA of biological material. It is usual to destroy the organic material with mineral acids after the irradiation. This is followed by a chemical separation step before the actual activity measurements are carried out. However, the complete destruction of the organic material is often time-consuming and may consequently prevent the determination of short-lived nuclides. A fast separation of the activity from the organic matrix by extraction would in such cases be valuable. This form of separation may, moreover, be of interest for the determination of any of the elements inasmuch as the analysis is facilitated.

Fast destruction of organic materials has been performed by fusing the samples in a mixture of sodium peroxide and sodium hydroxide. In this case, however, there might be a risk of loss of activities due to volatilization.

Let us now consider a biological specimen as an example for a Szilard-Chalmers separation procedure. In this connection it is necessary to have some knowledge about the chemical binding of the atom concerned[13.60]. In biological systems certain elements such as sodium and chlorine, present in about equivalent amounts in blood serum, occur in an ionic state. Further, the metabolism of bromine in most living organism is associated with the bromide ion. On the other hand, iron is mainly bound to a protein, such as porphyrin, or as in haeme-type compounds. Vitamin B_{12} contains cobalt, molybdenum is associated with the flavoprotein enzymes, and phosphorus occurs mainly as esterified P, as lipid P and as different nucleotides. In muscle tissue, about 30% of the magnesium present occurs as a free ion, while about 70% is bound in different complex forms. It might thus be exptected that the elements occuring as free ions together with the corresponding activities would be mostly extractable from the irradiated sample in a suitable medium. Further, the activities formed in the organically bound elements ought also to be extractable to some extent owing to the Szilard-Chalmers effect, as has also been proved[13.61].

Magnesium can suitably be determined using the Szilard-Chalmers separation process before performing the γ-spectrometric assay. With this technique magnesium has been determined in needle biopsy specimens from living human muscles (^{27}Mg $t_{1/2}$ = 9.5 min)[13.62].

13.7 References

13.1   D.A. Becker and P.D. La Fleur, J. Radional. Chem. 19, 155 (1974).

13.2   T. Westermark and B. Sjöstrand, Int. J. Appl. Rad. Isotopes 9,1 (1960).

13.3   T.T. Gorsuch, Analyst 87, 112 (1962).

13.4   D. Behne and P.B. Bratter, 4th Symp on the Recent Developments in NAA, 4-7 Aug, 1975, Churchill College, Cambridge.

13.5   R.M. Parr, Workshop on Chromium Analysis, Colombia, Mo, USA, 1-2 May, (1974).

13.6   D. Brune, Radiochim. Acta 5, 14 (1966).

13.7   T. Tackeuchi et al. J. Radioanal Chem. 53, 1 (1979).

13.8   J.M.J. Versieck and A.B.H. Speecke, Proc. of a Symp. on Nucl. Activation Techniques in the Life Sciences 1972, IAEA, Vienna 1972.

13.9   P.O. Wester, Acta Med. Scand. Suppl. 439, (1965).

13.10 D. Brune, Sci. Total Environ. 2, 111 (1973).

13.11 L. Bochirol et al., CEA-R 2514 (1964).

13.12 H. Meissner et al., Euronuclear, 2, 277 (1965).

13.13 N. Berglund et al., Nucl. Instr. Meth. 75, 103 (1969).

13.14 D. Brune and O. Landström, Radiochim. Acta, 5, 228 (1966).

13.15 K. Samsahl, Sci. Total Environ. 1, 65 (1972).

13.16 J. Kučera, Radiochem. Radioanal. Letters 24(3), 215 (1976).

13.17 Y. Kusaka and W.W. Meinke, NAS-NS 3104 (1961).

13.18 A.K. Lavrukhina et al., Chemical Analysis of Radioactive Materials. The Chem. Rubber Co., Ohio (1967).

13.19 G.J. Lutz (Ed), NBS, Techn. Note 467, parts 1 and 2 (1971).

13.20 F. Girardi, Modern Trends in Activation Analysis. NBS, Special Publ. 312, Vol.1, 577 (1969).

13.21 P. Ahlberg, Proc. of the Radioactivation Analysis Symposium, Vienna, June, (1959). Butterworths, London (1960).

13.22 R.L. Blanchard and G.W. Leddicotte, ORNL 2620 (1959).

13.23 K. Samsahl, AE-54, AE-56 (1961).

13.24 J. Hoste et al., Instrumental and Radiochemical Activation Analysis. Butterworths, London 42 (1971).

13.25 L.O. Plantin, Proc. of a Symp. on Nucl. Activation Techniques in the Life Sciences. IAEA, Bled, Yugoslavia, 10-14 April, 73(1972).

13.26 P.S. Tjioe et al., J. Radioanal. Chem. 16,153 (1973).

13.27 D. Comar, Article in Advances of Activation Analysis. Editors J.M.A. Lenihan and S.J. Thomson. Academic Press 163(1969).

13.28 K.A. Kraus and F. Nelson, Proc. Intern conf. peaceful uses of atomic energy. Geneva, 7, 113 (1955).

13.29 D. Brune et al, Clin. Chim. Acta 13, 285 (1966).

13.30 P.O. Wester, Scand. J. Lab. Clin. Invest. 17, 357 (1965).

13.31 F. Girardi and E. Sabbioni, J. Radioanal. Chem 1, 169 (1968).

13.32 M.P. Menon and R.E. Wainerdi, Proc. of the 1965 conf. on Modern Trends in Activation Analysis. Texas A and M University, College Station 152(1965).

13.33 W. Boch-Werthman and H. Schultze, Proc. IAEA Conf. on Nuclear Activation Techniques in the Life Sciences. Amsterdam, IAEA, Vienna, 173 (1967).

13.34 K. Rengan et al., J. Radioanal Chem. 54, 341(1979).

13.35 G.H. Morrision and R.A. Nadkarni, J. Radioanal. Chem. 18, 153 (1973).

13.36   J. Stary, The Solvent Extraction of Metal Chelates,
        Pergamon Press (1964).

13.37   Y. Marcus and A.S. Kertes, Ion Exchange and Solvent Ex-
        traction of Metal Complexes. John Wiley and Sons (1969).

13.38   S. Siekierski and B. Kotlinska, Atomnaya Energiya 8, 160
        (1959).

13.39   E.K. Hulet, J. Inorg. Nucl. Chem. 26, 1721 (1964).

13.40   I.H. Qureshi et al., Proc. of the 1968 Int. Conf. Modern
        Trends in Activation Analysis. NBS, Gaithersburg, Mary-
        land U S, Oct 7-11, (1968). NBS special publ. 312, 1,
        666.

13.41   G.D. Goode et al., Analyst, Sept, 728 (1969).

13.42   W. Rieman and R.S. Walton, Ion Exchange in Analytical
        Chemistry, Pergamon Press (1970).

13.43   J.R. DeVoe et al., Talanta 3, 298 (1960).

13.44   C.K. Kim and J. Silverman, Anal. Chem 37, 1616 (1965).

13.45   D. Brune and K. Jirlow, Radiochim. Acta 8, 161 (1967).

13.46   D. Monnier and E. Loepfe, Anal. Chim. Acta 41, 475 (1968).

13.47   T. Braun and A.B. Farag, J. Radioanal. Chem. 25, 5 (1975).

13.48   H.J.M. Bowen, J. Chem. Soc. Section A, (1970).

13.49   K. Ljunggren et al., Proc. of a Symp. on Nuclear Techni-
        ques in Environmental Pollution. Sanlsburg 26-30 Oct,
        (1970). IAEA, Vienna (1971) 373.

13.50   P. Strohal and D. Huljev, Ibid, 439.

13.51   R.E. Wainerdy et al., Ibid, 459.

13.52   A.R. Byrne et al., Ibid, 415.

13.53   E. Holm and R.B.R. Persson, In: Advances in Radiation
        Protection Monitoring IAEA, Vienna, 1978.

13.54   E. Holm and R.B.R. Persson, In: The Natural Radiation
        Environment III. Rice University, Houston, USA (1978).

13.55   J. Ruzicka and J. Stary, Talanta 10, 287 (1963).

13.56   J. Ruzicka and C.G. Lamm, Analyst, Sept., 157 (1969).

13.57   K. Kudo and N. Suzuki, J. Radioanal. Chem. $\underline{26}$, 327 (1976).

13.58   H.A. Das et al., J. Radioanal. Chem. $\underline{24}$, 383 (1975).

13.59   T. Shigematsu and K. Kudo, J. Radioanal. Chem. $\underline{59}$,63 (1980).

13.60   D.A. Phipps, Metals and Metabolism, Oxford Chemistry series, Clarendon Press (1975).

13.61   D. Brune, Anal. Chim. Acta $\underline{34}$, 447 (1966).

13.62   D. Brune, H. Sjöberg, Anal. Chim. Acta $\underline{33}$, 570 (1965).

14 Nuclear Chemical Analysis in Biological and Medical Research

14.1 Biological significance of trace elements

It has long been known that certain elements occurring in trace quantities in living tissues, such as copper and zinc which are associated with enzymes, hormones and vitamin systems, play a fundamental role in the process of biosynthesis. Similarly, it was discovered that cobalt is a constituent of vitamin B_{12}, that selenium is associated with vitamin E and that cobalt, copper and manganese are involved in the synthesis of hemoglobin. These elements are all easily determined using NAA technique, which thereby becomes a valuable tool for biomedical investigations[14.1, 14.2].

In a variety of diseases it has been shown that the levels of certain of these trace elements in blood and tissues exhibit differences as between the pathological and normal cases. For example conditions of fever, schizophrenia and chronic and acute leukaemia have been shown to be associated with an increase in the copper content of the blood.

To reach reliable conclusions on such matters requires considerable and painstaking effort, not least as regards the performance of the actual trace-element analysis. Straightforward correlation of results obtained from a few sick and healthy individuals is renderd difficult by the observation that where a variety of analytical techniques has been applied considerable differences can be found in the concentration of a given element in a specific tissue. To some extent these differences can be accounted for in terms of biological variation produced by diet, geographical condition, life history including age of individual etc. It therefore becomes necessary to establish the extent of the normal variation and the concentration of a given element before conclusions regarding the supposedly related pathological condition can be drawn. This, in its turn, requires a large number of chemical analyses to be performed on a wide variety of individuals followed up by statistical analysis of the results. Investigations of this type have demonstrated that trace elements in biological specimens follow different statistical distributions. Gaussian distributions occur frequently as do skewed distributions. At this point, however, it should be mentioned that a number of the earlier analysis are undoubtedly in error as a result of contamination (see secs. 13.2, 13.3).

One of the principal advantages of activation analysis in the field of medical-biological research is quite simply that contamination can be readily avoided. The specimen chosen for analysis is irradiated first and then treated with reagents. If these reagents should contain traces of elements capable of producing contamination effects, their presence is of no significance since they were absent at the irradiation stage. Thus, due to the high sensitivity and the possibility of analysis without interference, NAA is particular suitable for the characterization of trace elements in tissues, in biological fluids such as blood, urine and cerebro-spinal fluids, of biopsy specimens and even in cellular fractions.

14.2 NAA as an aid in medical diagnosis

With NAA technique as a future medical diagnostic tool in mind, interesting results were obtained in a comparison of the concentrations of the elements arsenic, bromine, copper, gold, iron, molybdenum, selenium and zinc in the blood of normal individuals and of those suffering from uraemia. It was expected that in the pathological case, where kidney activity is reduced, the blood would exhibit an abnormal concentration of arsenic and other elements, which are normally eliminated in the urine. In point of fact, the results demonstrated an appreciably higher arsenic concentration in the uraemic blood, namely an increase by a factor of ten over that in normal blood. Furthermore, molybdenum was found to occur in uraemic blood at a concentration of twice that observed in normal blood[14.3].

Similarly it has been shown that the arsenic content of various tissues in the human body increases by a factor of 2 to 3 in connection with the onset of uraemia[14.4].

Activation analysis has also demonstrated its power as a diagnostic aid in connection with the infant disease "Cystic Fibrosis", which is hereditary and relatively uncommon. The condition is accompanied by an abnormally high concentration of electrolyte in the sweat (NaCl). It is accordingly possible to detect the onset of this illness by performing sweat tests. In the meantime high concentrations of sodium have been demonstrated in the nail tissue of infants who exhibit this pathological condition, so that it is now possible to detect the illness at an early stage by a routine analysis of nail pairings. The technique has been applied in this way in screening examinations carried out at various university hospitals[14.5]. Table 14-1 gives the concentration of sodium in different parts of free nail edges of a control group of 5 cases as well as 5 cases of cystic fibrosis according to Kollberg and Land-

Table 14-1

Sodium concentration of different parts of the free nail edges[14.5], CF = Cystic Fibrosis.

	I		II		III		IV		V		
	R	U	R	U	R	U	R	U	R	U	Mean
Control											
Right	575	612	504	456	676	660	645	615	722	602	621
Left	810	445	870	730	650	680	604	520	473	571	
CF											
Right	4620	3970	4620	4440	4480	2530	5460	6480	10690	5930	5713
Left	4440	4020	8260	7830	4350	4150	5540	5330	9270	7620	

R = radial, U = ulnar

Values in μg/g

ström[14.5]. In the CF cases the mean sodium content is about 9 times higher.

Great care must be shown if such samples are cleaned prior to analysis. The ^{24}Na-activity induced may easily be extracted in water from nail-clippings, as shown in fig. 14-1[14.5].

The nuclear technique can also be applied as a diagnostic tool for the detection of lung diseases initiated by hard metal dust. In one instance a number of workers in a hardmetal industry were afflicted by the lung disease pneumoconiosis, which was attributed to the inhalation of hardmetal dust. Activation analysis of lymph node tissue from the lungs of a number of individuals was performed in a search for cobalt, tantalum and tungsten which are the common constituents of hardmetal alloy. The diseased tissue together with lymph-node tissue from normal individuals, and samples of various hardmetal alloys were irradiated simultaneously. The activities in the specimens were compared. The results clearly showed that a high level of tungsten was associated with the diseased tissue, while the samples of tissue from individuals showing no signs of pneumoconiosis were free of this metal[14.6].

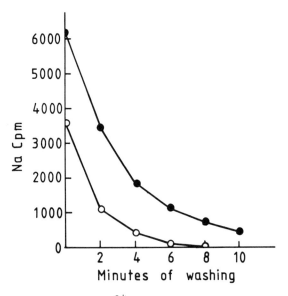

Fig. 14-1.   Extraction of ^{24}Na in water from nail clippings refering to the CF case and of controls respectively[14.5].

478

Table 14-2
Trace element distribution in lung and other tissues[14.10].

Tissue	Mean Concentration (ppm Wet Weight)						
	Sb	As	Cd	Hg	Cu	Mn	Zn
No of cases	15	9	8	4	5	4	6
Right lung							
Apex	0.084	0.014	0.33	0.094	1.26	0.135	10.3
Middle	0.038	0.011	0.40	0.070	1.29	0.169	9.3
Base	0.033	0.009	0.28	0.062	1.11	0.109	10.9
Left lung							
Apex	0.087	0.015	0.30	0.046	1.19	0.146	9.8
Base	0.035	0.011	0.21	0.068	1.04	0.104	8.9
Lymph glands	0.258	0.038	0.16	0.056	1.18	0.732	24.8
Right Paratracheal hilar	0.429	0.066	0.20	0.078	1.42	0.553	21.2
Left Paratracheal hilar	0.339	0.039	0.19	0.090	1.31	0.565	19.3
Visceral pleura	0.037	0.016	0.16	0.147	3.00	0.260	13.0
Trachea	0.007	0.008	0.09	0.050	0.77	0.089	7.3
Pulmonary artery	0.006	0.010	0.15	0.056	1.84	0.089	16.3
Pulmonary vein	0.007	0.012	0.13	0.040	1.49	0.064	14.1
Tongue	0.007	0.009	0.16	0.053	1.24	0.151	28.1

In a joint WHO/IAEA research programme, studies of trace elements in relation to cardiovascular diseases in several countries has been investigated[14.7-14.9]. Thus various trace elements such as Cr, Cu, F, Mn, Si, V, Zn, Cd, Co and Pb were studied in connection with the cardiovascular diseases. According to Plantin[14.8], low contents of calcium, cobalt and sodium of heart tissue in the cardiovascular disease material compared to the control material were observed. The sodium content of the liver was higher and content of manganese lower in the cardiovascular disease material than in the controls. The selenium and chromium content in kidney cortex was lower in the cardiovascular disease material and iron and cobalt contents were higher in this tissue compared to the controls. The effects of different geographical location, industrial activities, diets etc being considered in this investigation.

Molokia and Smith[14.10] studied the distribution of seven
essential and non-essential trace elements in lung tissue using
an NAA technique. Characteristic modes of distribution were
observed, (table 14-2). Cancerous specimens were also examined.

The diagnosis of exocrine pancreas insufficiency in patients with chronic pancreatitis is improved by measurement of
Zn in the duodenal juice sampled during the secretin-pancreo-
zymin-test. The concentrations of Zn and other elements, Fe,
Rb, Co, Cr, Se, Sb, Sc, Cs and Ag were determined by means of
instrumental neutron activation analysis. But the measurement
of the other elements had no significance according to Persig-
hel and Löffler[14.11].

14.3 Determination of O, N, P, Si or K in biological samples

The nuclear reactions for the determination of O, N, P,
Si or K in biological samples using fast neutrons are given
in table 14-3, (see sect 10.3). The determination of O by

Table 14-3

Nuclear reactions for the determination of O, N, P, Si or K
in biological samples using fast neutrons.

Nuclear reaction	$t_{1/2}$	E MeV	Bq/mg
$^{16}O(n,p)^{16}N$	7.4 s	6.13; 7.10	160
$^{31}P(n,\alpha)^{28}Al$	2.24 min	1.78	240
$^{28}Si(n,p)^{28}Al$	2.24 min	1.78	420
$^{29}Si(n,p)^{29}Al$	6.6 min	1.27; 2.43	4.1
$^{31}P(n,2n)^{30}P$	2.5 min	0.511	16
$^{14}N(n,2n)^{13}N$	10 min	0.511	7.6
$^{39}K(n,2n)^{38}K$	7.7 min	0.51; 2.21	2.1
$^{16}O(p,\alpha)^{13}N$	10 min	0.511	
$^{13}C(p,n)^{13}N$	10 min	0.511	

the (n,p) reaction need not to be corrected if the content
of fluorine is low. For the determination of P the Si content
must be considered.

By using the $^{14}N(n_{th}, p)^{14}C$ reaction the 585 keV protons
may be determined by a solid state track detector (cellulose-
nitrate plastic sheets)(see sec. 6.5.4). The protons cause
damage which are extended by etching and are visible in a
light-microscope as 'tracks' and counted. The cross section
for thermal neutrons of ^{14}N is σ_0 = 1.820 mb and that of ^{15}N
only σ_0 = 24 µb. The natural isotopic abundance of ^{14}N is
99.63% and that of ^{15}N 0.37% so there is practically no
contribution of ^{15}N atoms to the representation. By using a
suitable combination of ^{14}N and ^{15}N labelling, new possibilities
are opened for tracer investigations in biological systems,
through neutron-induced auto-radiography (NIAR).

14.4 Fluorine in tooth enamel

It has been generally recognized that the fluorine content,
as well as its distribution in human enamel is of primary importance for the resistance of the human population against caries.
The fluorine content and its depth distribution can be obtained by using a microanalytical method employing the resonance
nuclear reaction $^{19}F(p,\alpha)^{16}0$. The cross section for the fluorine reaction shows a strong resonance at 1375 keV and a weak
resonance at 1348 keV, see Möller and Starfelt[14.12]. Protons
from a 2 MeV proton Van de Graaff accelerator of variable
proton energy are directed to the enamel surface of human
tooth samples. The γ-rays of energies 6.1 or 7.1 MeV emitted in the reaction are detected either with NaI(Tl) or Ge(Li)
detectors. The depth resolutions varies between 0.2-0.6 µm
up to 10 µm depth.[e.g. 14.26]

14.5 In vivo nuclear chemical analysis

The development of in vivo nuclear chemical techniques has opened an area of research into elemental composition of human beings which can be compared to X-ray radiography for internal organs.

With the in vivo technique total body neutron activation (TBNAA) of a number of elements like Ca, P, Na and Cl have been measured.

By using prompt-γ neutron activation techniques Cd concentrations in kidney and liver are measured. The internal deposition of cadmium, lead and iodine can also be measured by using X-ray fluorescence methods. Total body potassium can be measured directly by means of whole-body counting. In table 14-4 the body elements which can be measured by whole body in vivo techniques are given. In table 14-5 elements which can be measured by partial body in vivo techniques are given.

14.5.1 HYDROGEN

The total body hydrogen content has been measured in vivo using neutron irradiation of the patient with an ^{241}Am-Be source and subsequent measurement of the thermal neutron-capture γ-radiation of 2.21 MeV, according to a study by Carlmark and Reizenstein[14.13].

14.5.2 NITROGEN

The method of in vivo neutron activation analysis routinely used for calcium measurements, is not so useful for nitrogen. The reasons are that (n,γ) reaction produced by capturing of thermal neutrons in ^{14}N leads after fast γ-deexcitation to the stable isotope ^{15}N. Fast neutrons, however, produce the radioactive isotope ^{13}N by a (n, 2n) reaction. This nuclide is a positron emitter like ^{15}O and ^{11}C which are also produced by (n,2n) reactions. By measuring the prompt γ-rays from the 10.8 MeV level of ^{15}N produced by neutron capture in ^{14}N it is, however, possible to determine nitrogen in vivo. Various irradiation geometries have been used.

Table 14-4

The elements of the body which can be analyzed by whole body in vivo techniques.

Element (% by weight in standard man)	Reaction	Induced nuclide	Measurement of γ- or X-radiation
Hydrogen (10%)	n_{th},γ	^2H	prompt γ(2.21 MeV)
Nitrogen (3%)	n(14 MeV), 2n n_{th},γ	^{13}N ^{15}N	delayed γ(0.51 MeV) prompt γ(10.8 MeV)
Oxygen (60%)	n(fast),p	^{16}N	delayed γ(6-7 MeV)
Calcium (1.5%)	n_{th},γ n(14 MeV),α	^{40}Ca ^{37}A	prompt γ(many); delayed γ(3.10 keV) delayed X-ray (2.6 keV)
Phosphorous (1%)	n(fast), α n_{th},γ	^{28}Al ^{32}P	delayed γ(1.78 MeV) prompt γ(0.08 MeV)
Sodium (0.15%)	n_{th},γ	^{24}Na	prompt γ(many); delayed γ(2.75 MeV)
Chlorine (0.15%)	n_{th},γ	^{38}Cl	prompt γ(many); delayed γ(1.6, 2.2 and 3.10 MeV)
Magnesium (0.05%)	n_{th},γ n(fast),α	^{27}Mg ^{24}Na	delayed γ(0.84 MeV) delayed γ(2.75 MeV)
Iron (0.006%)	n(fast),p	^{56}Mn	delayed γ(0.84 MeV)
Potassium (0.3%)	^{40}K-EC-β^+-decay	^{40}K	γ(1.46 MeV)

Table 14-5

Elements of the body which can be analyzed by _in vivo_ techniques irradiating only parts of the body.

Element	Reaction	Induced	Measurement of γ- or X-radiation
Fluorine	p,p'γ	^{19}F	prompt γ(110,197 keV)
Iodine	n_{th},γ	^{128}I	delayed γ(0.45 MeV)
	γ, K-X	^{127}I	K-X-ray 28 keV
Manganese	n_{th},γ	^{56}Mn	delayed γ(0.84 MeV)
Copper	n_{th},γ	^{64}Cu	delayed γ(0.51 MeV)
Cadmium	n_{th},γ	^{114}Cd	prompt γ(0.559 MeV)
	γ, K-X	^{113}Cd	K-X-ray
Lead	γ,K-X		K-X-ray

14.5.3 FLUORINE

The activation technique, which uses 3 MeV proton beams, allows _in vivo_ analysis of fluorine in teeth. The fluorine present in the dental enamel in proportions ranging between 0.01 to 0.1 % in weight gives rise to the following reactions:

$$^{19}F(p,p'\gamma)^{19}F \qquad\qquad E_\gamma = 110 \text{ keV}$$

$$^{19}F(p,p'\gamma)^{19}F \qquad\qquad E_\gamma = 197 \text{ keV}$$

The γ-rays are detected with a high resolution Ge(Li) detector. With a proton beam current of 20 nA the energy dissipated in the dental enamel is only 0.055 W. Each measurement takes about 60s which means about 3 J energy imparted[14.14].

The absorbed dose in the mouth area is less than 0.15 mrad (1.5 µGy) for each determination.

In conclusion this technique can be used safely without causing health problems. It can also be used as an aid in the study of the prevention of dental caries.

14.5.4 IODINE

Thermal neutrons are strongly attenuated by biological tissues (about a factor of 2.5 in passage through 2 cm of tissue) and it is evident that a uniform neutron irradiation cannot be obtained by simple means in the thyroid. Because of the marked absorption of neutrons in tissue and the differences in geometrical factors, the conditions required for accurate determinations are difficult to fullfil for in vivo irradiation of the human thyroid. Therefore Lenihan et al.[14.15] used ^{129}I as an internal standard. This nuclide posesses a half-life of $1.6 \cdot 10^7$ year and is readily introduced into the thyroid in amounts which may be measured on exposure to thermal neutrons. ^{129}I yields two radioactive nuclides, ^{130}I ($t_{1/2}$ = 12.5 h) and ^{130m}I ($t_{1/2}$ = 9.2 min).

On the macroscopic scale, the distribution patterns of ^{127}I and ^{129}I in the gland may be the same. Thus, the probability of activation (and of subsequent detection of the induced activity) will be the same for the two nuclides.

Difficulties of measurement produced by complicated geometrical effects were therefore avoided by using the radioactive nuclide ^{129}I as an internal standard. This nuclide, on injection into the body, is concentrated to the thyroid whose position can thereby be located by means of radioactivity measurements. In subsequent neutron activation, ^{129}I transforms to ^{130m}I which is then readily measured. Simultaneously a proportion of the natural iodine in the thyroid, comprising the isotope ^{127}I, undergoes transformation to the radionuclide ^{128}I, which is also readily measured. By using this simple procedure it accordingly becomes possible to determine the iodine content of the thyroid by means of the rule of three. Verification of the method was provided by an experiment carried out on sheep which were subjected to neutron irradiation in a special channel at the nuclear research centre at Saclay

Table 14-6

Total body calcium measured in normal men by total body neutron activation analysis[14.18].

Subject	Age (years)	Weight (kg)	Height (m)	Total-body calcium (g)
1	36	69	1.85	1361
2	23	70	1.85	1254
3	35	76	1.78	1206
4	40	71	1.77	1053
5	36	75	1.74	999
6	41	68	1.69	980
7	22	68	1.68	960
8	35	69	1.66	933
			Mean	1093
			Range	933 to 1361

in France. This method, however, is not recommended for use in human beings due to the radiation exposure involved.

The analysis of stable iodine in the thyroid is possible by registeration of the fluorescent KX-ray radiation emitted when the thyroid is irradiated by γ-radiation from an americium-241 source[14.16]. In vivo measurements of iodine by this method has also been used for studying the pharmaco-kinetics of X-ray-contrast media and to determine the kidney function from the elimination of iodinated compounds according to Ahlgren et al[14.17].

In patients undergoing urography it has been possible to follow iodine concentration by in vivo measurements on the finger-tips of patients with normal kidney function up to 6-10 hours after injection[14.17].

14.5.5 CALCIUM

Total-body calcium measurements by in vivo neutron activation analysis have found valuable applications in calcium ba-

Table 14-7

Total body calcium measured in male patients by total body
neutron activation analysis.

Age (years)	Weight (kg)	Height (m)	Calcium measured (g)	Calcium expected (g)	Deficit (%)	Disease
35	100	1.81	865	1190	27	Hand-Schuller-Christian
40	67	1.78	792	1120	29	Renal osteodystrophy transplant
38	66	1.75	700	1070	35	Osteoporosis
44	57	1.74	730	1050	30	Osteomalacia
52	72	1.63	664	860	23	Osteoporosis

lance studies in connection with changed pathological conditions.

A relationship between height of normal man and total body calcium content determined through NAA technique is given
in table 14-6, reported by Nelp and Palmer[14.18]. Table 14-7
gives the calcium deficiencies resulting from various diseases
as measured by total-body neutron activation analysis.

Also the in vivo method can be used for measurements of
the calcium content of a section of the lumbar spine as demonstrated by Al-Hiti et al[14.19]. In this study the subject was
irradiated with fast neutrons produced by bombardment of a thick
lithium target with 10.2 MeV protons. The neutrons were moderated to the thermal energy level by the light elements occuring
in the body before the appropriate nuclear reaction was initiated. Often the moderation of fast neutrons is accomplished
with water or polyethylene moderators surrounding the patient
(14.18-14.20).

In the NAA procedure the radionuclide ^{49}Ca, on which
the analysis is based, is produced through the reaction ^{48}Ca
(n, γ) ^{49}Ca with thermal neutrons. The neutrons are produced
either in a radionuclide neutron source, by means of a neutron
generator, or in a cyclotron[14.18-14.20].

Central problems in analysis of this kind are the construction of a workable neutron moderating system surrounding the patient, the accomplishment of uniformity of neutron flux density in the irradiated body and the achievement of optimum conditions for the neutron source and the subject geometry.

14.5.6 CADMIUM

Successful results from in vivo measurements of cadmium in liver tissue have been reported by Harvey et al[14.21]. The liver is irradiated with a narrow beam of pulsed neutrons produced in a cyclotron. The prompt γ-radiation following neutron capture in ^{113}Cd($\sigma \sim$ 20 000 b) is registrated in a semiconductor detector. Thus, using an anticoincidence technique liver-cadmium content has been studied in cadavers and in

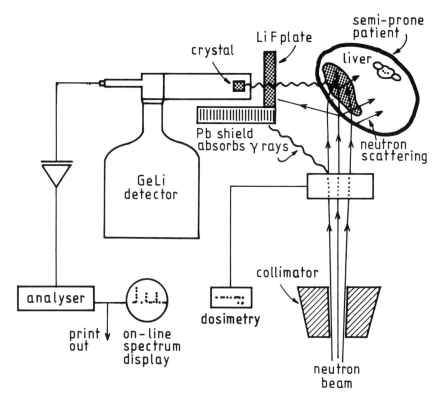

Fig. 14-2.   Experimental set up for in vivo measurement of cadmium[14.22].

four men with known or suspected cadmium poisoning. The patients
all showed high liver cadmium levels of between 35 and 200 ppm
compared with under 1 ppm in non-exposed subjects. The dose of
radiation used in clinical studies was 0.4 - 1.0 rem and the
detection limit of cadmium was 1.0 ppm[14.21]. The experimental
set up for the measurement is given in fig. 14-2.

Measurements of the cadmium concentration in the kidney
have been performed by in vivo X-ray fluorescence analysis[14.25].
A disc-shaped 11 GBq ^{241}Am-source was used to generate the characteristic X-rays of cadmium. The angle between primary and
measured radiation was 110° as seen in fig. 14-3. The colli-
mation of the radiation source and the detector was designed
so that about half the kidney cortex was studied.

Due to the attenuation of the characteristic X-rays of
cadmium and to the short distance between the radiation source-
sample and detector, the minimum detectable concentration is
strongly dependent on the thickness of the tissue layer between the detector and the kidney.

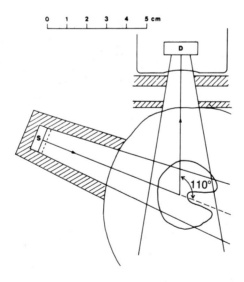

Fig. 14-3. Measurements of the cadmium concentration in renal
            cortex using X-ray fluorescence analysis. The right
            kidney is irradiated by a ^{241}Am source. The posi-
            tion of the Mo-filter is indicated. The angle bet-
            wenn incident and measured radiation is 110°[14.25].

Table 14-8

The mean absorbed dose and the energy imparted in in vivo X-ray
fluorescence analysis (XRF). Data for other type of radiological
examinations are given as a comparison[14.24].

Type of investigation	Mean absorbed dose mGy	Energy imparted mJ
XRF of lead in finger bones	2.5	0.01
Ordinary X-ray picture of a hand	0.65	0.10-0.25
XRF of cadmium in a kidney	0.6	0.4
Prompt-γ neutron activation analysis of cadmium in a kidney	0.57 (neutrons) 0.29 (photons)	1.1 0.6
Kidney function studied by XRF of the fingertip after injection of contrast medium	2.3	0.003
Urography		510
^{51}Cr-EDTA clearence		0.8

The detection limit for cadmium in a kidney located at
3 cm depth is 20 $\mu g \cdot g^{-1}$ and at 5 cm depth 110 $\mu g \cdot g^{-1}$.

14.5.7 LEAD

Since more than 90 % of the lead in the body is concentrated in bone in vivo measurement of the lead in the skeleton
should be a good method for estimating the body lead content.
The in vivo method for lead measurements are based on X-ray
fluorescence analysis by irradiating a finger according to Ahlgren and Mattsson[14.24].

The left forefinger was fixed in a perspex holder and about

1 cm^3 of the second phalanx was irradiated 40 minutes with two ^{57}Co collimated sources having a total activity of (20 mCi ~ 0.8 · 10^9 Bq).

The characteristic KX-rays from lead was recorded with a Ge(Li) spectrometer. The bone mineral content was estimated from the quotient of the coherent and incoherent scattered primary photons. The calibration factor for lead concentration was derived from measurements on a finger phantom made of silica paraffin wax and bone ash. The minimum detectable concentration of lead in forefinger bone was 20 µg for each g of bone tissue.

The mean absorbed dose and the energy imparted in in vivo X-ray fluorescence analysis (XRF) of lead, cadmium and iodine are given in table 14-8.

Data for other type of radiological examinations are also given for comparison purposes.

14.6 References

14.1   V. Valkovic, Trace Element Analysis, Taylor and Francis (1975).

14.2   D.A. Phipps, Metals and Metabolism, Oxford Chemistry Series Clarendon Press (1975).

14.3   D. Brune, K. Samsahl and P.O. Wester, Clin. Chim. Acta 13, 285 (1966).

14.4   N.A. Larsen et al., Proc. of Symp. Nucl. Activation Techniques in the Life Sciences 1972, IAEA, Vienna (1972).

14.5   H. Kollberg and O. Landström, Acta Pädiat. Scand. 63 (1974).

14.6   D. Brune et al., Proc. Int. Symp. Nucl. Activation Techniques in the Life Sciences 1972, IAEA, Vienna (1972).

14.7   M. Masironi, Ibid.

14.8   L-O. Plantin, Proc. Int. Symp. Nucl. Activation Techniques in the Life Sciences, IAEA, Vienna (1979).

14.9   R. Parr, IAEA, Vienna (personal communication) (1976).

14.10 M.M. Molokhia and H. Smith, Arch. Environm. Health 15,
       745 (1967).

14.11 M. Persigehl and A. Löffler, Proc. Int. Symp. Nucl. Acti-
       vation Techniques in the Life Sciences, IAEA, Vienna (1979).

14.12 E. Möller and N. Starfelt, Nucl. Instr. Meth. 50, 225
       (1967).

14.13 B. Carlmark and P. Reizenstein, In Vivo Neutron Activation
       Analysis, Proc. of a Symp. Vienna 17-21, April (1972).

14.14 I. Baijot Shoobants, F. Bodart, and G. Deconninck. Health
       Phys. 36, 423 (1979).

14.15 J.M.A. Lenihan et al., Nature 214, 1221, 5094 (1967).

14.16 P.B. Hoffer, et al. Radiology 90, 342 (1968).

14.17 L. Ahlgren, T. Grönberg, and S. Mattsson, In vivo X-ray
       fluorescence analysis for medical diagnosis. Adv. X-ray
       Anal. 23 (ed. J.R. Rhodes) Plenum (1980).

14.18 W.B. Nelp and H.E. Palmer, Ibid, 193.

14.19 K. Al-Hiti et al., Int. I. Appl. Rad. Isotop. 27, 97
       (1976).

14.20 S.H. Cohn et al., Ibid, 173.

14.21 T.D. Harvey et al., The Lancet, June 7, 1269 (1975).

14.22 B.J. Thomas et al., Proc. of Gasteiner Intern. Symp. Ra-
       dioaktive Isotope in Klinik und Forshung (ed. R. Höfer).
       Verlag H. Egerman (1976).

14.23 L. Ahlgren and S. Mattsson, Phys. Med. Biol. 26, 19 (1981).

14.24 L. Ahlgren and S. Mattsson, Phys. Med. Biol. 24, 136 (1979).

14.25 L. Ahlgren, J-O. Christoffersson, T. Grönberg and
       S. Mattsson: In X-ray Fluorescence (XRF and PIXE) in
       Medicine, p 61-74, Field Educational Italia, 1981.

14.26 U. Lindh and A.B. Tveit, J. Radioanal Chem. 59, 167
       (1980).

15 The Use of Nuclear Chemical Analysis in the Field of Criminology

15.1 Detection of gunshot residues

In criminological laboratories routine tasks include the
identification of the origin of narcotic drugs, powder residues on palms of hands in connection with shootings, hair
fibres, earth samples, blood smears etc. Microscopy, together
with sensitive chemical analytical techniques such as atomic
absorption spectrophotometry and NAA, are used in a routine manner in such investigations at several laboratories.

As regards NAA, a major effort has been directed to the
identification of powder residues on palms of hands in connection with firing. In this connection the determination of antimony and barium present in gunshot residues becomes the main
objective. Detection of gunshot residues on surfaces can further pinpoint the location of the firing.

In the commonly used paraffin test, soft wax is pressed
into the palms of the person suspected of having fired a gun.
The wax impressions are then analyzed for their content of
antimony and barium, which are regarded as specific for cart-
ridge propellants. Both antimony and barium are particulary
suitable for activation analysis and can be readily identified
by their respective nuclides, ^{122}Sb (2.8 d) and ^{139}Ba (83 min).
In connection with the NAA procedure, simple chemical separation steps are performed or an instrumental technique is used.
With instrumental analysis, $1 \cdot 10^{-3}$ µg of antimony and
$2 \cdot 10^{-2}$ µg of barium can be detected[15.1].

A technique described by Hoffman[15.2] depends upon the
absorption or powder residues on cotton swabs. For this purpose one end of a wooden rod is bound with a cotton wool pad
moistened with nitric acid solution. These rods are stored rea-

dy for use in plastic vials and are easy for police staff to handle. The palms of hands and clothes of a person suspected of having fired a shot are brought into contact with the cotton wool which is wiped over the crucial surfaces and then irradiated in a reactor. A person who has recently fired a shot may show antimony and barium amounts that are in the range of 0.3-6 µg and 0.2 - 1.4 µg respectively on the hand in contact with the weapon, compared to the respective amounts of 0.003 - 1 µg and 0.07 - 0.3 µg on the other hand. Different weapons yield different amounts of powder deposited on the hands[15.3, 15.4].

It should also be pointed out that in the event of a bullet passing completely through a body, use of the technique described above makes it possible to distinguish from which side of the body the bullet has entered, since the highest content of powder residues is clearly to be found around the point of entry. It has even been possible to find a connection between the concentration of powder residues on a body and the distance from which the shot was fired, according to studies by Baumgärtner et al.[15.5] and Krishnan[15.3].

15.2 Analysis of hair

Trace element analysis of blood smears, hair fibres, paint samples, etc found either at the site of the crime or on a suspect has been used for purposes of identification. In a number of trials in the United States evidence of this type has been accepted in court. The exactitude of such identification is open to doubt, however, partly in view of the great risk of contamination to which the samples are exposed and partly because of the considerable normal variation often exhibited by trace elements in samples of biological origin. An important evaluation of significant U S court cases involving forensic activation analysis has been presented by Guinn[15.6].

Identification of unknown persons by the analysis of hair strands has received a great deal of attention. It has often occurred that finds at the site of the crime included stands

of hair. These acquire great value if they can be positively
identified with a particular person. Microscopical techniques
are often employed in this type of work. In recent years, however, such identification has been possible by means of comparison analysis of a number of trace elements present in hair
strands. In this context, neutron activation analysis has proved to be of considerable value in view of its very high sensitivity and because it can be performed in a non-destructive
fashion leaving the samples to be presented as evidence in an
undisturbed condition. The principal disadvantage remains the
unreliability of evaluation referring to the known magnitude
of the normal variation of trace elements in hair. In addition, washing and surface contamination are capable of altering
the trace element contents while variations can occur as a
result of age etc.[15.7] Extensive forensic hair analysis for
individual characterization has been performed by e.g. Coleman[15.8].

In connection with a serious mercury accident in Iraq,
the mercury content of normal individuals and of those who
has ingested organic mercury compounds, i.e. methyl mercury,
through consumption of grains treated with fungicides has been
studied by Al-Shahristani and Al-Haddad[15.9]. They observed
that the hair mercury content of "normal" people in unconta-
minated and contaminated areas was from 0.1 - 4 and 1 - 12 ppm
respectively. People who ingested mercury compound but showed
no symptoms had hair mercury concentrations from a few ppm to
300 ppm. Mild symptoms appeared with 120 - 600 ppm hair mercury levels; moderate with 200 - 800 ppm, and severe with
400 - 1 600 ppm.

A comprehensive survey of the applications of nuclear
science in crime investigation has been given by Guinn[15.4].

The use of hair as a mirror of the environment with respect to heavy metal pollution has been surveyed by Lenihan
and Dale[15.10]. Trace element characterization is thus a
suitable indicator of human exposure to various pollutants
as described by Ryabukhin[15.11] (see sec. 16.3).

15.3 Detection of falsification of paintings

Authentication of paintings through a combination of NAA and autoradiography has been successfully demonstrated by Cotter[(15.12)] who irradiated whole paintings in the beam port of the Brookhaven Medical Research Reactor in thermal fluxes of about $10^9 n$ cm^{-2} s^{-1}. The uniformity of activation was checked by activating grids of pure iron wires.

It is well known that the pigments chosen for use in artist's colours have been subject to the influence of fashion and are thus characteristic of the time and even the locality of their manufacture. By measuring the γ-ray spectra of the activated paintings, the elements of many pigments can also be identified and their relative abundances within the paintings determined. Thus the combination of activation analysis with autoradiography can yield, non-destructively, much information concerning the pigments used as well as the distribution of a number of them[(15.12)].

The application of NAA techniques in the study of art objects and archaeological materials has further been surveyed by e.g. Sayre[(15.13)].

15.4 Identification of origin of oil discharge by neutron activation analysis

Various methods have been used for identifying the source of an oil discharge, e.g:

- Labelling of oil by the addition of radioactive compounds in trace amounts[(15.14)]

- Labelling of oil by means of minute objects, which can be identified microscopically, as for example balls of plastic or metal varying in size, colour or composition in accordance with a recognized code

- Characterization of elements such as vanadium which occur in oil in trace quantities.

Of these, the fingerprint method which is performed in accordance with possibility 3 above owes its application to the fact that cobalt, manganese or vanadium may exhibit variations of concentration in different oils by a factor of between 20 and 200. This technique has been applied with considerable success by Guinn and Bellanca[15.15].

Various analytical techniques such as infrared, fluorescence and mass spectrometry used in oil spill identification have been reviewed by Bentz[15.16].

15.5 References

15.1 W.B. Renfro and W.A. Jester, J. Radioanal. Chem. 15, 79 (1973).

15.2 C. Hoffman, Identification News, Oct 7 (1968).

15.3 S.S. Krishnan, J. Radioanal. Chem. 15, 165 (1973).

15.4 V.P. Guinn, Ann. Rev. Nucl. Sci., Vol.24, 561 (1974).

15.5 F. Baumgärtner et al., Z. Anal. Chem. 197, 424 (1963).

15.6 V.P. Guinn, J. Radioanal. Chem. 15, 389 (1973).

15.7 A. Gordus, J. Radioanal. Chem. 15, 229 (1973).

15.8 R.F. Coleman, J. Brit. Nucl. Eng. Soc. 6, 134 (1967).

15.9 H. Al-Shahristani and I.K. Al-Haddad, J. Radioanal. Chem. 15, 59 (1973).

15.10 J.M.A. Lenihan and I.M. Dale, Advisory Group on Applications of Nuclear Methods in Environmental Research, IAEA, Vienna 22-26 March (1976).

15.11 Yu.S. Ryabukhin, J. Radioanal. Chem. 60, 7 (1980).

15.12 M.J. Cotter, Am. Scientist 69, 17 (1981).

15.13 E.V. Sayre, in: Advances in Activation Analysis Vol.II, J.M.A. Lenihan, S.J. Thomson and V.P. Guinn (Eds.), Academic Press, 155 (1972).

15.14   S. Carlsson et al., in: Nuclear Techniques in environ-
        mental Pollution, IAEA, Vienna (1971).

15.15   V.P. Guinn and S.C. Bellanca, Proc. of the 1968 Int.
        Conf. on Modern Trends in Activation Analysis, NBS,
        Gaithersburg, Oct. 7-11, 1968, NBS Special Publ. 312,
        1, 93 (1969).

15.16   A.P. Bentz, Anal. Chem. 48, 455A (1976).

16 Nuclear Chemical Analysis in Environmental Sciences

16.1 Radionuclides in the environment

The radionuclides in the environment can be divided into two general classes:
- Natural radionuclides and
- Artificial radionuclides.

The natural radionuclides comprise two subclasses, i.e., "cosmogenic" and "primordial". Among the primordial radionuclides are the decay series headed by U-238 and Th-232, which have always existed in the earth's crust. The cosmogenic radionuclides are spallation products from the collision of high energy cosmic radiation with the earth's atmosphere. Examples are C-14, Be-7 and Na-22.

The environmental artificial radionuclides are products of the nuclear age, initiated in 1945 with the first nuclear weapon explosion. There are three subclasses of artifical radionuclides, "fission products", "activation products" and the "transuranium elements". A major part of the transuranium elements have been released from atmospheric nuclear explosions between 1959 and 1961[16.1].

The release of radioactive nuclides plays an important role in the environmental aspects of nuclear power production. The radioactive nuclides already present and released into the environment might, however, serve as tracers in biological, ecological and geological studies. Their presence is often used for age determination. By studying the ratio between various nuclides in the U- and Th-decay schemes, it is possible to date geological samples. Polonium-210 and cesium-137 are often used for age determination of sediment core samples. In archeology, the carbon-14 isotope is of fundamental importance.

16.2 Trace elements in the environment

In addition to radioactive pollution, the modern industrial world also contaiminates the environment with heavy metals,
of which Hg, Cd and Pb have been devoted special attention.
A large portion of such elements is discharged into the air from
fuel combustion, industrial processes and various types of
burning.

Nuclear techniques have been used for the analysis of
more than 15 of such elements. Neutron activation analysis
is in this respect a useful technique. There are, however,
also several other analytical techniques in current use.

The elements released into the air are deposited on the
ground and contaminate water and growing plants. There is a
great need for trace element analysis in a great variety of
sample matrixes. Nuclear chemical techniques have been shown
to be convenient for this type of studies.

16.3 Application of NAA to environmental studies

Applications of neutron activation analysis (NAA) in environmental studies involve both non-destructive instrumental
activation analysis technique and radiochemical separations.
A special procedure for using NAA of trace elements in environmental systems is the tracing of selected elements delibe-
rately added to the system under study. These elements are
subsequently detected by being activated. The addition of
such tracers makes it possible to study the movement of gas,
aerosols and suspended or dissolved material in natural waters
etc. Most environmental studies, however, involve the measurement of substances which are inherent to the system, or to
the source of pollution under investigation. Very high sensitivity is obtained with the elements In, Dy, La, Br and Co,
which can be detected through NAA in most environmental samples.

Special features of activation analysis techniques in
environmental studies include:

(i)   The ability to detect most of the chemical elements

at concentrations as low as to parts-per-million,i.e. 10^{-6} (ppm) and part-per-billion, i.e. 10^{-9}(ppb).

(ii)   The use of instrumental techniques which eliminate the need for tedious dissolution of samples, such as sediments, soils or biological materials, e.g. human tissues.

(iii)   The versatility NAA offers in terms of the nature of the sample matrix, i.e. there is no need to standardize the analytical procedure for a particular sample matrix pertinent to the concentration range.

(iv)   The ability to detect, identify and quantify a wide range of trace elements simultaneously in a single specimen.

16.4 Trace elements in hair

Determination of trace element composition in the hair has made a valuable contribution to environmental sciences and toxicology. Abnormal concentrations of toxic elements in hair can serve as evidence of exposure to abnormal quantities of toxic substances. Hair is a sensitive indicator of man's exposure to environmental pollutants, since it concentrates a number of trace elements. It also has a unique property to map the history of contamination since most peoples' hair is more than 10 cm, and that represents almost a year's growth.

In order to judge if a population has been exposed to abnormal amounts of trace element pollutants, it is essential to establish the background concentration of various trace elements in the hair of normal population. A number of authors have determined the trace element content of hair by using neutron activation analysis. A comprehensive compilation of the reported values has been made by Ryabukhin[16.2] (see also sec. 15.2).

It has also been shown that in certain cases, when a single exposure to a pollutant has taken place, the biological elimination of the element in man can be studied by following the concentration along the hair specimen. Fig. 16-1 shows a

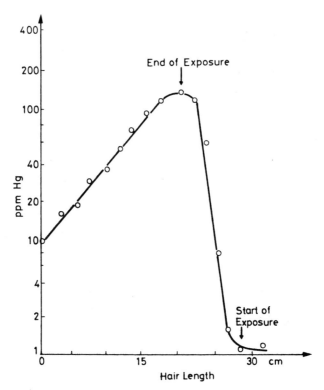

Fig. 16-1   Variation of Hg-concentration along a hair specimen[16.3].

typical variation of mercury concentration in a hair specimen. The mercury body burden, B_{Hg}, can be expressed by the equation (eq. 2.52).

$$B_{Hg} = \frac{I}{k} [1-\exp(-kt_e)]\exp(-kt_o) \qquad (16.1)$$

where I - daily intake, k - excretion rate constant, t_e - exposure time and t_o - excretion time after the end of the exposure. The time can be expressed in terms of hair length. Thus, if g is the hair growth rate, $X_i = t_e \cdot g$ is the hair length from the start to the end of the exposure and $X_d = t_o \cdot g$ is the hair length from the end of the exposure[16.3].

The curve shown in fig. 16-1 thus makes it possible to calculate the individual's hair growth rate (g) by dividing the length of the hair from the point of concentration increase to the root-end by the time elapsed from the onset of exposure unitil the date of sample collection. The excreation rate constant of the polluting element can be determined from the excretion part of the curve in a manner similar to radioactive decay half-life determination.

Examination of single hair strands has favourably been accomplished with PIXE-technique as demonstrated by Li Hong-Kou[16.12].

16.5 Mercury pollution

The advantages of using NAA for measuring environmental pollutants have been shown e.g. by Johnels and Westermark, who studied mercury levels in the feathers of wild birds in Scandinavia[16.4]. The incidents of environmental pollution of mercury in Japan, Sweden and Canada focused a great interest on the use of the NAA-technique, for exploring and documenting the extent of mercury contamination in the biosphere.

Neutron activation of mercury comprises the radionuclides $^{197}Hg (t_{1/2} = 65$ h) emitting 77 keV γ-rays and ^{203}Hg $(t_{1/2} = 46.6$ d) emitting 279 keV γ-rays. The highest analytical sensitivity can be obtained by using the 77 keV peak. The energy region of this peak can often contain significant contributions from interfering matrix nuclides. The 279 keV peak, however, can be measured after 5-10 days, when most of the interferences from matrix nuclides has decayed. Sensitivity of 1 ppm has been reported for instrumental NAA in the analysis of samples of soil, sediments, wild animals and food products. Desiring a higher sensitivity the accomplishment of chemical procedures are required.

Through comparison of the mercury levels in feathers from living birds with collections of feathers from museums, Johnels and Westermark demonstrated that there has been an extreme large increase in mercury content since 1930[16.4]. The increase started when the chlorine-alkali plants using the Dow mercury process were started.

In order to assess the extent of mercury intake by the victims of environmental mercury accidents in Japan, Ukita and Ohuchi (1966) used the NAA-technique to analyze hair samp-

les[16.5]. In Canada, large scale surveys of mercury content
have been carried out to define the extent of mercury pollution
resulting from chlorine-alkali plants and the agricultural use
of mercury fungicides and herbicide agents. The reference data
on mercury levels in normal healthy residents of North Ameri-
ca prior to 1964 had been measured using NAA technique[16.6].

16.6 Cadmium pollution

A widespread interest in environmental cadmium levels
arose after cadmium environmental toxicity had been demonstrated among Japanese women exposed to industrial cadmium pollution from consumption of vegetables.

Both photon and neutron activation methods could be used
in the measurements of ppb-levels of Cd in vegetablegrowing
soil, vegetation and in human tissues, e.g. hair. Instrumental
photon activation analysis can be applied to environmental samples, whereas NAA-technique requires radiochemical separations.
Most environmental samples contain less than 50 ppb of cadmium.
Higher values are found in the vicinity of smelteries and in
chemical fertilizers plants[16.7].

16.7 Vanadium and energy production

Vanadium, usually occurring at trace levels, is an element
of increasing interest in the field of environmental pollution.
It occurs naturally in fossil fuels as a porphyrin complex at
levels ranging from 10-14 000 $mg \cdot kg^{-1}$ (0.2 - 300 $mmol \cdot kg^{-1}$).
The rapid increase in the production of electrical power by
coal- and oil-fired power plants and domestic heating mobilizes
considerable amounts of vanadium into the environment, since
these processes require the combustion of enormous amounts of
fuel. In addition, high concentrations of this element have
been found in urbain air originating solely from industry, since there are no natural sources[16.8].

As a consequence, man is exposed to continuously increasing amounts of vanadium, and investigations of the vanadium
behaviour in various ecosystems, including animal and man, are
well motivated and urgent. Neutron activation analysis appears
to be a particularly suitable analytical technique for vanadium
determination in biological samples. This technique has now been
suggested both as a reference method and to be used in the development of vanadium reference materials.

16.8 Arsenic pollution

Arsenic is a ubiquitous element occurring in many parts
of the lithosphere and biosphere in trace quantities. Although
arsenic is present in air and water, and local areas, soils
seem to be the major receptor.

Comprehensive reviews of typical levels of arsenic in the
environment and its dispersal pathways and mechanisms are given
in three recent treatises by the US EPA, the US National Aca-
demy of Sciences and the National Research Council of Canada,
respectively. The technique used for determination of trace
amounts of arsenic in environmental samples are neutron activation analysis, atomic absorption spectrophotometry and fluorescence spectrophotometry.

Use is made of γ-ray spectrometry to determine ^{76}As
$(t_{1/2} = 26.3$ h) after 3 days following irradiation of the
sample in the thermal neutron flux. With Ge(Li) detectors,
the ^{76}As-peak (559.2 keV) is well resolved from the ^{82}Br-peak
(555.3 keV). If low levels of silver are present, the ^{110m}Ag
(657.8 keV) does not contribute to the 657.0 keV ^{76}As peak.
The detection limit for the determination of arsenic with
NAA is about 5 ng (60 picomol). The processing of arsenic-
containing ores for the recovery of metals, such as gold,
copper or lead, can cause both occupational health hazards
for smelter workers and an environmental health problem to
persons living in the vicinity of the refineries. Surface
soil is the most sensitive indicator of local arsenic conta-

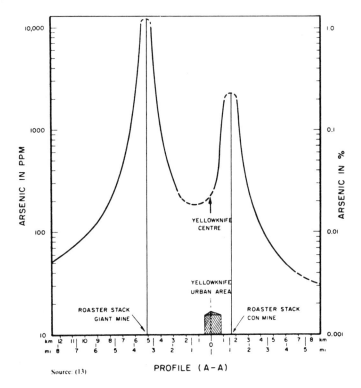

Fig. 16-2   Arsenic concentrations in soil, in the vicinity of
Yellowknife smelter[16.9].

mination. An example of the distribution of soil arsenic in
the vicinity of two smelteries is shown in fig. 16-2. Activation analysis of environmental samples and of scalp hair from
this area permitted assessment of the potential risk to local
residents and to workers exposed to arsenic emitted from the
gold smelteries[16.9].

16.9 Heavy elements in tissues of industrially exposed
workers

Various human populations, particularly certain occupational groups, are exposed to substances from air which represent potential health risks.

Knowledge about the various heavy metal levels in the internal working environment is not readily available. It is likely that workers exposed in the past have accumulated such elements to a certain extent in their internal tissues.

Valuable information might consequently be obtained from analysis of tissues from diseased exposed workers as well as non-exposed control groups living in the same districts.

A study performed in the vicinity of the metal refinery Rönnskärsverken in North Sweden indicates increased levels of a great number of elements in the exposed worker population[16.10].

Another study dealing with manganese in the hair of welders resulted in a geometric mean manganese content of 6.1 ± 2.5 ppm, compared to 1.2 ± 2.9 ppm in the control group. Fig. 16-3 shows the distribution of the contents of both groups. Beyond the intersection point of the distribution curves chosen as threshold, 7% of the reference persons and 75% of the welders are found[16.11].

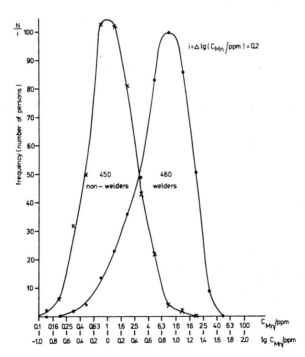

Fig. 16-3   Distribution of manganese concentration in hair.

16.10 References

16.1   M. Eisenbud, Environmental Radioactivity 2nd ed., Academic Press Inc. New York (1973).

16.2   Yu. Ryabukhin, Report on the Co-ordinated Research Programme. Neutron Activation Analysis of Pollutants in Man using Research Reactor. IAEA/RL/41 A Vienna (1977).

16.3   D. Al-Shahristani, K.M. Shihab and M. Jahl. In Nuclear Activation Techniques in the Life Sciences. IAEA-SM-227/7, Vienna (1979).

16.4   A.G. Johnels and T. Westermark. On Chemical Fallout (eds.M.W. Mills and G.G. Berg, Chas.) C. Thomas Publ., Springfield, Ill. pp.221-241 (1969).

16.5   T. Ukita and A. Ohuchi. In Proceed. 7th Japan Conf. on Radioisotopes, Japan Atomic Industr. Forum, Paper 8/4-2 (1966).

16.6   R.E. Jervis and B. Tiefenbach. In Proceed. 1st Int. Conf. on Nucl. Method in Environmental Res. (ed. J.R. Vogt) Am. Nucl. Soc. pp. 188-196 (1971).

16.7   A.K. Perkons and R.E. Jervis, J. Forensic Sc. 11:50-63 (1966).

16.8   R. Cornelius, L. Mees, J. Hoste, J. Ryckebusch, J. Versieck and F. Barbier. In Nuclear Activation Techniques in the Life Sciences, IAEA, SM-297/25, Vienna (1979).

16.9   R.E. Jervis and B. Tiefenback. In Nuclear Activation Techniques in the Life Sciences, IAEA-SM-227/67, Vienna (1979).

16.10   D. Brune, G.F. Nordberg and P.O. Wester. Sci. Total Environm. $\underline{16}$, 13 (1980).

16.11   W. Wiesener, W. Goerner and S. Niese. In Nuclear Activation Techniques in the Life Sciences. IAEA-SM-227/73, Vienna (1979).

16.12   Li Hong-Kou. Doctoral Dissertation, University of Lund, oct. 28. 1983.

17 Geology and Mineral Exploration

17.1 Introduction

NAA is a common technique to determine the elemental composition of minerals[17.1, 17.2]. A major objective in the NAA of minerals has been the determination of trace element partitioning between coexisting minerals and a silicate melt. Such data provide partition coefficients which are used for studies of element abundance variations[17.3, 17.4].

NAA of minerals and inclusions separated from rocks have been utilized to elucidate the physical and chemical processes taking place during the formation of rocks.

Instrumental and radiochemical neutron activation analyses have been extensively used to study meteorites and lunar rocks[17.5, 17.6, 17.7]. Table 17-1 shows the contents of various elements determined through thermal and epithermal activations combined with measurements with NaI(Tl) and Ge(Li) detectors according to Geissler[17.8].

17.2 Field applications using neutron activation and gamma ray spectrometry

Characteristic data of various radionuclide neutron sources are given in tables 17-2 and 17-3.

Using a 3.0 mCi ^{241}Am-Be-neutron source, Zaghloul et al.[17.10] determined the iron content of ore using thermal neutron capture γ-ray analysis.

Of the various radionuclide neutron sources (tables 17-2 and 17-3), the ^{252}Cf-source has attracted special interest. In this source, the γ-radiation, following the emitted neutrons of approximate fission spectrum energy, is low. A source can vary from 100 μg to about 10 mg[17.11].

Table 17-1
Elemental levels in lunar materials[17.8].

Element	Luna 16	Luna 20, Probe A	Luna 20, Probe B
		Content, ppm	
Na	3040 ± 80	2420 ± 100	2450 ± 140
		2810 ± 110	2790 ± 90
K	1080 ± 200		560 ± 80
Sc	55 ± 0.9	18.0 ± 0.2	19.0 ± 0.2
Cr	1860 ± 30	1190 ± 20	1450 ± 20
Mn	1660 ± 40	732 ± 39	751 ± 28
		809 ± 36	872 ± 25
Fe,%	12.7 ± 0.1	6.13± 0.04	6.61± 0.14
Co	27.0 ± 0.5	25.5 ± 0.6	45.9 ± 0.9
As	<7		<4
Rb	<38	<16	<19
Mo		5.0 ± 0.7	3.5 ± 0.4
Sb	<0.6		<0.5
Cs	<0.6	<0.7	<0.8
La	11.7 ± 0.4	8.38± 0.16	7.42± 0.16
Ce	48 ± 2	22.5 ± 1.4	20.4 ± 0.8
Sm	9.0 ± 0.3	3.9 ± 0.1	3.6 ± 0.1
Eu	2.0 ± 0.05	0.96± 0.04	0.93± 0.04
Tb	2.3 ± 0.4	1.4 ± 0.6	1.6 ± 0.6
Yb	7.2 ± 0.5	4.2 ± 0.2	3.9 ± 0.1
Lu	0.65± 0.04	0.39± 0.02	0.40± 0.02
Hf	7.4 ± 0.2	3.1 ± 0.4	2.9 ± 0.3
W	16 ± 4	42 ± 3	38 ± 2
		39 ± 2	46 ± 2
Th	1.3 ± 0.3	1.0 ± 0.1	0.7 ± 0.1

After californium-252 neutron sources became available in 1972, they have been extensively used for borehole logging. Neutron activation γ-ray spectrometer systems can be divided into two basic parts, i.e. the neutron source and the γ-ray detector. A shadow shield of tungsten, lead or bismuth is generally used between the neutron source (Cf-252) and the detector (HPGe). The shield is used to attenuate the direct

Table 17-2

Yield of 1-Curie radionuclide neutron sources[17.9].

Source	Type	Average neutron energy	Yield, $n \cdot s^{-1}$
^{124}Sb-Be	γ, n	24 keV	$1.6 \cdot 10^6$
^{239}Pu-Be	α, n	4.5 MeV	$2.0 \cdot 10^6$
^{241}Am-Be	α, n	4 MeV	$2.0 \cdot 10^6$
^{210}Po-Be	α, n	4.3 MeV	$2.5 \cdot 10^6$
^{238}Pu-Be	α, n	4 MeV	$2.8 \cdot 10^6$
^{244}Cm-Be	α, n	4 MeV	$3 \cdot 10^6$
^{242}Cm-Be	α, n	4 MeV	$4 \cdot 10^6$
^{226}Ra-Be	α, n	3.6 MeV	$1.5 \cdot 10^7$
^{252}Cf	Spontaneous Fission	2.3 MeV	$4.4 \cdot 10^9$

Table 17-3

Characteristic data for various radionuclide neutron sources[17.9].

Source	Half-life	Gamma exposure rate R/h at 1 meter	Heat generation watts	Volume[b] cm^3
^{124}Sb-Be	60 d	$4.5 \cdot 10^4$	20	200
^{210}Po-Be	138 d	2.0	640	200
^{242}Cm-Be	163 d	0.3	600	2
^{252}Cf	2.65 y	2.9	0.8	<1
^{244}Cm-Be	18.1 y	0.2	600	70
^{238}Pu-Be	89 y	0.4	550	350
^{241}Am-Be	458 y	2.5	750	$2.2 \cdot 10^4$

Normalized to $5 \cdot 10^{10}$ neutrons per second.

[b]Not including void space for helium from α-decay.

Table 17-4

Detection limits[1] for various elements in a sea bed[(17.14)].

Element	Detectable concentration	Element	Detectable concentration
Aluminium	0.1 - 1%	Magnesium	0.1 - 1%
Antimony	1 - 10%	Manganese	0.01 - 0.1%
Arsenic	0.1 - 1%	Mercury	1 - 10%
Barium	0.1 - 1%	Molybdenum	0.01 - 0.1%
Cadmium	1 - 10%	Nickel	1 - 10%
Chromium	4 - 40%*	Niobium	1 - 10%
Cobalt	0.1 - 1%	Scandium	1 - 10 ppm
Copper	0.1 - 1%	Selenium	1 - 10 ppm
Dysprosium	10 - 100 ppm	Silver	10 -100 ppm
Fluorine	0.01 - 0.1%	Strontium	0.1 - 1%
Gold	10 - 100 ppm*	Tin	3%
Hafnium	1 - 10 ppm	Titanium	10 -100 ppm
Indium	1 - 10 ppm	Tungsten	0.1 - 1%
Iron	50 - 100%*	Vanadium	10 -100 ppm
Lantanum	0.01 - 0.1%		

[1] Based on 2 min irradiation and 2 min counting (1 mg ^{252}Cf source)
* Sensitivity increases almost linearly with irradiation and counting time.

γ-radiation from the source and to direct neutrons from the source into the rock of the wall. Both the source and the detector with cryostat and electronics are mounted in a single hermetically sealed sonde, that can be moved, within the borehole.

17.3 Prospecting copper ores by means of ^{252}Cf-sources

Various copper minerals from Peru, consisting of malach-
ites, azurites and pyrites, have been analyzed by standard wet
chemical and neutron activation techniques. The results show
that a small ^{252}Cf-source can be used for the analysis of copper concentration as low as 1%, with a precision of 5 to 10 %.

The irradiation facility consisted of a 660-μg ^{252}Cf-
source surrounded by a circle of polyethylene bottles of heavy
water. The flux in each irradiation position was measured by
irradiating pure copper foils and comparing the activity to
that of a standard ^{22}Na-positron emitter. Copper was determined using the ^{63}Cu (n,γ) ^{64}Cu-reaction. The counting system
consisted of a simple coincidence system with a resolving time
of 100 ns. The overall coincidence efficiency was estimated to
be 5%[17.11].

A borehole logging system for in-situ assay of uranium in
ores based on activation with a ^{252}Cf-source and measurements
of delayed neutrons has been developed by Steinman et al.[17.18]
This system is mounted on a vehicle and includes the downhole
sonde, logging winch, data recording and display equipment. It
has source shielding allowing logging speeds from stationary
to 8 meters/min, with the capacity for determining 0.01% U_3O_8-
concentration at 1.5 meters/min.

Borehole neutron γ-ray spectroscopy has also been used
to determine uranium, sulphur, silicon, aluminium, iron and
tungsten[17.19].

Because of its high hydrogen content, coal is an ideal
matrix for using thermal neutron capture techniques[17.20].

The californium-252 progress report describes various
applications of this source in detail. The compactness and
portability of the californium neutron source make it attractive for in situ applications. On the ocean floor, the technique can be used to identify manganese nodules. Wiggins et
al.[17.12] surveyed the possibility of using such a technique
for locating the materials on the bottom of the sea. For studies of the elemental composition in bore-holes, either on land
or on the continental shelf, a sonde consisting of a ^{252}Cf-

source and a high resolution germanium semiconductor detector is useful. With a device described by Tanner et al.[17.13], neutron capture γ-rays, as well as γ-rays of induced radionuclides, can be measured for establishing the elemental composition in drill holes.

Detection limits for several elements in the sea-bed have been established by Wogman et al.[17.14] The results of their study are given in table 17-4.

The ^{252}Cf-source has also been applied in environmental studies. In heavy-metal polluted water, mercury and cadmium have been identified at a lower level limit of 20 ppm with a 10 μg ^{252}Cf-source, as reported by Handley and DeCarlo[17.15].

Borehole γ-ray spectrometry can also be applied without any neutron source for measuring naturally radioactive compounds, such as potassium-40 and uranium and thorium daughters.

A special advantage of the method is that several radionuclides in the Th- and U-series can be measured simultaneously, so that the state of disequilibrium can be determined in different part of the ore body[17.16].

17.4 Delayed neutron measurements of uranium and thorium

During fission, a small fraction of the neutrons emitted is delayed by a fraction ranging from one second up to 55 s. The formation of these nuclides has been discussed by e.g. Rudstam et al.[17.21].

Table 17-5

Properties of delayed neutrons and neutron yields[17.22].

| Delayed neutron i group | Half-life | Mean energy keV | Yield per fission % β_i | | | |
			^{235}U thermal	^{233}U thermal	^{239}Pu thermal	^{238}U fast
1	55·0	250	0·026	0·023	0·007	0·021
2	20·6	460	0·136	0·079	0·063	0·215
3	5·0	405	0·122	0·066	0·045	0'254
4	2·1	450	0·246	0·074	0·069	0·610
5	0·61	-	0·072	0·014	0·018	0·355
6	0·28	-	0·026	0·009	0·009	0'119
		$\Sigma \beta_i$: =	0·658%	0'265%	0·211%	1·574%

For practical reasons, nuclear technology divides these so-called delayed neutrons into six groups according to their half-lives (see table 17-5).

From an analytical point of view, the measurement of delayed neutrons can be based on the determination of fissionable nuclides, such as ^{235}U and ^{232}Th. This analytical technique was introduced by Amiel[17.23].

The delayed neutron counting technique is used advantageously in routine applications of uranium analysis in ores. Typical samples amount to 1-4 g.

The whole analytical set-up according to Andersson[17.24] is shown in fig. 17-1. The pneumatic transfer system allows transport of the sample from the irradiation position of the detector within one second. The precision rates obtained for various amounts of uranium in geological samples are given in

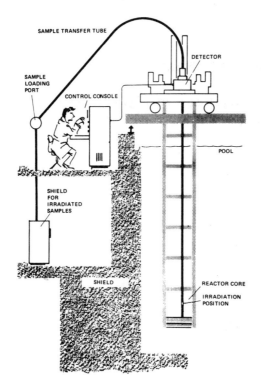

Fig. 17-1   Analytical set-up for uranium and thorium analysis[17.24].

table 17-6. The neutrons are recorded with a number of convential BF_3-counters.

If the sample also contains thorium, which will also undergo fission, some of the delayed neutrons registered will be due to the fissions in the thorium. This effect can interfere with the uranium analysis. The influence of the thorium can, however, be largely eliminated by the choice of a good thermal irradiation position in the reactor.

Table 17-6

Typical values for the precision of a routine analysis[17.24].

Amount of uranium in the sample	1 µg	3 µg	5 µg	25 µg	100 µg	1000 µg
Precision	20 %	6 %	4 %	2 %	1.5 %	1 %

17.5 Analysis of thorium

An estimation of the thorium content can also be made with a delayed neutron technique. The method is based on a second irradiation of the same sample, in a different irradiation position, using fast fission. The neutron signal from uranium is then suppressed and the thorium contribution will play a relatively large role. By making one measurement in each of the two irradiation positions, the uranium and thorium contributions can be distinguished from each other. By this method, the uranium content can be accurately determined, and the thorium content can be estimated.

17.6 References

17.1   R. Gijbels, Mineral.Sci. Engng. 5, 304 (1973).

17.2   R. Gijbels and J. Hertogen, Pure and Appl. Chem. 49, 1555
        (1977).

17.3   J.G. Arth, J. Res. U.S. Geol. Survey 4,41(1976).

17.4   F.A. Frey, Rev. Geophys. Space Phys. 17, 803 (1979).

17.5   L. Grossman, R. Ganapathy, and A.M. Davis, Geochim.
        Cosmochim. Acta 41, 1647 (1977).

17.6   L. Grossman, R. Ganapathy, R.L. Methot, and A.M. Davis,
        Geochim. Cosmochim. Acta 43, 817(1979).

17.7   A.A. Smales, The Place of Activation Analysis in Geoch-
        emistry and Cosmochemistry in Activation Analysis in
        Geochemistry and Cosmochemistry (eds. A.O. Brunfelt, E.
        Steinnes), Proc. NATO Advanced study Institute Kjeller
        1970, Universitetsforlaget, Oslo, 1971.

17.8   M. Geisler, J. Radioanal. Chem. 28, 209(1975).

17.9   W.C. Reinig and A.G. Evans, Proc. of a Symp. Modern Trends
        in Activation Analysis, Gaithersburg, Md. Oct. 7-11, 1968.
        NBS-Special Publ. 312, Vol. II, 953.

17.10   R. Zaghloul, M. Obeid, and R. Henkelman, Radiochem. Ra-
        dional. Letters, 17, 7(1974).

17.11   Californium-252 Progr. (US ERDA).

17.12   P.F. Wiggins, D. Duffy, and A.A. El Kady, Anal. Chim.
        Acta 61, 421(1972).

17.13   A.B. Tanner et al., Nucl. Instr. Meth. 100, 1 (1972).

17.14   N.A. Wogman et al., J. Radioanal. Chem. 15, 591(1973).

17.15   T.H. Handley and V.A. DeCarlo, J. Radioanal. Chem. 11,
        265(1972).

17.16   A.B. Tanner, R.M. Moxham, and F.E. Senftle, U.S.
        Geol. Survey Open-File Report 77-571, 22 pp., 1977

17.17   S.S. Nargolwalla and H. Seigel, Can.Min.J. 98, 75(1977).

17.18   D.K. Steinman et al., Meeting on Cf-252 utilization,
        Fontenai-Aix-Roses, France, 26-28 April, 1976.

17.19   F.E. Senftle, Field Studies of borehole gamma-ray spec-
        trometer methods for mineral exploration. A selected
        bibliography. U.S. Geological Survey. Open-File Report,
        (1980).

17.20   F.E. Senftle, A.B. Tanner, P.W. Philbin, G.R. Boynton
        and C.W. Schram. Mining Engineering (AIME) 30, 666
        (1978).

17.21   G. Rudstam, S. Shalev, and O.C. Jonsson, Nucl. Instr.
        Meth. 120, 333(1974).

17.22   D. Jakeman, Physics of Nuclear Reactors, The English
        University Press, London, 1966.

17.23   S. Amiel, Anal. Chem. 34, 1683 (1962).

17.24   T.L. Andersson, AB Atomenergi, Studsvik, Sweden (priva-
        te communication), 1976.

18 Radiation Protection

18.1 Radiation protection standards

18.1.1 PRINCIPLES OF RADIATION PROTECTION

All ionizing radiation from accelerators or radioactive material is potentially hazardous. Radiation sources outside the body give rise to external radiation of personnel, and radioactive materials which enter the body irradiate its organs and tissues. Scientists working with ionizing radiation and radioactive materials must therefore be capable of preventing and controlling these hazards. There is a golden rule which is always applicable:

"As many exposures may involve some degree of risk, any unnecessary exposure must be avoided, and all doses must be kept as low as is readily achievable, economic and social consideration being taken into account".

There are three principles which can be applied to prevent or control the exposure of personnel to radiation hazards:

- Remove the hazard
- Guard the hazard
- Guard the worker

These principles imply that working places are properly designed and that appropriate equipment and shielding are provided to ensure the maximum amount of protection. The last principle refers to the requirements of periodic measurements of the radiation levels in the working environments and continous personal monitoring[18.1]. These principles are summarized in table 18-1.

Table 18-1
Principles of radiation protection a perspective

Radiation worker   Non-Radiation worker

Minimize the hazard

(1) Keep the amount of radioactive material required to a minimum

(2) **Choose radioactive material presenting the least possible hazard**

(3) Choose the safest and most practicable procedures

(4) Dispose of radioactive waste safely

(5) Restrict the movement of radioactive materials to a minimum

Guard the hazard (containment)

General

Prevent hazardous release of radioactive material to the environment

Local	Local
(1) Partial(fume cupboards)	(1) Sealed transport containers
(2) Total (glove boxes)	
(3) Temporary (sealed containers)	

Shielding

General	General
(1) Reduce radiation level outside the controlled area to well within permissible levels	(1) Reduce radiation level in public access position to well below non-occupational levels

Local	Local
(1) Permanent (hot cell)	(1) Shielded transport containers
(2) Temporary (lead bricks, lead coffins)	

Guard the worker
(Planning and instruction)

(1) Choice of material, instruments and facilities	(1) Transport regulations
(2) House rules	(2) Emergency procedures
(3) Operating instructions	
(4) Emergency instructions	

Monitoring
(Radiation and contamination)

(1) Personnel	(1) Transport containers
(2) Equipment	
(3) Area and site	
(4) Biological	

Protective clothing

(1) General types for routine operations

(2) Special types for emergency use

Note: In all processes involving radioactive materials (storage experiment or process, handling of materials, transport and the disposal of active waste) consideration must be given to the appropriate methods of control.

18.1.2 RADIATION UNITS AND STANDARDS

Radiation measurements and investigations of radiation effects require various degrees of specification of the radiation field at the point of interest. The most elementary quantities associated with the radiation field deal either with particle number or energy and this is denoted in their names, e.g., particle flux or energy flux, etc. The word particle can be replaced by the more specific term for the considered entity, e.g., neutron flux, electron fluence, etc.

The "particle number" N, is the number of particles emitted, transferred, or received, and the (particle) flux, $\dot{N}$, is the quotient of dN by dt, where dN is the increment of the particle number in the time interval dt.

$$\dot{N} = \frac{dN}{dt} \tag{18.1}$$

Unit: s^{-1}

The (particle) fluence, ϕ, is the quotient of dN by da, where dN is the number of particles incident on a sphere of cross-sectional area da.

$$\phi = \frac{dN}{da} \tag{18.2}$$

Unit: m^{-2}

The (particle) fluence rate, φ, is the quotient of dϕ by dt, where dϕ is the increment of the particle fluence in the time interval dt.

$$\varphi = \frac{d\phi}{dt} = \frac{d^2N}{da\ dt} \tag{18.3}$$

Unit: $m^{-2}\ s^{-1}$

The term particle flux density is also used as the name for this quantity. As the word density has several connotations, the term particle fluence rate is preferable.

When ionizing radiation passes through matter it interacts with the atoms and molecules in the medium it traverses, producing ionizations and excitations. Depending on the medium, the absorbed energy may give rise to observable effects, for example, ionization, photographic effects, viological effects

and heating. The energy imparted may be expressed in joule
per kg, which has led to the concept of absorbed dose. The
definition of the absorbed dose, D, is the quotient of $d\bar{\epsilon}$ by
dm, where $d\bar{\epsilon}$ is the mean energy imparted by ionizing radiation
to the matter in a volume element and dm is the mass of the
matter in that volume element.

$$D = \frac{d\bar{\epsilon}}{dm}$$ (18.4)

The special unit of absorbed dose is the 'gray' abbreviated Gy.

1 Gy = 1 J/kg (=100 rad)

The earlier special unit was the rad.

From the biological point of view, evidence has accumulated that the effects of the various types of ionizing radiation
are not the same. One can assume that radiation can bring about
a change in an organism only by virtue of the energy that is
actually absorbed. A biological effect, however, may also depend
upon the spatial distribution of the energy released along the
track of the ionizing particle. It will therefore depend on the
type and quality of the radiation, and equal energy imparted by
different types of radiation may not produce the same biological effects. For radiation protection purposes, a separate quantity, the dose equivalent, is therefore used for comparison of
risks of biological effects of radiation from different types
of sources. The dose equivalent, H, is defined as the product
of the absorbed dose, D, and a quality factor, Q, which is different for various types of radiation:

$$H = D \cdot Q$$ (18.5)

The special SI-unit for the dose equivalent is the sievert,
when the absorbed dose is given in gray$^{(18.2)}$:

1 J/kg = 1 sievert = 1 Sv (=100 rem)

The earlier unit was rem, when the absorbed dose was given in
rad.

The linear energy transfer or collision stopping power, L_∞,
can be used to specify the radiation quality. The relationship
between L_∞ and the quality factor, Q, which is recommended to
be used for radiation protection purposes, is given in table 18-2.

Table 18-2

The relationship between the collision stopping power, L_∞, and the quality factor[18.2].

L_∞ in water keV/µm	Quality factor Q
< 3.5	1
7.0	2
23	5
53	10
>175	20

For β-, γ- and X-rays, the quality factor, Q, is equal to 1. For α-particles in nature and multi-charged particles (and particles of unknown charge) and α-particles of unknown energy, the quality factor Q is equal to 20. For neutron radiation, the quality factor varies with energy according to the following table 18-3.[18.3]

For neutrons, protons and singly-charged particles of rest mass greater than one atomic mass unit of unknown energy the quality factor is equal to 10.

18.1.3 BIOLOGICAL EFFECTS OF IONIZING RADIATION

The biological effects of ionizing radiation in humans depend on the fact that energy is imparted to the tissue. This results in chemical transformations of biologically important molecules like DNA. The physio-chemical transformations take

Table 18-3

The quality factor for neutrons of various energies[18.3].

Neutron energy MeV	Quality factor Q
thermal	2.3
0.1	8
1	10.5
10	6.5
100	4.5

place within fractions of a second, while the biological effects appear after hours, days and even years after the exposure.

If the transformation of the DNA of a cell is extensive, the cell will die. If it survives, it can be transformed to an out-law cell, which can cause cancer. If the testes or ovaries are irradiated, transformation of the genetic material can be transferred to the next generation. This is called genetic effects, as it does not affect the exposed individuals but rather their progeny. The biological effects of ionizing radiation which effect the exposed individual are called somatic effects. The nature of these effects can be morbid or even fatal (i.e., cancer).

Irradiation during a short period of time with high absorbed dose might take place in accidents with accelerators or other nuclear installations. In table 8-4, the acute effects of whole body irradiation after a short period of time with various levels of absorbed dose are given[18.1]. One must however, bear in mind that the individual variations are great and that the effects are reduced when only part of the body is exposed. In medical radiation for tumour treatment, the total absorbed dose to a part of a body is in the order of 10-60 Gy, delivered in daily fractions of about 2 Gy during a period of several weeks.

18.1.4   RISKS AT LOW RADIATION DOSES

An interesting question in radiation protection is the risk for biological effects at very low absorbed dose levels. Our knowledge of the biological effects at low levels of radiation exposure is still very diffuse and limited. This is mainly due to the fact that, at low absorbed dose levels, the effects are so rare that it is extremely difficult to get significant experimental data. One has, thus to extrapolate from experience at absorbed dose levels above 1 Gy down to the mGy-level.

We don't know for certain that the dose/effect-relationship is linear down to zero, but this is the assumption used by the International Commission on Radiological Protection for establishing their guidelines for permissible levels[18.2].

Table 18-4

The biological effects of whole-body exposure to ionizing radiation[18.1].

Absorbed dose Gy	Probable effects
< 0.25	No detectable clinical effects. Probably no delayed effects on individuals.
~ 0.50	Slight transient blood changes. No other clinically detectable effects. Delayed effects possible, but serious effects on average individual very improbable.
1.00	Nausea and fatigue with vomiting possible above 1.25 Gy. Marked changes in blood pictures with delayed recovery. Shortening of life expectancy.
2.00	Nausea and vomiting within 24h. Following latent period of about one week, epilation, loss of appetite, general weakness and other symptoms, such as sore throat and diarrhoea. Possible death in 2-6 weeks for a small number of the individuals exposed. Recovery likely, unless complicated by poor previous health, superimposed injuries or infections.
4.00	Nausea and vomiting after 1-2h. After a latent period of about one week, beginning of epilation, loss of appetite, and general weakness accompanied by fever. Severe inflamation of mouth and throat in the third week. Symptoms such as pallor, diarrhoea, nose bleeds and rapid emaciation in about the fourth week. Some deaths in 2-6 weeks. Eventual death can occur in about 50% of the exposed individuals.
6.00	Nausea and vomiting in 1-2h. Short latent period following initial nausea. Diarrhoea, vomitting, inflammation of mouth and throat toward end of the first week. Fewer, rapid emaciation and death as early as the second week for all exposed individuals.

Table 18-5

Incidence risk as number of cases per 10 000 individuals exposed to 1 Sv each[18.2].

Organ	Genetic	Fatal somatic	Morbid somatic
Gonads	40	-	-
Breast	-	25	50
Red Marrow	-	20	-
Lung	-	20	-
Thyroid	-	5	100
Bone	-	5	-
Other organ	-	50	-
Skin	-	1	100
Total	40	126	250

In table 18-5, the linear extrapolated risk factors for different tissues are given.

18.1.5 RECOMMENDED DOSE-EQUIVALENT LIMITS

In order to prevent non-stochastic effects, and to limit
the occurrence of stochastic effects to an acceptable level,
the International Commission on Radiological Protection makes
the following recommendations:
- The dose-equivalent limit to all single tissues except the eye lens is 0.5 Sv(50 rem) in a year.
- The dose-equivalent limit to the lens of the eye is
  0.3 Sv (30 rem) in a year.
- The annual dose-equivalent limit for uniform irradiation
  of the whole body is 0.05 Sv (50mSv = 5 rem).
- In case of non-uniform irradiation, the dose limitation is based on an assumption of equal risk, i.e.,
  that the whole body is irradiated uniformly.

This condition will be met if:

$$\sum_{T} \omega_T H_T \lesseqgtr H_{wb,L} \tag{18.6}$$

Table 18-6

Weighting factors, ω_T, representing the proportion of the stochastic risk resulting from tissue (T) to the total risk, when the whole body is irradiated uniformly[18.2].

Tissue	ω_T
Gonads	0.25
Breast	0.15
Red bone marrow	0.12
Lung	0.12
Thyroid	0.03
Bone surfaces	0.03
Other organ	0.30

where ω_T is a weighting factor representing the proportion of the stochastic risk resulting from tissue (T) to the total risk, when the whole body is irradiated uniformly; H_T is the dose-equivalent in tissue (T) in a non-uniform irradiation, and $H_{wb,L}$ is the recommended annual dose-equivalent limit for uniform irradiation of the whole body, namely 50 mSv (5 rem). The values of ω_T recommended by the ICRP are given in table 18-6[18.2].

18.2 Radiation shielding

18.2.1 HEAVY CHARGED PARTICLES

The range of heavy charged particles like protons and α-particles, with energy of a few MeV, is very short. As an example, it can be mentioned that an α-particle with the energy of 8 MeV has a range of only 50 µm in soft tissue, which corresponds to the length of a cell. A proton of 2 MeV energy has the same range according to the relation (see sec. 5.1):

$$R_p(T_p) = R_\alpha(4T_p) \tag{18.7}$$

527

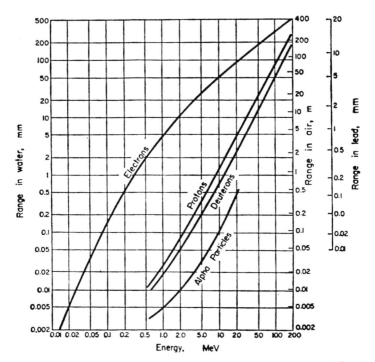

Fig. 18-1. Particle ranges in water, air and lead[18.3].

At higher energies, however, the range can be considerable.
In fig. 18-1 the ranges of α-particles, protons and electrons
in soft tissues are shown[18.3].

18.2.2 ELECTRONS AND BETA PARTICLES

The range of electrons is not as well defined as that of
heavy particles, partly because of larger straggling and bremsstrahlung production (sec. 5.2). In soft tissue the range of
electrons is shown in fig. 18-1. It is also possible to calculate the range at various energies according to the following
empirical relationships:

$$R(0.15 < T < 0.8) = 0.407 \cdot T^{1.38} \; (g/cm^2) \qquad (18.8 \text{ a})$$

$$R(T > 0.8 \text{ MeV}) = 0.542T - 0.133 \quad (g/cm^2) \qquad (18.8 \text{ b})$$

Radiation shielding barriers for high energy electrons
are preferably built of concrete, which has low bremsstrahlung
production. The bremsstrahlung produced must also be considered
in the dimensioning of the shielding barrier.

18.2.3 PHOTON RADIATION

The fluence of photons or the photon fluence rate from a
point source is inversely proportional to the square of the
distance to the source:

$$\varphi(x) = \frac{\varphi(1)}{x^2} \qquad (18.9)$$

Thus, the distance itself rather effectively reduces the
irradiation. This so-called "inverse square law" is only theoretically valid for point sources, but in practice it is used
in those cases where the irradiated object is much larger than
the source itself.

Attenuation of monoenergetic photons, in absorbing material
of thickness 'd' m, follows an exponential law (see eq. 5.26).

$$\varphi(d) = \varphi(0)e^{-\mu_\ell d} \qquad (18.10)$$

where μ_ℓ is the linerar attenuation coefficient.

Due to the exponential law, it is possible to give a half-
value thickness of the material:

$$\varphi(HVT) = \frac{1}{2}\,\varphi(0)$$

$$\frac{1}{2} = e^{-\mu_\ell \cdot HVT}$$

$$\ln2 = \mu_\ell \cdot HVT$$

$$HVT = \frac{\ln2}{\mu_\ell} = \frac{0.693}{\mu_\ell} \qquad (18.11)$$

In the case of broad-beam irradiation of a shielding barrier,
one often gets contribution from Compton-scattered photons and
annihilation quanta. The fluence rate of this secondary radia-

tion is φ_s. Thus, the total fluence rate behind the shield is the sum of the primary φ_p and secondary φ_s photon fluence rates:

$$\varphi = \varphi_p + \varphi_s = \varphi_p \ [1+\frac{\varphi_s}{\varphi_p}] \qquad (18.12a)$$

as

$$\varphi_p = \varphi_0 \ \exp(-\mu_\ell d) \qquad (18.12b)$$

we get

$$\varphi = \varphi_0 \ \exp(-\mu_\ell d)[1+ \frac{\varphi_s}{\varphi_p}] = \varphi_0 B_\Phi(\mu_\ell d)\exp(-\mu_\ell d) \qquad (18.12c)$$

where

$$B_\Phi(\mu_\ell d) = [1+ \frac{\varphi_s}{\varphi_p}] \qquad (18.12d)$$

The expression in the parenthesis is always larger than 1 and increases with the thickness of the absorber, $\underline{d}$. This expression is called the build-up factor and is written $B_\Phi(\mu_\ell d)$.

The general expression for the total photon fluence rate at distance $\underline{x}$ from a point source behind a radiation shield of thickness $\underline{d}$ is the following:

$$\varphi_d(x) = \frac{\varphi_0(1)}{x^2} \ B_\Phi(\mu_\ell d)e^{-\mu_\ell d} = \varphi_0(x)B_\Phi(\mu_\ell d)e^{-\mu_\ell d} \qquad (18.13)$$

where

$\varphi_0(1)$ = photon fluence rate at 1 m without radiation shield

$\varphi_d(x)$ = photon fluence rate at distance $\underline{x}$ m with radiation shield of thickness $\underline{d}$.

In radiation shielding discussions the transmission of photon radiation through a shielding barrier is defined as the ratio between the absorbed dose in air, recorded at the same point and under the same conditions with and without the barrier. Thus the transmission can be given by the expression:

$$D_d(x) = \frac{D_0(1)}{x^2} \ B_D(\mu_\ell d)e^{-\mu_\ell d} = D_0(x)B_D(\mu_\ell d)e^{-\mu_\ell d} \qquad (18.14)$$

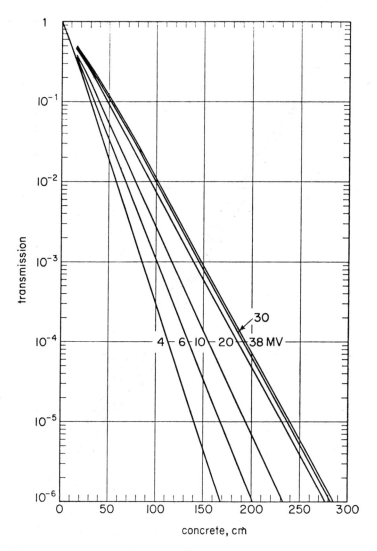

Fig. 18-2   Broad-beam transmission of X-rays through concre-
             te, density 2.35 g/cm³. 4MV: linear accelerator;
             1 mm gold target followed by 20 mm aluminium
             beam flattener. 6-38 MV: Betatron; target and
             filtration not stated. The 38 MV curve may be
             used up to 200 MV[18.4].

In fig. 18-2, broad-beam transmission curves of X-rays through
concrete are given. The high-energy X-rays are generated in li-

531

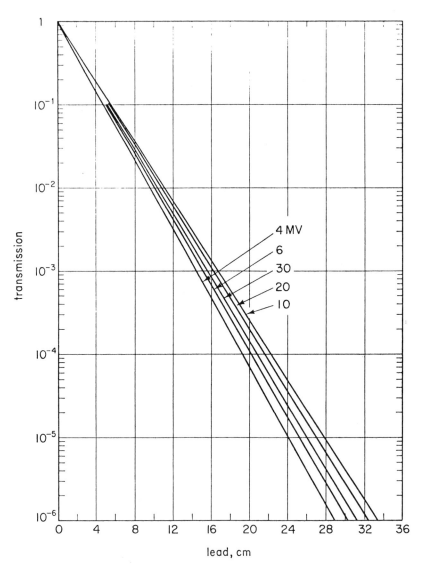

Fig. 18-3   Broad-beam transmission of X-rays through lead, density 11.35. Betatron; platinum wire target 2 mm · 8 mm; no beam filtration[18.4].

near accelerators and betatrons. Corresponding curves for lead[18.4] are given in fig. 18-3.

18.2.4 NEUTRON RADIATION

The physical processes involved in neutron attenuation are:

- elastic scattering
- inelastic scattering
- nuclear radiation with emission of secondary particles (n,2n), (n,p), (n,α)
- capturing processes (n,γ) of slow neutrons, mainly thermal.

The final capturing of the neutrons which are attenuated in a shielding barrier, takes place by (n,α)- and (n,γ)-processes. Radiation protection shields for high energy neutrons must contain materials that reduce by scattering the kinetic energy of the neutrons and then capture them. Hydrogen is such a material and thus materials containing a lot of hydrogen, like water or paraffin, are often used in neutron shields. Hydrogen also has a high cross section for neutron capturing at thermal energy. Other materials containing high-capture cross sections for thermal neutrons are also often added to the shield e.g. cadmium and boron.

The attenuation of a narrow beam of fast neutrons in hydrogen-containing material, or in combinations of hydrogen and rich amounts of heavy materials (high-Z), can be described rather well by an exponential equation:

$$\varphi = \varphi_0 e^{-\Sigma_r d} \qquad (18.15)$$

where φ = the fluence rate of primary neutrons behind the absorbing material
   φ_0 = the fluence rate at the same point without absorbing material
   d = the thickness of the absorbing material
   Σ_r = the macroscopic cross section for attenuation of fast neutrons in the absorbing material, also called the 'removal'- cross section.

$$\Sigma_r = \sigma_r \cdot 10^{-28} \frac{N_A \varphi}{A} = 0.06 \frac{\varphi}{A} \quad m^{-1} \qquad (18.16)$$

σ_r = the microscopic cross section in barn, 1 barn = $10^{-28} m^2$

N_A = Loschmidts or Avogadros number i.e. $6.023 \cdot 10^{26}$ atoms per kg $\cdot$ atom

A = The atomic weight

ρ = The density kg $\cdot$ m^{-3}

Due to scattering the fast neutrons are thermalized. They are then captured in cadmium, boron or lithium-rich materials placed in thin sheets behind the barrier, or simply mixed in the concrete.

Cadmium has a very high capture cross section (25 406 barn) but in the capture process, γ-radiation of energies between 3-5 MeV is emitted. From shielding point of view it is better to use boron even if its capture cross section is lower (760 barn), as the γ-radiation emitted has an energy of 0.48 MeV. Lithium has still lower capturing γ-energies, but the capture cross section is only 71 barn. It is also rather expensive. Thus boron is the material mostly used to absorb thermal neutrons in radiation shields.

The quantity one wants to estimate for radiation protection purposes is the dose equivalent, which must be below 2.5 mrem $\cdot$ h^{-1} or 25 μSv $\cdot$ h^{-1} for radiological personnel.

The dose equivalent at distance $\underline{x}$ m from a "point source" behind a radiation shielding barrier of thickness $\underline{d}$ m can be written as:

$$H'_d(x) = \frac{\varphi_0(1)}{x^2} \; e^{-\Sigma_r d} \; (\frac{H'}{\varphi}) B_H \qquad \qquad \mu Sv \cdot h^{-1} \qquad (18.17)$$

where

$H'_d(x)$ = dose equivalent rate behind the radiation protection barrier μSv h^{-1}

$\varphi_0(1)$ = the neutron fluence rate of primary neutrons at 1 m distance from the point source, without any radiation protection barrier n $\cdot$ m$^{-2} \cdot$ s^{-1}

$(\frac{H'}{\varphi})$ = dose equivalent rate/neutron fluence rate μSv $\cdot$ h^{-1}/n $\cdot$ m$^{-2} \cdot$ s^{-1}

This factor can be derived from the "conversion factor" (neutrons/cm$^2 \cdot$s per mrem $\cdot$ h^{-1}) in table 8-7:

$$(\frac{H'}{\varphi}) = \frac{10^{-3}}{\text{conversion factor}} \; \mu Sv \cdot h^{-1}/n \cdot m^{-2} \cdot s^{-1} \qquad (18.18)$$

B_H = dose equivalent build-up factor depending on the
primary neutron energy, the absorption material
and thickness. Values of B_H are given in table 18-8.

Table 18-7

Conversion factors and effective quality factors for neutrons[18.4].

Neutron energy, MeV	Conversion factor[a], neutrons/cm^2.s per mrem/h	Effective quality factor[b], Q
$2.5 \cdot 10^{-8}$ (thermal)	260	2.3
$1 \cdot 10^{-7}$	240	2
$1 \cdot 10^{-6}$	220	2
$1 \cdot 10^{-5}$	230	2
$1 \cdot 10^{-4}$	240	2
$1 \cdot 10^{-3}$	270	2
$1 \cdot 10^{-2}$	280	2
$1 \cdot 10^{-1}$	48	7.4
$5 \cdot 10^{-1}$	14	11
1	8.5	10.6
2	7.0	9.3
5	6.8	7.8
10	6.8	6.8
20	6.5	6.0
50	6.1	5.0
$1 \cdot 10^2$	5.6	4.4
$2 \cdot 10^2$	5.1	3.8
$5 \cdot 10^2$	3.6	3.2
$1 \cdot 10^3$	2.2	2.8
$2 \cdot 10^3$	1.6	2.6
$3 \cdot 10^3$	1.4	2.5

a Calculated at maximum of depth-dose equivalent curve.
b Maximum dose equivalent divided by the absorbed dose at
the depth where the maximum dose equivalent occurs.

Figure 18-4 shows the reduction of fluence rate of neutrons of various energies penetrating water and concrete[18.3].

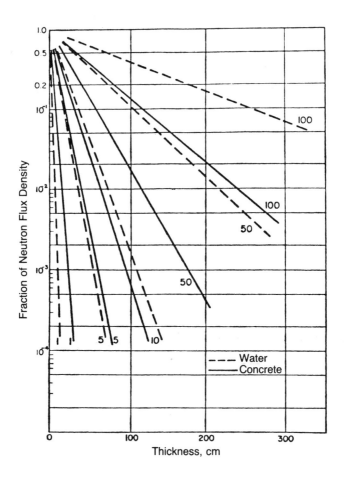

Fig. 18-4   Approximate broad beam absorption of neutrons in water and concrete (concrete density 2.37 g/cm^3). Note that the curve shows the reduction of neutron flux density, not of dose rate[18.3].

Table 18-8

The macroscopic cross section and the dose equivalent build-up factor for some commonly used materials in neutron shielding barriers[18.5].

Neutron energy MeV	Water Σ_r m^{-1}	Water B_H	Paraffin Σ_r m^{-1}	Paraffin B_H	Concrete Σ_r m^{-1}	Concrete B_H	Iron Σ_r m^{-1}	Iron B_H
3	11	1.0	18	1.5	6.4	1.0		
~5	10	1.0	12	1.3	9	1.2		
6-8	-	-	-	-	8.3	-	17	-
14-15	5	1.0	5.7	1.25	6.1 5.9	1.2* 2.3**	7.1	1.0
Fission	10.3				8.9		15.8	

* Up to 0.3m.

** Up to 0.6-1.0 m.

18.3 Safe handling of radioactive materials

18.3.1 RADIOTOXICITY

In the handling of radioactive material, one must consider both the risk of external irradiation and the risk of internal contamination. The various radionuclides are therefore divided into different classes according to the following properties:

- the type of decay: α-, β-, or γ-rays.
- the physical half-live.
- the efficiency of incorporation and excretion by the human body.
- the concentration pattern in the human body.

Group 1: Very high radiotoxicity - α-emitting radionuclides.

Group 2: High radiotoxicity -radionuclides which are concentrated to specific organs or tissues (^{131}I, ^{90}Sr).

Group 3: Moderate radiotoxicity - radionuclides which are relatively evenly distributed in the body and relatively rapidly excreted.

Group 4: Low radiotoxicity - those radionuclides which emit very low energy radiation, or which have a very short half-life.

In table 18-9 examples are given of the most common radionuclides, classified as to group of radiotoxicity[18.6].

In dealing with unsealed radioactive materials, special attention has to be given to internal radiation hazards. The extent of the precautions to be taken will depend upon such factors as the nature of the operation, the total activity, the specific activity, the radiotoxicity, the chemical composition and toxicity in exceptional cases, and other chemical and physical properties of the radionuclides or radioactive compounds. The radiotoxicity classification in table 18-9 is based primarly on the inhalation risk. It serves in turn as the basis for the standards for the design of laboratories required for a wide range of levels of unsealed radionuclides handled in different ways, as shown in table 18-10[18.6].

18.3.2 WORKING CONDITIONS

The various types of working places or laboratories required for using a wide range of levels of unsealed radionuclides handled in different ways are classified in Type C, Type B and Type A. Type C is conventional chemical laboratory of good quality. Type B is a specially-designed radionuclide laboratory with absorbing, easily cleaned surfaces and adequate separate ventilation with a special fume hood. Type A is a specially-designed laboratory with glove boxes and shielding barriers in which large activities of highly radiotoxic materials can be handled.

Modifying factors can be applied to the quantities of activity indicated in table 18-10 according to the complexity of the procedures to be followed[18.6].

The following factors are suggested, but due regard should be paid to the special circumstances in individual cases.

538

Table 18-9

Radionuclides classified according to relative radiotoxicity per unit activity (18.6).

Group 1: very high radiotoxicity

^{210}Pb	^{226}Ra	^{227}Th	^{231}Pa	^{233}U	^{238}Pu	^{241}Pu	^{243}Am	^{244}Cm	^{249}Cf
^{210}Po	^{228}Ra	^{228}Th	^{230}U	^{234}U	^{239}Pu	^{242}Pu	^{242}Cm	^{245}Cm	^{250}Cf
^{223}Ra	^{227}Ac	^{230}Th	^{232}U	^{237}Np	^{240}Pu	^{241}Am	^{243}Cm	^{246}Cm	^{252}Cf

Group 2: high toxicity

^{22}Na	^{56}Co	^{95}Zr	^{125}Sb	^{131}I	^{144}Ce	^{181}Hf	^{207}Bi	^{228}Ac	^{198}Au	^{231}Th
^{36}Cl	^{60}Co	^{106}Ru	^{127}Te^{m}	^{133}I	^{152}Eu(13 yr)		^{210}Bi	^{230}Pa	^{199}Au	^{233}Pa
^{45}Ca	^{89}Sr	^{110}Ag^{m}	^{129}Te^{m}	^{134}Cs	^{154}Eu	^{182}Ta	^{211}At	^{234}Th	^{197}Hg	^{239}Np
^{46}Sc	^{90}Sr	^{115}Cd^{m}	^{124}I	^{137}Cs	^{160}Tb	^{191}Ir	^{212}Pb	^{236}U	^{197}Hg^{m}	
^{54}Mn	^{91}Y	^{114}In^{m}	^{126}I	^{140}Ba	^{170}Tm	^{204}Tl	^{224}Ra	^{249}Bk	^{203}Hg	
		^{124}Sb							^{200}Tl	
									^{201}Tl	
									^{202}Tl	
									^{203}Pb	
									^{206}Bi	
									^{212}Bi	
									^{220}Rn	
									^{222}Rn	

Group 3: moderate toxicity

^{7}Be	^{48}Sc	^{65}Zn	^{91}Sr	^{103}Ru	^{125}Te^{m}	^{140}La	^{153}Gd	^{187}W
^{14}C	^{48}V	^{69}Zn^{m}	^{90}Y	^{105}Ru	^{127}Te	^{141}Ce	^{159}Gd	^{183}Re
^{18}F	^{51}Cr	^{72}Ga	^{92}Y	^{105}Rh	^{129}Te	^{143}Ce	^{165}Dy	^{186}Re
^{24}Na	^{52}Mn	^{73}As	^{93}Y	^{103}Pd	^{131}Te^{m}	^{142}Pr	^{166}Dy	^{188}Re
^{38}Cl	^{56}Mn	^{74}As	^{97}Zr	^{109}Pd	^{132}Te	^{143}Pr	^{166}Ho	^{185}Os
^{31}Si	^{52}Fe	^{76}As	^{93}Nb^{m}	^{105}Ag	^{130}I	^{147}Nd	^{169}Er	^{191}Os
^{32}P	^{55}Fe	^{77}As	^{95}Nb	^{111}Ag	^{132}I	^{149}Nd	^{171}Er	^{193}Os
^{35}S	^{59}Fe	^{75}Se	^{99}Mo	^{109}Cd	^{134}I	^{147}Pm	^{171}Tm	^{190}Ir
^{41}A	^{57}Co	^{82}Br	^{96}Tc	^{115}Cd	^{135}I	^{149}Pm	^{175}Yb	^{194}Ir
^{42}K	^{58}Co	^{85}Kr^{m}	^{97}Tc^{m}	^{115}In^{m}	^{135}Xe	^{151}Sm	^{177}Lu	^{191}Pt
^{43}K	^{63}Ni	^{87}Kr	^{97}Tc	^{113}Sn	^{131}Cs	^{153}Sm	^{181}W	^{193}Pt
^{47}Ca	^{65}Ni	^{86}Rb	^{99}Tc	^{125}Sn	^{136}Cs	^{152}Eu(9.2 h)	^{185}W	^{197}Pt
^{47}Sc	^{64}Cu	^{85}Sr	^{97}Ru	^{122}Sb	^{131}Ba	^{155}Eu		^{196}Au

Group 4: low toxicity

^{3}H	^{58}Co^{m}	^{71}Ge	^{87}Rb	^{97}Nb	^{103}Rh^{m}	^{131}Xe^{m}	^{135}Cs	^{191}Os^{m}	^{232}Th	^{238}U
^{15}O	^{59}Ni	^{65}Kr	^{91}Y^{m}	^{96}Tc^{m}	^{113}In^{m}	^{133}Xe	^{147}Sm	^{193}Pt^{m}	NatTh	NatU
^{37}A	^{69}Zn	^{85}Sr^{m}	^{93}Zr	^{99}Tc^{m}	^{129}I	^{134}Cs^{m}	^{187}Re	^{197}Pt^{m}	^{235}U	

Table 18-10

Limitation on activities in various types of working place or laboratory*.[18.6]

Radiotoxicity of radionuclides	Minimum significant quantity (μCi)	Type of working place or laboratory required		
		Type C	Type B	Type A
1. Very high	0.1	10 μCi or less	10 μCi - 10 mCi	10 mCi or more
2. High	1.0	100 μCi or less	100 μCi -100 mCi	100 mCi or more
3. Moderate	10	1 mCi or less	1 mCi - 1 Ci	1 Ci or more
4. Low	100	10 mCi or less	10 mCi - 10 Ci	10 Ci or more

1μCi = 0.037 MB$_q$; 1 mC$_i$ = 37MB$_q$; 1C$_i$ = 37000 MB$_q$

* Type C, Type B and Type A have the meanings normally used in the classification of laboratories for handling radioactive materials. Type C is a good quality chemical laboratory. Type B is a specially designed radioisotope laboratory. Type A is a specially designed laboratory for handling large activities of highly radioactive materials. In the case of a conventional modern chemical laboratory with adequate ventilation and fume hoods, as well as polished easily cleaned, nonabsorbing surfaces, etc., it would be possible to increase the upper limits of activity for Type C laboratories towards the limits for Type B laboratories for toxicity groups 3 and 4.

Procedure	Modifying factor
Storage (stock solutions)	• 100
Very simple wet operations	• 10
Normal chemical operations	• 1
Complex wet operations with risk for spills (liquid extraction)	• 0.1
Simple dry operations	• 0.1
Dry and dusty operations (grinding)	• 0.01

The handling tools and equipment used should be placed in nonporous trays and pans on absorbent disposable paper, which should be changed frequently. Pipettes, stirring rods and similar equipment should never be placed directly on the bench or table. After use, all vessels and tools should be set aside for meticulous cleaning.

18.3.3 CONTROL OF THE EXTERNAL IRRADIATION

There are four conditions which determine the amount of irradiation man is exposed to during work with radionuclides:

- type and energy of the emitted radiation.
- distance to the source.
- duration of the operation.
- shielding.

The α-emitters should always be handled in glove boxes. The gloves and the window of the box give enough shielding for external α-irradiation.

The β-irradiation is also rather easy to control due to its short range. In water, tissue and plastic materials, the range is about 0.5 cm for β-radiation with maximum energy of about 1 MeV, but one must always take the bremsstrahlung into consideration when high-density material is used.

The γ-radiation is more penetrating than both α- and β-radiations, and therefore one has to take all the above mentioned factors into consideration in order to obtain an optimal and economic shielding against γ-irradiation.

When planning an operation involving a γ-emitting radionuclide, it is necessary to estimate the exposure of the personnel.

This can often be done with sufficient accuracy by using the equation:

$$\underline{\overline{X}} = \Gamma \frac{A \cdot t}{x^2} T_d \qquad mR \qquad (18.19)$$

where

$\underline{\overline{X}}$ = the exposure mR

Γ = the specific γ-constant $\frac{mRm^2}{mCih}$.

A = the activity in the source mCi

t = the time estimated to perform the preparation h

x = the distance between the source and the personnel m

T_d = the transmission through the shielding barrier of thickness $\underline{d}$ m, as given in figs. 18-5 and 18-6.

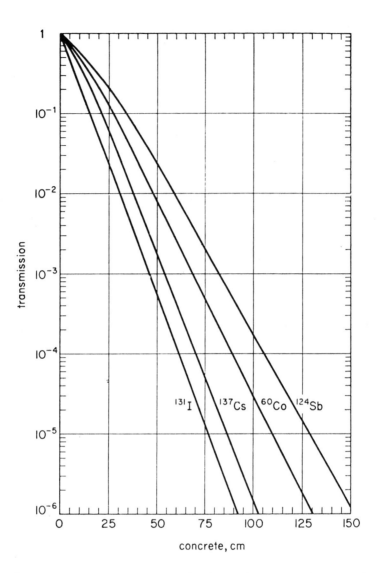

Fig. 18-5.   Broad-beam transmission of γ-rays from various
radionuclides through concrete, density
2.35 g/cm$^{3(18.4)}$.

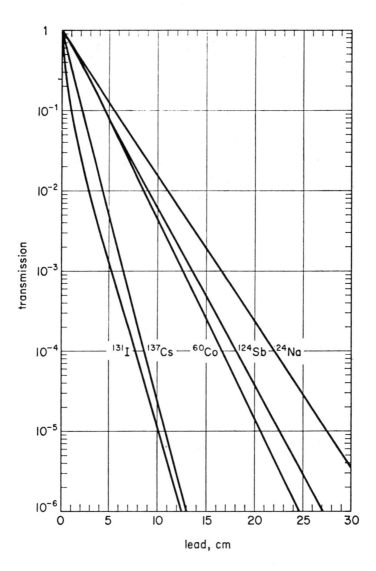

Fig. 18-6. Broad-beam transmission of γ-rays from various radionuclides through lead, density 11.35 g/cm^3 (18.4).

Values of specific γ-ray constants for some usual radionuclides are given in table 18-11.

Table 18-11

Output of γ-ray sources[18.4].

Nuclide	Half-life	Principal γ-ray energies, in MeV, and per cent photons per disintegration	Exposure rate R/h at 1 m from 1 Ci
^{24}Na	15.0 h	γ: 1.37(100%) 2.75(100%)	1.84
^{60}Co	5.24y	γ: 1.17(100%) 1.33(100%)	1.30
^{124}Sb	60d	γ: 0.60 to 2.09	0.98
^{131}I	8.05d	γ: 0.08 to 0.72 x: 0.005(0.6%) 0.03(5%)	0.22 (and 0.025 due to x-rays)
^{137}Cs	30y	γ: 0.66(85%)	0.32
^{182}Ta	115d	γ: 0.07 to 1.23	0.60
^{192}Ir	74d	γ: 0.30 to 0.61	0.48
^{198}Au	2.70d	γ: 0.41 to 1.09 x: 0.009(1%) 0.07(3%)	0.23 (and 0.014 due to x-rays)
^{226}Ra and daughters	1620y	γ: 0.047 to 2.4	0.825

18.4 References

18.1   IAEA Safety Series No. 38, Radiation Protection Procedures, IAEA, Vienna 1973.

18.2   ICRP Publication 26, Recommendations of the International Commission on Radiological Protection, Pergamon Press 1977.

18.3   ICRP Publication 4, Protection Against Electromagnetic Radiation Above 3 MeV and Electrons, Neutrons and Protons, Pergamon Press 1963.

18.4   ICRP Publication 21, Data for protection against ionizing radiation from external sources, Pergamon Press 1973.

18.5   NCRP Report No. 38, Protection Against Neutron Radiation, National Council on Radiation Protection and Measurements, Washington D.C. 1971.

18.6   IAEA Safety Series No. 1, Safe Handling of Radionuclides, IAEA, Vienna, 1973.

Index

spectrum, 46

transistor characterisation, 48-49, 59-61

binding energy, per nucleon, 23-24, 54, 82, 102

binomial distribution, 258-259

biological effects, 523-527
  ionizing radiation, 523-525
  dose equivalent, definition (Sv), 522

blank observation (x_b), 269-274

blocking, 172-173

Boltzmann's konstant (k) 107, 207

bore-hole logging 347-348, 429-430, 513-514

boron trifluoride (BF_3) counter, 194, 296, 409, 421, 516

Bowen's kale standard, 331-332

branching, decay, 42-43, 58, 60, 65-66
  transformation, 76-78

breeder reactor, 147-148

Breit-Wigner formula, 99, 101-103, 110-111, 116-117

bremsstrahlung, 121-125, 144-146, 170-171, 391, 394, 419, 435-439, 529
  activation yield, 121-122, 406-408, 410-411
  broad beam (γ,X) transmission 531-532, 542-543

bubbel chamber, 213-215

build up factor, 64, 530

C

cadmium pollution, 504

cadmium threshold (T_{cd}), 109-111, 336-339

capture γ-rays, 161, 428

capture reactions, 33, 84, 100-102, 105-196, 108-117, 147-148, 174-178, 298, 352-354, 360-362

carbon analysis, 413-417
  of steel, 368-371

carriers, 454-455

centre-of-mass system, 86, 88-89, 96

Cerenkov counter, 205-206

cesium-137, 303

chain decay, 67-68

chain reaction, 146

channelling, 172-173, 365, 387

characteristic X-rays 391-393, 437-439, 489

charge collection
  in ion chambers, 194-196
  in semiconducters, 211

charge independence, 83

chemical separations, 454-471
  electroplating, 467
  group separation, 457-48
  ion exchange 461-464
  isotope and amalgam exchange, 464-467
  removal of sodium, 457
  solvent extraction, 458
  substoichiometry, 468-469
  Szilard-Chalmers effect 469-471

chi-square distribution (χ^2), 286-288

cloud chamber, 213-215, 297

cobalt-60, 305

coincidence techniques, 319-323

cold irradiation, 452-453

collision stopping power (L_∞), 522-523

comparative life-time ($ft_{1/2}$), 48-49, 59-60

comparator technique, 329-331

H

hair analysis, 494-496, 501-503
  arsenic pollution, 505-506
  cadmium pollution, 504
  mercury pollution, 503-504
half-life $(t_{1/2})$, 48, 61-62, 66, 69-72
hard core in nucleons, 83
Heisenberg uncertainty relation, 97
high flux reactor, 326
hole mobility, 208-209
HPGe (high purity germanium), 212-213

I

independent activities, mixture, 64-65
industrial pollution, 503-507
inelastic scattering, 84, 95, 103, 106, 112
integral discriminator, 233
integral non-linearity, 239
integrating circuit, 232
internal conversion, 49-52, 60-61
internal standard, 229-231
in vivo analysis, 482-491
  cadmium, 488
  calcium, 486
  fluorine, 484
  hydrogen, 482
  iodine, 485
  lead, 490
ionization chamber, 194-196, 313-314
ion source, neutron generator, 135-139
isobar
  decay, 26, 58, 60, 66
  definition, 19

isomer
  decay, 58, 60, 76-78
  definition, 19
  production, 413, 423-426
isotone, definition, 19
isotope, definition, 19

K

K-capture, 47-48
knock-out reaction, 84
K-shell ionization, 391-393, 431-436

L

laboratory system, 85-89
L-capture, 47-48
Leachman equation, 111
least squares fitting, 274-280, 282-284, 291
linac, 145, 157-160
linear amplifiers, 230-231
linear attenuation coefficient (μ_ℓ), 179-180, 189, 431, 529-530
linear energy transfer (L_∞), 522
linear momentum conservation law, 85
liquid scintillators, 202
lithiumdrifted detectors, 211-212, 305-307
live-time, 239
log decrement (ξ), 175-177
log ft-value, 48-49, 59-61
logic symbols, 241
Lorentz line, 118
L-shell ionization, 393, 435

M

macroscopic cross section (Σ), 92, 177, 179, 432-433, 533
magic numbers, 23, 28-29, 35, 411
manganese in hair, 507

Watt's equation, 111
wave length (λ), 97-99
wave number (k), 97-99
Wilkinson ADC-converter, 235-236
working conditions, 538-540

X

X-ray
  emission, particle induced
  (PIXE), 390-402
  energy measurements, 307
  fluorescence, 407, 431-440

Y

yield (Y)
  bremsstrahlung reaction,
     121-125, 406-408, 411
  maximum (Y_{max}), 116-117
  reaction, 93, 116-117,
     362-363

Z

zinc sulphide, ZnS(Ag),
  201-202